Recent Advances in Biotechnology

(Volume 5)

Advances in Assisted Reproduction Technologies

Edited by

Islam M. Saadeldin

Research Institute of Veterinary Medicine
Chungnam National University
34134 Daejeon
Republic of Korea

College of Veterinary Medicine
Chungnam National University
34134 Daejeon
Republic of Korea

&

Department of Physiology
Faculty of Veterinary Medicine, Zagazig University
44519 Zagazig
Egypt

Recent Advances in Biotechnology

Volume # 5

Advances in Assisted Reproduction Technologies

Editor: Islam M. Saadeldin

ISSN (Online): 2468-5372

ISSN (Print): 2468-5364

ISBN (Online): 978-981-5051-66-7

ISBN (Print): 978-981-5051-67-4

ISBN (Paperback): 978-981-5051-68-1

Published by Bentham Science Publishers Pte. Ltd. Singapore. All Rights Reserved.

First published in2022.

need for a court order if at any point you breach any terms of this License Agreement. In no event will any delay or failure by Bentham Science Publishers in enforcing your compliance with this License Agreement constitute a waiver of any of its rights.

3. You acknowledge that you have read this License Agreement, and agree to be bound by its terms and conditions. To the extent that any other terms and conditions presented on any website of Bentham Science Publishers conflict with, or are inconsistent with, the terms and conditions set out in this License Agreement, you acknowledge that the terms and conditions set out in this License Agreement shall prevail.

Bentham Science Publishers Pte. Ltd.
80 Robinson Road #02-00
Singapore 068898
Singapore
Email: subscriptions@benthamscience.net

BENTHAM SCIENCE

CONTENTS

PREFACE

More than 4 decades have passed since the birth of the first in vitro fertilized baby in 1978. The use of assisted reproductive technology (ART) to overcome infertility has increased exponentially with the simultaneous increase in the number of fertility centers in every part of the world. The use of ART continues to increase around the world, due to ever-increasing global access to infertility clinics in the treatment of different forms of infertility (like tubal disease, ovarian aging, or ovarian dysfunction).

This book captures the current and recent advances in assisted reproduction technology in humans and livestock in an easy and comprehensive way for the non-expert and the junior embryologists through simplifying each ART tool by providing definition and explanation, and how the methods are practiced and how to overcome troubleshooting, and showing the ARTs uses and significance.

This book will be an ideal reference for junior embryologists as it provides take-home messages for the current and recent ARTs. It contains ten chapters with a detailed explanation of normal reproductive physiology, male ARTs such as sperm retrieval and freezing, female ARTs such as oocyte activation, and cryopreservation, and finally embryo ARTs that include assisted hatching and cloning techniques.

Islam M. Saadeldin
Research Institute of Veterinary Medicine
Chungnam National University
Daejeon
Republic of Korea

ii

List of Contributors

Anuja Gopalakrishnan	Department of Reproductive Medicine, Chennai Fertility Center and Research Institute, Chennai, Tamil Nadu, India
Deepa Onnarakatt	Department of Reproductive Medicine, Chennai Fertility Center and Research Institute, Chennai, Tamil Nadu, India
Elham K. Eltahawy	Chemistry Department, Faculty of Science, Helwan University, Cairo, Egypt
Eman A. Hussien	Biotechnology Program, Faculty of Science, Helwan University, Cairo, Egypt
Firas Al-Rshoud	Faculty of Medicine, Hashemite University, Zarqa, Jordan
Islam M. Saadeldin	Research Institute of Veterinary Medicine, Chungnam National University, 34134 Daejeon, Republic of Korea College of Veterinary Medicine, Chungnam National University, 34134 Daejeon, Republic of Korea Department of Physiology, Faculty of Veterinary Medicine, Zagazig University, 44519 Zagazig, Egypt
Jayesh Parasharam Shinde	ART Fertility Clinic PVT LTD, B 404 Aishwaryam Comfort, Akrudi, Pune-411019, Maharashtra, India
Krishnavignesh Lakshmanan	Department of Reproductive Medicine, Chennai Fertility Center and Research Institute, Chennai, Tamil Nadu, India
Manar M. Hozyen	Ganin Fertility Center, Cairo, Egypt
Mohammad Ibrahim	Department of Theriogenology, Faculty of Veterinary Medicine, Alexandria Governorate 5424041, Egypt
Mohamed M. Z. Hamada	Department of Physiology, Faculty of Veterinary Medicine, Zagazig University, 44519 Zagazig, Egypt
Mohamed Fadel El Mohr	Dr. Faris Medical Center for Infertility and Human Reproduction, Heliopolis, Cairo, Egypt Dar El Om for Infertility and Human Reproduction, Mokatam, Cairo, Egypt
Mohamed M. Omran	Chemistry Department, Faculty of Science, Helwan University, Cairo, Egypt
Ömer Uçar	Muğla Sıtkı Koçman University, Milas Veterinary Faculty, Division of Reproduction & A.I., Milas-Muğla, Turkey
Poongothai Muthuswamy	Department of Biotechnology, Dr. NGP Arts and Science College, Coimbatore, Tamil Nadu, India
Smisha Sridev	Department of Reproductive Medicine, Chennai Fertility Center and Research Institute, Chennai, Tamil Nadu, India
Sara A. Mekkawy	Biotechnology Program, Faculty of Science, Helwan University, Cairo, Egypt
Tamara Darwish	Faculty of Medicine, Hashemite University, Zarqa, Jordan

Thomas Ebner   Department of Gynecology, Obstetrics and Gynecological Endocrinology, Johannes Kepler University, Altenberger Strasse 69, 4040 Linz, Austria

V. M. Thomas   Department of Reproductive Medicine, Chennai Fertility Center and Research Institute, Chennai, Tamil Nadu, India

CHAPTER 1

Physiology of the Reproductive System

Mohamed M. Z. Hamada[1] and **Islam M. Saadeldin**[1,2,3,*]

[1] *Department of Physiology, Faculty of Veterinary Medicine, Zagazig University, 44519 Zagazig, Egypt*

[2] *Research Institute of Veterinary Medicine, Chungnam National University, Daejeon, Republic of Korea*

[3] *College of Veterinary Medicine, Chungnam National University; 34134 Daejeon, Republic of Korea*

Abstract: The reproductive system of the living organism is the biological system made up of all the anatomical organs involved in sexual reproduction. This system involves the interaction of several fluids and hormones to regulate the functions of the reproductive system. The ultimate goal of the reproductive system is to successfully produce gametes (sperms and oocytes) to attain a combination of genetic material between two individuals, which allows for the possibility of greater genetic fitness of the offspring. In this chapter, we introduce the physiological process of gonadal development, male, and female reproductive system, embryo formation, and development to give the reader the basic concepts for application in the field of assisted reproductive techniques.

Keywords: Oocyte, Ovary, Physiology, Reproduction, Sperm, Testis.

GONADAL DEVELOPMENT AND SEX DETERMINATION

Gonadal Development

The gonads represent a unique embryological situation in that: the rudiments of all body organs except the gonads can normally differentiate into only one type of organ. For example, a lung rudiment can become only a lung, and a liver rudiment can develop only into a liver. On the other hand, the gonadal rudiment has two normal options. When it differentiates, it can develop into either an ovary or a testis. The path of differentiation taken by this rudiment determines the future sexual development of the organism. Before this decision is made, the mammalian gonad first develops through a bi-potential (indifferent) stage, during which time

* **Corresponding author Islam M. Saadeldin:** Research Institute of Veterinary Medicine, Chungnam National University, 34134 Daejeon, South Korea; Tel: 00821024817666; E-mails: islamms@cnu.ac.kr and islamsaad82@gmail.com

Islam M. Saadeldin (Ed.)
All rights reserved-© 2022 Bentham Science Publishers

it has neither female nor male characteristics (Figs. **1** and **2**). The indifferent gonads consist of several components:

1. Coelomic epithelium, which is the precursor of Sertoli cells in males and granulosa cells in females.

2. Mesenchymal stromal cells, which are the precursor of Leydig cells in males and theca cells in females.

3. Germ cells that have migrated there from the yolk sac endoderm.

This assembly is organized into the indifferent gonads into two layers, cortex and medulla, and proceeds as follows:

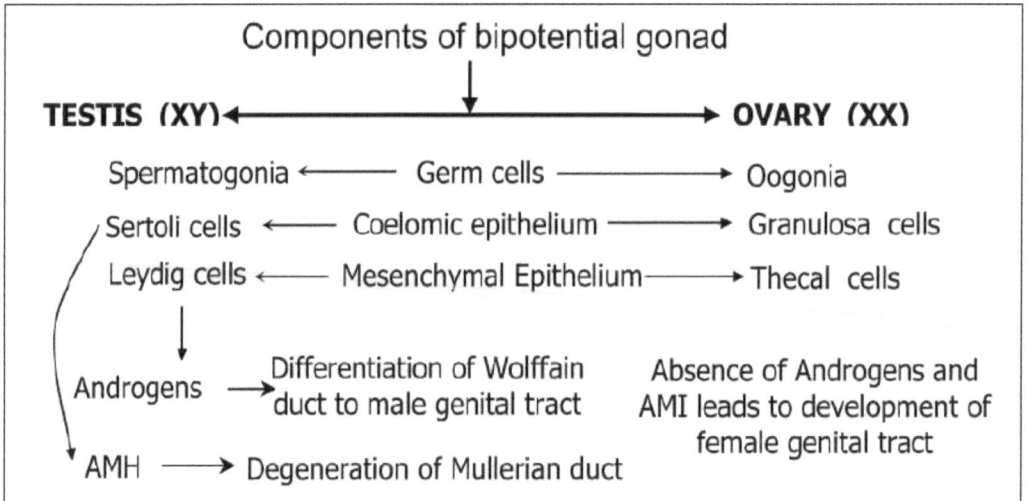

Fig. (1). Differentiation of the indifferent gonad components to their analog in both testes and ovary.

In a Male Fetus

Spermatogenic tubules begin to be formed at 6 weeks. This is followed by differentiation of the Sertoli cells at 7 weeks and Leydig cells at 8 to 9 weeks. At this point, the testes are structurally recognizable, and testosterone secretion begins.

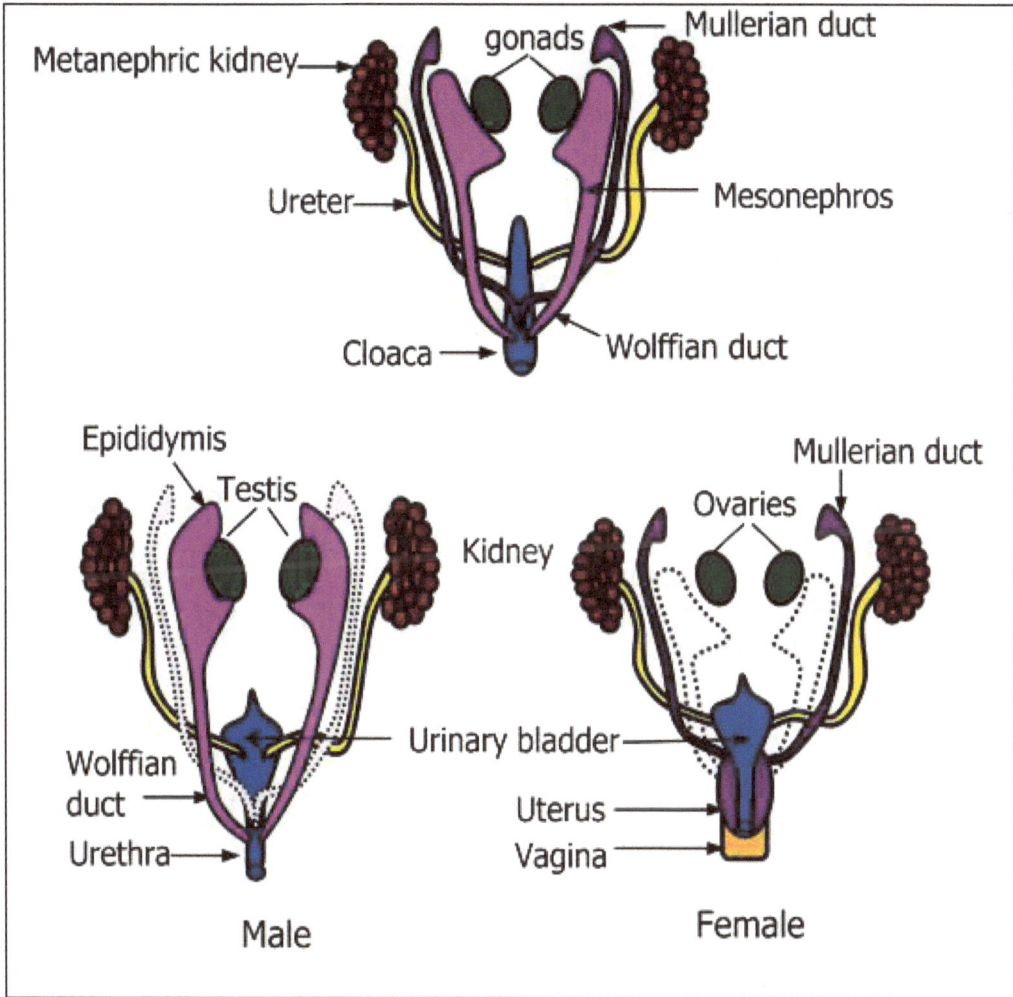

Fig. (2). Development of the gonads and their ducts in mammals. The upper figure represents the undifferentiated gonads and the presence of both male and female ducts. In the lower figures the male and female development is due to gonadal differentiation.

The germ cells become enclosed within the medulla, whereas the cortex is regressed. No known hormonal influences are required for the differentiation of the indifferent gonad into a testis till that stage. The urogenital groove (sinus) is the progenitor of the external genitalia. The Wolffian duct differentiates into the epididymis and the vas deferens.

In a Female Fetus

Differentiation of the indifferent gonad into an ovary does not start until 9 weeks gestation. At this time, the activity of both X chromosomes within the germ cells is essential. The germ cells begin to undergo mitosis, giving rise to daughter cells called oogonia, which continue to proliferate. Shortly thereafter, meiosis is initiated in some oogonia, and each is surrounded by differentiating granulosa cells and precursor theca cells to form follicles.

The germ cells, now known as primary oocytes remain in the first stage, or prophase of meiosis until they are activated hormonally at puberty.

In contrast to the male arrangement of gonadal zones, the cortex, which contains the follicle, predominates whereas the medulla regresses. The primitive ovary begins to synthesize estrogen which contributes to the latter ovarian differentiation by blocking androgen actions.

Sex Determination

The genital system is generally composed of primary sex organs including the gonads, and secondary sex organs, which include the rest of the sex organs like internal and external sex organs.

Primary Sex Determination

The determination of the gonads, in mammals primary sex determination is strictly genetic and is not influenced by the environment. In most cases, the female is XX and the male is XY, every individual must have at least one X chromosome. Since the female is XX, each of her eggs has a single X chromosome. The male, being XY, can generate two types of sperms: half bear the X chromosome, half the Y. If the egg receives another X chromosome from the sperm, the resulting individual is XX, the ovary is formed, and is female; if the egg receives a Y chromosome from the sperm, the individual is XY, testes are formed, and becomes male. Thus, Y chromosome is an important factor for determining sex in mammals.

<u>*Mechanisms of Primary Sex Determination*</u>

Several genes have been found essential for normal sex differentiation these are:

<u>*Sex Determining Region of the Y Chromosome (SRY gene)*</u>

The organization of indifferent gonad into the characteristic spermatogenic tubules of the male is directed by a segment on the short arm of the Y chromosome known as the sex-determining region of the Y chromosome (SRY gene). SRY has different effects on converting the bipotential gonads into the testis, as follows:

a) It works directly to convert the coelomic epithelium into male-specific Sertoli cells.

b) SRY encodes another gene called testes determining factor (TDF). This TDF stimulates the development of primary sex cords to a seminiferous tubule.

Steroidogenic Factor I (SFI)

SFI is another protein activated by SRY, it is essential in masculinizing both the Leydig and Sertoli cells:

a) In Sertoli cells, SFI works in collaboration with Sox9 to elevate the levels of antimullerian hormone (AMH) transcription.

b) In Leydig cells, SFI activates the genes encoding the enzymes that facilitate the synthesis of testosterone.

Sox9 gene (Autosomal Sex Reversal)

It is one of the autosomal genes involved in sex determination. It is essential for testis formation since it is expressed only in the male genital ridge.

Gene Encoding H-Y antigen

It is identical to or closely linked to the SRY or TDF genes. H-Y antigen is a glycoprotein present on all male cells except diploid germ cells. This H-Y antigen causes virilization of the cells of the indifferent gonad.

Autosomal Genes for Androgen Receptors

These genes are responsible for the formation of androgen receptors in target organs for sensitizing the genital ducts and external genitalia to the masculinizing effect of testosterone and dihydrotestosterone.

DaxI (Ovary-determining Gene on X Chromosome)

It is a potential ovary-determining gene present on the X chromosome.

It is expressed in the genital ridges of the mouse embryos, shortly after SRY expression. Indeed, in XY mice, SRY and DaxI are expressed in the same cells; DaxI appears to antagonize the function of SRY, and it down-regulates SFI expression. Thus DaxI is probably a gene involved in ovary determination.

WNT4 (Ovary-determining Gene on an Autosome)

The WNT4 gene is another gene that may be critical in ovary determination. This gene is expressed in the mouse genital ridge while it is still in its bipotential stage. WNT4 expression then becomes undetectable in XY gonads, whereas it is maintained in XX gonads as they begin to form ovaries (Fig. **3**).

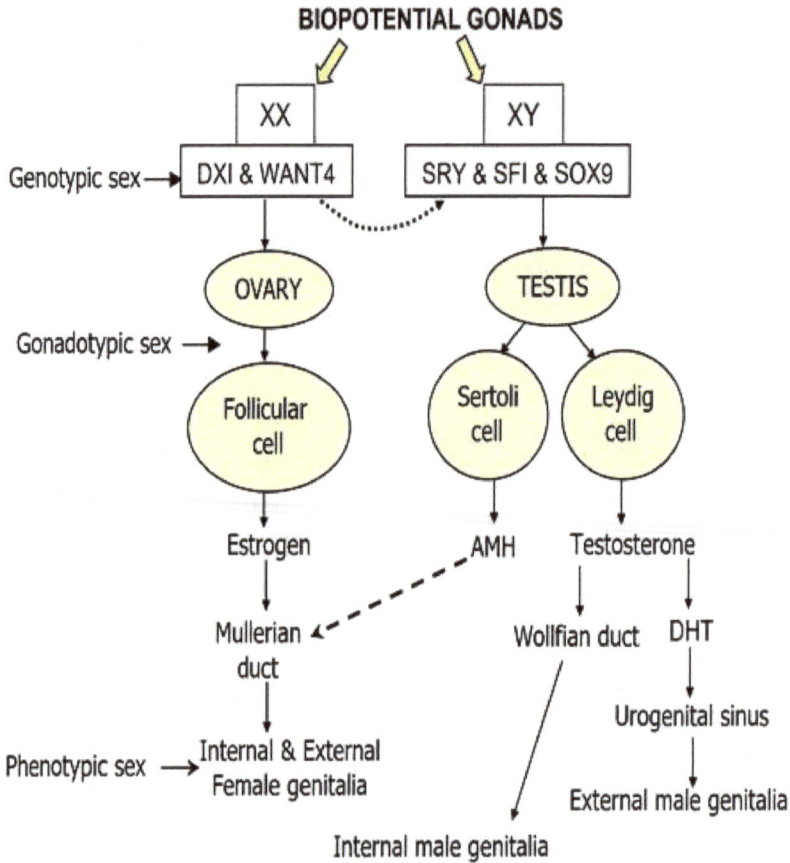

Fig. (3). Summary of mammalian sex determination. Dashed arrows mean inhibition.

Secondary Sex Determination

Secondary sex determination in mammals involves the development of the female and male phenotypes in response to hormones secreted by the gonads. Both female and male secondary sex determination has two major temporal phases, the first occurs within the embryo during organogenesis and the second occurs during adolescence.

Male Phenotype Determination

The formation of the male phenotype involves the secretion of two testicular hormones; AMH and androgens (testosterone and dihydrotestosterone).

Antimullerian Hormone (AMH)

It is a 560-amino acid glycoprotein secreted from the Sertoli cells and causes degradation of the Mullerian duct. AMH bind to the mesenchyme cells surrounding the Mullerian duct causing these cells to secrete a paracrine factor that induces apoptosis in the Mullerian duct epithelium.

Testosterone and Dihydrotestosterone (DHT)

Testosterone secreted from fetal Leydig cells is responsible for the differentiation of the Wolffian duct into the epididymis, vas deferens, and seminal vesicles. DHT is produced in some target organs like the urogenital sinus and swellings by reduction of testosterone under the effect of 5 α-reductase enzymes. DHT is responsible for the differentiation of urogenital sinus and swellings into the male external genitalia including the scrotum and penis. In absence of this enzyme, the male external genitalia will be a female one.

Female Phenotype Determination

The absence of TDF, testosterone, and AMH allows the following changes:

a) Indifferent gonads differentiate into the ovary.

b) Wolffian ducts degenerate.

c) Mullerian ducts develop into the oviducts, uterus, and cervix.

d) Tissue around the urogenital sinus becomes the clitoris, labia, and vagina.

Thus, the development of the female phenotype depends on the absence of androgens during early development.

MALE REPRODUCTIVE SYSTEM

Introduction

The male reproductive system is made up of several individual organs acting in concert to produce spermatozoa and deliver them to the reproductive tract of the female. This concerted effort involves both the neuro-endocrine (hypothalamus and pituitary gland) and the genital system.

The gonad consists of two testes, each suspended within the scrotum by a spermatic cord and external cremaster muscle. During embryogenesis, the testes develop retro-peritoneally on the posterior wall of the abdominal cavity. As they descend into the scrotum, they carry with them a portion of the peritoneum. This peritoneal out-pouching, the tunica vaginalis, forms a serous cavity that partially surrounds the anterolateral aspect of each testis, permitting it some degree of mobility within its compartment in the scrotum [1].

Spermatozoa produced by the seminiferous tubules of testes enter a duct system including short straight ducts, tubuli recti, which connect the opened end of each seminiferous tubule to the rete testis, a system of labyrinthine spaces housed within the mediastinum. The spermatozoa leave the rete testis through 10-20 short tubules, vasa efferentia, which eventually fuse with the epididymis, which is connected with ductus deferens. In addition, there are accessory sex glands that play an important role, for instance providing the seminal fluid. These glands include paired ampullae, paired seminal vesicle, a prostate gland, and paired bulbourethral glands (Cowper's gland) [2].

The Male Gonad (Testes)

Each testis is surrounded by a capsule of dense, irregular collagenous connective tissue known as tunica albuginea. Immediately deep into this layer is a highly vascularized loose connective tissue, the tunica vasculosa, which forms the vascular capsule of the testis.

From the tunica albuginea, connective tissue septa radiate to subdivide each testis into intercommunicating compartments called testis lobules.

In all domestic animals except stallions, these septa units near the center of the testis form a fibrous cord called the mediastinum testis. Each testis lobule has one to four blindly ending seminiferous tubules, which are surrounded by a richly innervated and highly vascularized loose connective tissue derived from tunica vasculosa. Dispersed throughout this connective tissue are small conglomerations of endocrine cells, the interstitial cells of Leydig, which are responsible for testosterone production [3].

Seminiferous Tubules

Seminiferous tubules are highly convoluted hollow tubules, 30 to 70 cm long and 150 to 250 µm in diameter, it is surrounded by extensive capillary beds. About 1000 seminiferous tubules are present in the testes, for a total length of nearly 0.5 km, dedicated to the production of spermatozoa (Fig. **4**).

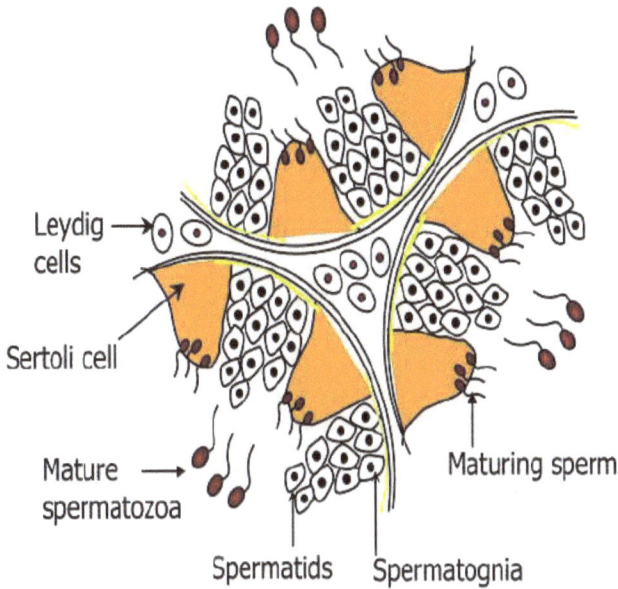

Fig. (4). Cross-section of Mammalian Testis.

The seminiferous epithelium is composed of two types of cells.

1. Spermatogenic cells produce spermatozoa a process called spermatogenesis.

2. Sertoli cell which is has many functions either in the male reproductive system or during spermatogenesis.

Spermatogenic Process

Spermatogenesis is the production of sperms from the primordial germ cells (PGCs) lining the seminiferous tubules. It is divided into:

(1) Spermatocytogenesis. (2) Meiosis. (3) Spermiogenesis (Spermateliosis).

Spermatocytogenesis

Definition: It is the process in which the spermatogonia lining the seminiferous tubules differentiate into primary spermatocytes.

Mechanism: The spermatogonia lie on the basal lamina of the seminiferous tubules and, after puberty, become influenced by testosterone to enter the cell cycle to produce sperms.

Meiotic Division

Definition: Meiotic Division is composed of the first and second divisions. In the first meiotic division, the diploid number of chromosomes in each primary spermatocyte is reduced to haploid number forming two secondary spermatocytes. In the second meiotic division, each secondary spermatocytes divide into two spermatids of the same haploid number of chromosomes. Thus each primary spermatocyte (diploid number, 2n) gives rise by meiotic division to four spermatids (haploid number, n) as shown in Fig. (**5**).

Spermiogenesis

Definition: It is the process by which, the immotile spermatids are differentiated into motile sperms without division.

Mechanism: The newly formed spermatid has a central nucleus, well-developed Golgi body, endoplasmic reticulum, mitochondria, and a pair of centrioles. The sequence of spermatid transformation to sperm is as follows:

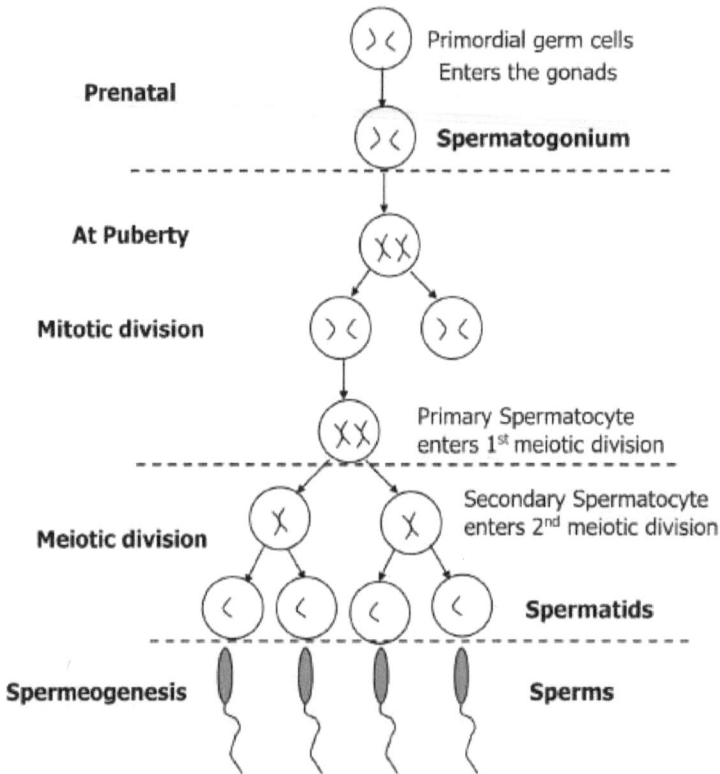

Fig. (5). Different divisions during the stages of spermatogenesis.

1. Hydrolytic enzymes are formed on the rough endoplasmic reticulum, modified in the Golgi apparatus, and packed as small, membrane-bound vesicles. These small vesicles fuse, forming an acrosomal vesicle, which enlarges to its final size known as the acrosome or acrosomal cap.

2. The nucleus becomes condensed and flattened and most of the spermatid cytoplasm is removed.

3. One centriole becomes attached to the nucleus and the other will elongate to form the tail of the spermatozoon.

4. Mitochondria location is shifted and becomes associated with the proximal portion of the developing flagellum. They form the mitochondrial sheath, which constitutes the middle piece of the mature spermatozoon (Fig. **6**).

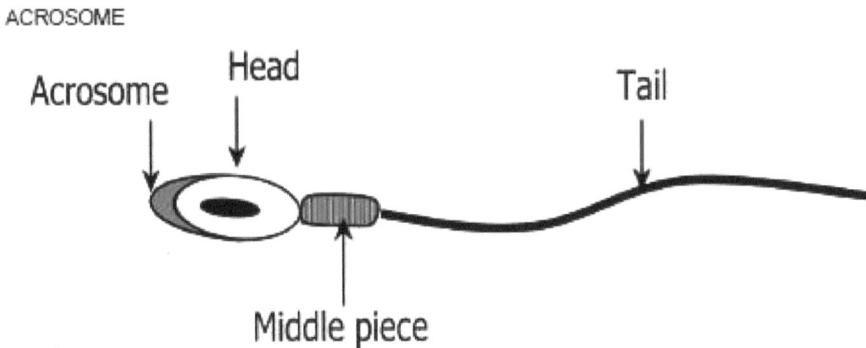

Fig. (6). Schematic representation of the sperm.

In mice, the entire spermatogenesis process takes 34.5 days, the spermatogonial stages last 8 days, meiosis lasts 13 days, and spermiogenesis takes up another 13.5 days. In most domestic animals, this process takes about 60-65 days.

The differentiation of mammalian sperm is not completed in the testes. After being expelled into the lumen of the seminiferous tubules, the sperm are stored in the epididymis, where they acquire the ability to move. Motility is achieved through changes in the ATP-generating system as well as changes in the plasma membrane that make it more fluid.

The sperm released during ejaculation can move, yet they cannot yet bind to and fertilize an egg. The final sages of sperm maturation (called capacitation).

Sperm Capacitation

Definition: It is the set of changes that allow the sperm to be able to fertilize the ova. This process does not occur until the sperm has been subjected to the fluid of the female reproductive tract for a certain period. The process involves the removal of a protective substance termed glycerophosphocholine (GPC), present on the acrosome of the sperm that has a protective purpose and thus interferes with the fertilizing efficiency of the sperm. The acrosome reaction involves the release of a series of hydrolytic enzymes from the acrosomal cap like hyaluronidase that digests the hyaluronic acid that binds the granulosa cells covering the oocyte. Another enzyme called **acrosine** has also been released that digest the cellular coating around the oocyte. Detachment of the acrosomal cap is triggered by the rapid influx of calcium ions that induce disruption of the acrosome and release of its hydrolytic enzymes. Detachment of the acrosome exposes a special protein on the head of sperm called **fertilin** that is important for fusing the sperm head with the denuded zona pellucida. The acrosome reaction also results in vigorous flagellar tail movement of the sperm that allow penetration of the oocyte. Once sperm is attached to the oocyte its membrane is depolarized preventing entry of other sperms inside (polyspermy) [4].

Sertoli Cells

The lateral cell membranes of adjacent Sertoli cells form occluding (tight) junctions with each other, thus subdividing the lumen of the seminiferous tubules into two isolated compartments:

1. The basal (outer) compartment, located basal to the tight junctions and contains the spermatogonia only.

2. The adluminal (inner) compartment: it is wider and begins from the tight junction till the lumen of the seminiferous tubules. It contains the primary and secondary spermatocytes, spermatids, and spermatozoa.

Thus the tight junctions of these cells establish a blood-testis barrier that isolates the adluminal compartment from the interstitial tissues, thereby protecting the developing gametes from the immune system.

Because spermatogenesis begins after puberty, the newly differentiating germ cells, which have different chromosome numbers as well as express different surface membrane receptors and molecules, would be considered foreign cells by the immune system. If there is no tight junction of Sertoli cells for the isolation of germ cells from the interstitial tissue surroundings, an immune response would be mounted against them.

Functions

1. They give physical and mechanical support for germ cells.

2. They secrete a fructose-rich medium that nourishes the sperms and facilitates the transport of the sperm to the genital ducts.

3. They establish a blood-testis barrier which has very important functions:

   i. It protects the spermatocytes, spermatids, and sperms from damaging substances present in the bloodstream.
   ii. It prevents the entry of anti-sperm immunoglobulin molecules into the lumen of seminiferous tubules
   iii. It prevents the passage of antigens, produced during the differentiation of germ cells, from returning back and mixed with blood. Thus it prevents the autoimmune reactions leading to the death of sperms.
   iv. Its secretion is rich in K^+, HCO_3^- and ABP, this secretion, in the inner compartment of the seminiferous tubules, provides a driving force to expel the sperm to the duct system.
   v. Its secretion is rich in K^+, HCO_3^- and ABP, this secretion, in the inner compartment of the seminiferous tubules, provides a driving force to expel the sperm to the duct system.

4. They synthesize and release very important hormones and other proteins:

   i. Androgen-binding protein (ABP): It is a macromolecule that facilitates an increase in the concentration of testosterone in the seminiferous tubule by binding to it and preventing it from leaving the tubule.
   ii. It has the ability to convert the androgens to estrogens by the aromatase enzyme, in addition to its ability to synthesize estrogen.
   iii. Inhibin hormone: It is a hormone that inhibits the release of follicle-stimulating hormone (FSH) by the anterior pituitary.
   iv. During embryogenesis Sertoli cells synthesize and release the anti-Mullerian duct hormone (AMH), which suppresses the formation of the Mullerian duct (the precursor of the female reproductive tract) and thus establishes the maleness of the developing embryo.

Leydig Cells

They are small collections of endocrine cells dispersed in a richly vascularized loose connective tissue, the tunica vasculosa, which fill the spaces between the seminiferous tubules. These cells secrete the male sex hormone (androgen).

Androgen

Androgens are steroid hormones that exert masculinizing effects, and they promote protein anabolism and growth. Testosterone from the testes is the most active androgen, and adrenal androgens have less than 20% of its activity [5].

Secretion of the adrenal androgens is controlled by ACTH, but not by gonadotrophins (FSH and LH). Gonadotrophins are important regulators of steroidogenesis in the gonads. In the testes, LH acts on the Leydig cells whilst FSH acts on the Sertoli cells. About 98% of testosterone in plasma is bound to proteins: 65% to a β-globulin called gonadal steroid-binding globulin (GBG) or sex steroid-binding globulin, and 33% bound to albumin. A small amount of circulating testosterone is converted to estradiol, but most of it is converted into 17-ketosteroids, principally androsterone.

The secreted androgens are metabolized in the liver by oxidation and reduction. The reduced metabolites are conjugated to glucuronides and sulfates and are excreted in the urine. About two-thirds of the urinary 17-ketosteroids are of adrenal origin, and one-third are of testicular origin [6].

Functions

Testosterone diffuses to the target cells, wherein many sites, it is reduced to dihydrotestosterone (DHT) by the 5α-reductase enzyme. Testosterone receptor binds have a greater affinity for DHT than for testosterone. Thus, DHT formation is a way of amplifying the action of testosterone in target tissues.

I. During embryogenesis

1. Testosterone-receptor complexes are responsible for the maturation of Wolffian duct structures and consequently for the formation of male internal genitalia during development.

2. DHT-receptor complexes are needed to form the male external genitalia from the urogenital sinus and structures.

II. At Puberty

a) Metabolic effects

Androgens are anabolic hormone it increases the synthesis and decreases the breakdown of protein so has a positive nitrogen balance. It stimulates bone growth during puberty, but it ultimately halts linear growth by closing the epiphyseal growth centers. It causes enlargement of the muscle mass of males by increasing the size of muscle fibers. It also increases red blood cell mass by stimulating erythropoietin synthesis and by directly affecting the maturation of erythroid precursors.

b) On secondary sex characters

i. It enlarges the larynx, thickens the vocal cords, and thereby deepens the voice.
ii. It is required for the growth of secondary sex organs including the penis, scrotum, and prostate, and stimulates prostatic secretions.
iii. It is responsible for the typical body conformation of males (broad shoulders, horn shape); male aggressive behaviors; and expression of sexual desire (libido).
iv. It is essential for maintaining the spermatogenesis process.
v. It stimulates the hair follicles to produce the typical masculine hair distribution. It stimulates the growth of sebaceous glands and their production of sebum.

CONTROL OF TESTIS FUNCTIONS

Neuroendocrine Control

Regulation of the reproductive axis begins at the level of the hypothalamus, where neurosecretory cells synthesize and release GnRH in a pulsatile fashion into the hypothalamic-hypophysial-portal circulation.

In response, gonadotropes in the anterior pituitary synthesize and release the gonadotropins (FSH and LH), that control the gonadal function. This is known as the Hypothalamic-Pituitary-Gonadal Axis (Fig. **7**).

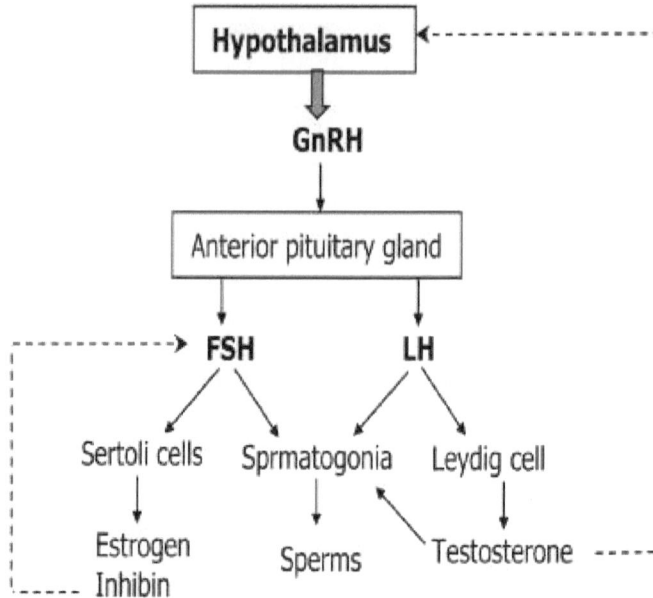

Fig. (7). Neuro-Endocrine Control of Testes Function (Hypothalamo-Hypophysial-Testicular Axis). Dashed arrows mean inhibition.

The pulsatility of GnRH and gonadotropins (FSH/LH) actions on their target organs and the production of high concentrations of testosterone are essential components in spermatogenesis. FSH, LH, and testosterone coordinate with local estradiol, inhibin, and activin as well as prolactin and GH in the regulation of spermatogenesis.

An adult male is unlike a female in that FSH and LH act on different cell types whose secretions have separate negative feedback effects on the secretion of both gonadotropins.

a. GnRH stimulates the gonadotropes to secrete FSH/LH. Because GnRH is secreted in a pulsatile manner, FSH and LH also are secreted in a pulsatile manner.
b. FSH stimulates Sertoli cells to produce inhibin, which in turn inhibits FSH secretion; LH stimulates Leydig cell to produce testosterone, which in turn inhibits LH secretion.
c. Inhibin has a negative feedback effect on the pituitary gonadotropes, whereas testosterone exerts a negative feedback effect primarily in the hypothalamus.
d. When plasma testosterone drops; the frequency of pulsatile GnRH secretion

increases. Thus plasma FSH and LH are maintained in a dynamic equilibrium with plasma inhibin and testosterone respectively.

Pituitary Gonadotropins

We mentioned before, that FSH acts exclusively on Sertoli cells, whilst LH acts exclusively on Leydig cells. Now, we can discuss, in some detail, the effect of each one on its target cell.

Effects of FSH on Sertoli Cells

2. FSH stimulates the Sertoli cell's production of activin, which modulates mitochondrial changes that occur as spermatogonia enter meiosis and become primary spermatocytes. Thus FSH, acting at least in part *via* Sertoli cells, enhances the early stages of sperm production.

3. FSH stimulates Sertoli cells to synthesize ABP, inhibins and stimulates estrogen synthesis from testosterone provided by Leydig cells.

4. Other products of FSH actions on Sertoli cells are providing energy sources, such as lactic acid, to the germ cells and facilitating the expulsion of spermatozoa into the lumen of the tubule. The release of FSH is inhibited by the inhibin hormone produced by Sertoli cells.

Effect of LH on Leydig Cells

The physiologic role of LH is to maintain testosterone secretion by Leydig cells. In addition, LH produces a locally high concentration of androgen in the testes, and this maintains spermatogenesis. The release of LH is inhibited by increased levels of testosterone and DHT.

Androgen

High intracellular testosterone levels are essential for normal spermatogenesis. Testosterone is concentrated in the tubule by ABP. The stages from spermatogonia to spermatids appear to be androgen-independent. The maturation from spermatids to spermatozoa depends on androgen acting on Sertoli cells in which the spermatozoa are embedded. FSH acts, also on the Sertoli cells to facilitate the last stages of spermatid maturation.

Inhibin and Activin

Inhibins and activins are disulfide-linked dimeric glycoproteins. Inhibin is dimers of α subunit linked to either a $β_A$ or $β_B$ subunit to generate inhibin A (α $β_A$) and inhibin B (α $β_B$). Dimerization of β subunits alone gives rise to three forms of activins referred as activin A ($β_A$ $β_A$), activin B ($β_B$ $β_B$), and activin AB or activin C ($β_A$ $β_B$).

Inhibins and activins were isolated from gonads mainly from Sertoli cells in males and granulosa cells of ovarian follicles in females. They can modulate the pituitary FSH secretion as activin stimulates FSH whereas inhibin inhibits FSH secretion through negative feedback.

Testicular Temperature

In domestic animals, normal testicular function, especially normal spermatogenesis, is temperature-dependent and requires an environmental temperature 3C to 5C lower than core body temperature.

Hence, in normal domestic males, the testes are located outside the abdominal cavity, in the scrotum. Failure of one or both of the testes to descend into the scrotum is known as cryptorchidism.

Mechanism Of Testicular Temperature Regulation

Role of Testicular Venous Pampiniform Plexus

The vascular supply of each testis is derived from the testicular artery, which descends with the testis into the scrotum accompanying the ductus deferens. The testicular artery forms several branches before it pierces the capsule of the testis to form the intra-testicular vascular elements.

The capillary beds of the testes are collected into several veins forming the pampiniform plexus of veins, which are wrapped around the testicular artery.

Blood in the pampiniform plexus of veins, which is cooler than that of the testicular artery, acts to reduce the temperature of the arterial blood, thus forming a countercurrent heat exchange system.

In this fashion, it helps keep the testis's temperature a few degrees lower than that of body temperature. At this cooler temperature, spermatozoa develop normally; at body temperature, spermatozoa that develop will be sterile.

Role of the Scrotum

Each testis is suspended within the scrotum by a spermatic cord and external cremaster muscle. The scrotum, along with the cremaster muscle and the vascular anatomy of the testicular arteries and veins, protects and thermo-regulates the testes in the following way:

a. The scrotum is essentially a hairless skin pouch with a subcutaneous fibroelastic and muscular layer called tunica dartos, and without subcutaneous fat.
b. Contraction and relaxation of the tunica dartos and cremaster muscles occur with changes in ambient temperature as well as in response to other tactile stimuli. Contraction of these muscles pulls the testes toward the abdomen to decrease the loss of temperature (*e.g.* during cold weather), while relaxation keeps the testes hanging away from the abdomen to increase the heat loss by radiation (*e.g.* during hot weather).
c. The scrotum of some species such as the horse contains numerous sweat and a sebaceous gland that further contribute to the thermoregulatory mechanism.

Cryptorchidism

Definition: It is the failure of one or both of the testes to descend into the scrotum. Although the cryptorchid testis is still capable of producing androgens, it is incapable of producing normal spermatozoa. Consequently bilaterally cryptorchid males would be sterile. The cryptorchid testis is more prone to torsion of the spermatic cord and 10 times more likely to be neoplastic. Cryptorchidism appears to be genetic, although the exact mechanism is not completely understood and may vary between species. It is most common in boars, dogs, and stallions and least common in bulls, rams, and bucks.

DUCT SYSTEM

The genital ducts can be subdivided into two categories:

The Intra-testicular ducts connect the seminiferous tubules to the epididymis. They include tubuli recti and rete testis and extra-testicular ducts, which are associated with each testis, including ductuli efferentes (vasa efferentia), the epididymis, ductus (vas) deferens, and the ejaculatory duct.

Intra-testicular Ducts

a. Tubuli Recti: They are short, straight tubes continuous with the seminiferous tubules and deliver the spermatozoa formed by the seminiferous tubules into the rete testis. These short tubules are lined by Sertoli cells in their first half, near the seminiferous tubule, and by the cuboidal cell in their second-half near the rete testis. The presence of Sertoli cells in the tubuli recti is an essential role to get rid of the remnant sperm cytoplasm.

b. Rete Testis: They are multiple narrow tubules forming anastomosing channels continuous with the tubuli recti. They are lined with cuboidal cells that have short microvilli with a single flagellum to deliver the immotile sperm to the ductuli efferentes.

Extra-testicular Ducts

a. **Ductuli efferentes** (vasa efferentia); composed of 10-20 short tubules attached by vascular connective tissue. They are lined with two types of cells:
   i. High columnar ciliated cells, which are responsible for the movements of the immotile sperms to the epididymis.
   ii. Non-ciliated low columnar cells, which are responsible for reabsorption of luminal fluid secreted by the Sertoli cells.

Thus, one of the functions of the efferent duct (vasa efferentia) cells is to absorb about 90% of the water from the lumen of the rete testis. This concentrates the sperm, giving them a longer life span and providing more sperm per ejaculate. This absorption of water is regulated by estrogen secreted by Sertoli cells.

If estrogen or its receptor is absent in mice, this water is not absorbed, and the mouse will be sterile. While blood levels of estrogen are higher in females than males, the levels of estrogen in the rete testis are even higher than that in female blood.

Epididymis

It is a single, thin, long highly convoluted tubule of considerable length (from 2 m in the cat to 80 m in the stallion). Anatomically, it is divided into three segments: head, or caput; body, or corpus; and tail, or cauda. The epididymis is not only a conduit for spermatozoa, but also provides a special environment in which spermatozoa are concentrated, undergo maturation, and acquire a fertilizing capacity.

Spermatozoa entering the caput from the rete testis are immotile and incapable of fertilization. Only after they undergo migration and maturation (through the caput and corpus), both motility and the fertilization capacity are achieved in the cauda epididymis.

The cauda epididymis serves as a storage depot for mature sperm. The media of cauda epididymis is slightly acidic to suppress the sperm activities, so the sperm can live for months in the epididymis. However, the lumen of the epididymis is lined by two types of cells: basal cells that function as stem cells, regenerating themselves, as well as the principal cells which have different functions including:

  i. Reabsorb the luminal fluid.
 ii. Phagocytose is the remnants of sperm cytoplasm that were not removed by Sertoli cells.
iii. Manufacturing glycerophosphocholine (GPC), which inhibits sperm capacitation until the sperm enters the female genital tract.

Ductus Deferens (Vasa Deferentia)

Each ductus deferens is a thick-walled muscular tube that conveys the spermatozoa from the cauda epididymis to the ejaculatory duct, which in turn deliver the sperms and seminal fluid into the prostatic urethra. The ductus deferens passes through the inguinal rings into the abdomen and connects the cauda epididymis with the pelvic urethra.

In most species, the terminal portion of the ductus deferens enlarges to form prominent ampullae such as those found in the bull and stallions. In other species, the ampullae are either absent or are anatomically indifferent from the ductus deferens. The ampullae serve as an additional storage depot for sperms, and in some species, such as the bull, stallion, and dog, the ampullary glands add to the ejaculate. In stallion, the secretion of ampullary glands is rich in ergothioneine, which acts as an anti-oxidizing agent protecting susceptible chemicals in semen from oxidation.

Accessory Genital Glands

Grossly, the main components of semen are the spermatozoa produced by the seminiferous tubules; and the seminal fluid produced by the accessory genital ducts. The seminal fluid (seminal plasma) produced by the accessory genital glands provides:

a. Substrates for conveying the sperms to the female genital tract and ensuring the maturation of the sperm.
b. It supplies an energy source (fructose) and protects from changes in osmotic pressure especially in stallion (citrate), and oxidizing agents (ergothioneine).
c. It also contains clotting factors, in the form of fibrinogen or a gel that is present in seminal vesicle fluid in the stallion, that forms a partial clot in semen, helping to minimize the escape of semen from the female genitalia.

Seminal Vesicle

Lies laterally to the terminal parts of each ductus deferens. In ruminants, it is a compact lobulated gland, in the stallion and boar, it is a large pyriform glandular sacs. In dog and cat, a seminal vesicle is absent. The duct of the seminal vesicles and the ductus deferens share a common ejaculatory duct that opens into the urethra. The secretions of the seminal vesicles are the terminal portions of the ejaculate.

a. These secretions are rich in fructose, which is considered an important oxidative substrate for the spermatozoa; as well as large quantities of prostaglandins and fibrinogen.
b. Prostaglandins are thought to play a role in fertilization by making the cervical mucous more receptive to the sperm movement; and by enhancing the antiperistaltic contractions of the uterus and fallopian tubes, helping the movement of sperms toward the ovary.
c. The clotting enzymes present in the prostatic secretion act on the fibrinogen of seminal vesicle secretions to form a weak fibrin coagulum, which holds up the semen in the vagina.
d. Moreover, the seminal vesicle secretions contain Ca^{2+}, zinc, LH, FSH, prolactin, testosterone, estradiol, inhibin, oxytocin, endorphin, and a variety of enzymes. The exact source and the role of these components remain obscure.

Prostate Gland

Present in all animals, It is intimately associated with the pelvic urethra, but among different species, it varies in size and appearance. Just before ejaculation (during emission), secretions of the prostate are discharged into the urethra meeting the sperms coming from ductus deferens. The motility of spermatozoa is suppressed in the epididymis due to its acidic medium, being rich in citric acid and metabolic end-products of spermatozoa, in addition, the vaginal secretions are also acidic in nature.

a. The prostatic secretion is slightly alkaline, containing Ca^{2+}, citrate (in the stallion, the prostatic secretion is rich in citric acid), clotting enzymes, and a profibrinolysin. This alkalinity of prostatic secretion plays a role in neutralizing the acidic medium of the seminal fluids, and the vagina and thus enhances the motility of the sperms.

b. Fibrinolysin formed from prostatic profibrinolysin, dissolves the coagulum, which is formed by the action of prostatic clotting enzymes on the seminal vesicle fibrinogen, and thus sperms can be released.

c. In stallion, prostatic secretion is rich in citric acid, which is known to act as an osmotic pressure buffering system. It maintains the osmotic pressure in the seminal fluid within the sperm heads, thus preventing the potential bursting of the sperm heads due to an influx of water from the surrounding fluids.

Bulbourethral Gland (Cowper's Gland)

These glands are dorsal to the urethra near the termination of its pelvic portion. They are present in all animals except dogs.

The secretions of these glands are the first secretions produced during sexual excitation. These secretions are clear, serous, and slightly alkaline. These secretions help to clean out the urine and bacteria collected in the urethra prior to ejaculation. The secretions also act as lubricants easing the passage of sperm.

PHYSIOLOGY OF MALE SEXUAL ACT

Sexual intercourse, mating, copulation, or coitus are all synonymous with male and female interaction to facilitate the deposition of male gametes into the female reproductive tract for fertilization. The sexual responses are mediated by the coordinated activity of the autonomic (Parasympathetic and sympathetic) and somatic innervation. Although these reflexes differ in detail in males and females, basic similarities allow the two sexes to be considered together, not only in humans but also in mammals generally. The relevant autonomic effects include:

a. Mediation of the vascular dilation which causes a penile or clitoral erection.

b. Stimulation of prostatic or vaginal secretions

c. Smooth muscle contraction of the vas deferens during ejaculation or rhythmic vaginal contraction during orgasm in both sexes.

Parasympathetic Activity

The reproductive organs of both males and females receive preganglionic parasympathetic innervation from the sacral segments (S2-S4) of the spinal cord and reach the target organs *via* the pelvic nerves. The activity of postganglionic neurons causes dilation of penile or clitoral arteries, and a corresponding relaxation of the smooth muscles of venous (cavernous) sinusoids, which leads to expansion of sinusoidal spaces. As a result, the amount of blood in the tissue is increased, leading to a sharp rise in the pressure and an expansion of cavernous spaces, that is to say, erection.

The parasympathetic activity also provides excitatory input to the vas deferens, seminal vesicles, and prostate in males and vaginal glands in females. The chemical Mediators of Parasympathetic in Penile Tissue:

i. Acetylcholine and Vasoactive intestinal peptide (VIP) as contransmitters since parasympathetic fibers supplying penile tissue contain both acetylcholine and VIP. Thus these fibers release both acetylcholine and VIP as chemical transmitters when stimulated.
ii. Nitric Oxide (NO): Some parasympathetic nerve fibers supplying deep cavernous tissues of the penis secretes NO synthase, the enzyme that catalyzes the formation of NO from arginine. NO activates guanylyl cyclase resulting in increased production of cGMP, and cGMP is a potent vasodilator. Thus it seems clear that NO plays a prominent role in the production of erection.

Sympathetic Activity

In contrast, the sympathetic activity causes vasoconstriction and sub-subsequently loss of erection. Emission means the movement (first part of ejaculation) of semen into the urethra. Emission is a sympathetic response as a result of afferent pathways from touch receptors in the gland penis and reaches the spinal cord through the internal pudendal nerves.

The preganglionic sympathetic innervation originates from the thoracolumbar segments (T11-L2) and reaches the target organs *via* the corresponding sympathetic chain ganglia and the inferior mesenteric and pelvic ganglia. The response of sympathetic activity leads to contraction of smooth muscles in vas deferens and seminal vesicles, and subsequently propulsion of semen into the urethra.

Somatic Activity

The somatic component of reflex sexual function arises from α-motor neurons in the lumbar and sacral spinal cord segments. These neurons provide excitatory innervation to the bulbocavernosus and ischiocavernosus muscles, which are active during ejaculation proper in males and mediate the contractions of the perineal (pelvic floor) muscles that accompany orgasm in both males and females.

Ejaculation proper means the propulsion of semen outside the urethra. Thus the sexual function can be summarized into three events:

a. Erection as a response to parasympathetic activity.
b. Ejaculation: which can be divided into:
   i. Emission as a result of sympathetic activity.
   ii. Ejaculation proper as a result of parasympathetic and somatic activities.

FEMALE REPRODUCTION

Physiology of the female reproductive system includes the following items.

Ovarian Cycle

The ovarian cycle includes three main phases:

I. **Follicular phase** during which two main events affecting each other take place in parallel:
   a. Oogenesis process by which mature oocyte is produced.
   b. Folliculogenesis process by which mature follicles are produced and secrete estrogen, the female sex hormone.
II. **Ovulation**, is a phase in which the dominant follicle (Graafian follicle) ovulates releasing mature secondary oocyte and the remaining cellular part of the Graafian follicle changes to corpus luteum.
III. **Luteal phase** or Secretory phase, during which the corpus luteum secretes progesterone mainly in addition to a smaller amount of estrogens and androgens.

Follicular Phase

Oogenesis

It is the differentiation of the ovum as shown in Fig. **8**. The gamete formed by oogenesis contains all the materials needed to initiate and maintain metabolism and development. In addition to forming a haploid number of chromosomes, oogenesis also builds up a store of cytoplasmic enzymes, mRNAs, organelles, and metabolic substrates that are essential for future embryo development.

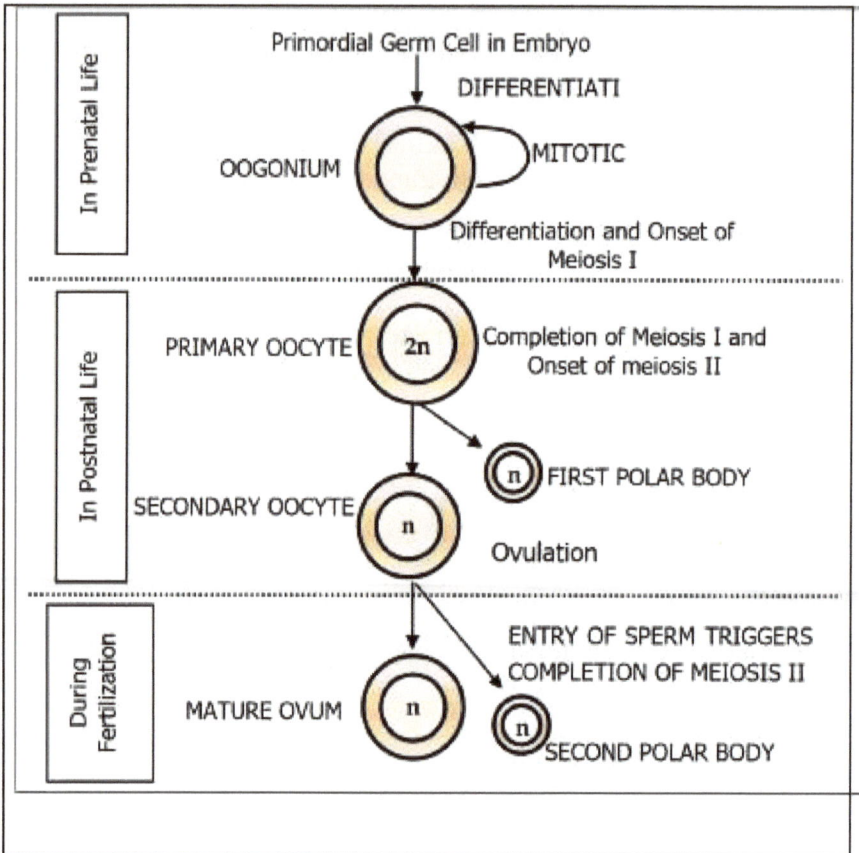

Fig. (8). Stages of mammalian oogenesis: Note that the first meiotic division starts during the prenatal period and is completed after puberty (during ovulation); while the second meiotic division is completed after fertilization.

Mechanism of Oogenesis

It varies among species since the reproductive patterns vary greatly among species, only a few eggs are produced during the lifetime of the animal. The oogonia divide to form a limited number of egg precursor cells as follows:

a. In the higher primates and human embryos, the oogonia divide rapidly from the second to the seventh month of gestation to form roughly 7 million germ cells. After the seventh month of embryonic development, the number of germ cells drops suddenly. Most oogonia die during this period, while the remaining oogonia enter the first meiotic division. These latter cells are called primary oocytes it progress through the first meiotic prophase until the diplotene stage, at which point they are maintained until puberty.

b. With the onset of adolescence, groups of oocytes periodically resume meiosis. Thus the first part of meiosis begins in the embryo, and the signal to resume meiosis is not given until puberty.

c. At Ovulation, the first meiotic division is completed producing two daughter cells: one of these cells contains hardly any cytoplasm and is called the first polar body, whereas the other cell has nearly the entire volume of cellular constituents and is the secondary oocyte.

d. During ovum fertilization by sperms, the second meiotic division is completed producing also two daughter cells, (second polar body and mature ovum). During the second meiotic division, most of the cytoplasm is retained by the mature egg (ovum) [4].

Folliculogenesis

You should remember that the female is born with ovaries containing a limited number of ova. In mammalian ovaries, folliculogenesis is the first event toward the generation of the offspring and is defined as the formation of Graafian follicle (mature preovulatory follicle) from a pool of primordial (non-growing) follicles.

Types of Ovarian Follicles

a. Primordial Follicles, present in the earliest stages and are the most abundant in number (100,000-400,00/female domestic animals). The primordial follicle constitutes a centrally located oocyte and a single layer of flattened granulosa cells associated with the oocyte. The cell cycle of oocytes at this stage is arrested at the prophase of the first meiotic division, and oocytes are not released from the arrest until they have passed through the growing phase (Fig. **9**).

b. Once the oocyte starts to grow and granulosa cells change morphology from a flattened to cuboidal shape, the follicles are considered to have entered the **primary-follicle stage** which is followed by a rapid proliferation of granulosa cells, resulting in the construction of the multi-laminar structure of the follicles (**secondary follicles**) [1].

c. During this course of development, the basement membrane is formed to wrap the outermost layer of granulosa cells, and the exterior of the membrane becomes covered by a thecal layer organized by ovarian stromal cells. As the follicle and oocyte reach a certain size a fluid-fill the space and the antral follicles are formed.

d. Only one antral follicle is selected and dominated over the other follicles to reach the ovulatory size and become the Graafian follicle.

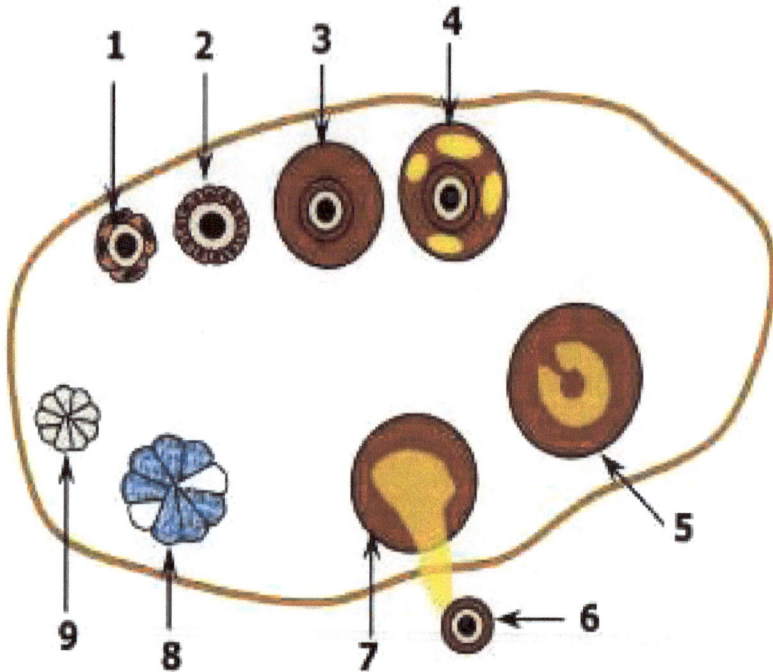

Fig. (9). Development of ovarian follicles. (1) Primordial Follicle, (2) Primary Follicle, (3) Secondary Follicle, (4) Formation of Antral follicle, (5) Graafian Follicle, (6) Free Ova (Ovulation) (7) Ruptured Follicle (8) Corpus Luteum (9) atretic follicle.

Development beyond the early antral follicle stage is dependent upon gonadotropins; antral follicles are sometimes referred to as tertiary follicles. Well-developed antral follicles are often referred to as Graafian follicles (Fig. **10**). However, after the preovulatory gonadotropin surge, they are called a preovulatory follicle.

The formation of a follicular antrum divides the population of granulosa cells into two main groups: (1) cumulus cells associated with oocytes; and (2) mural granulosa cells lining the follicular wall. The mural granulosa cells nearest the antrum are called periantral granulosa cells.

Life History of Follicles

A fixed number of primordial follicles are endowed during early life, and most of them are maintained in a resting state.

Growth of some of these dormant follicles is initiated before (**Initial Recruitment**) and throughout the reproductive life (**Cyclic Recruitment**).

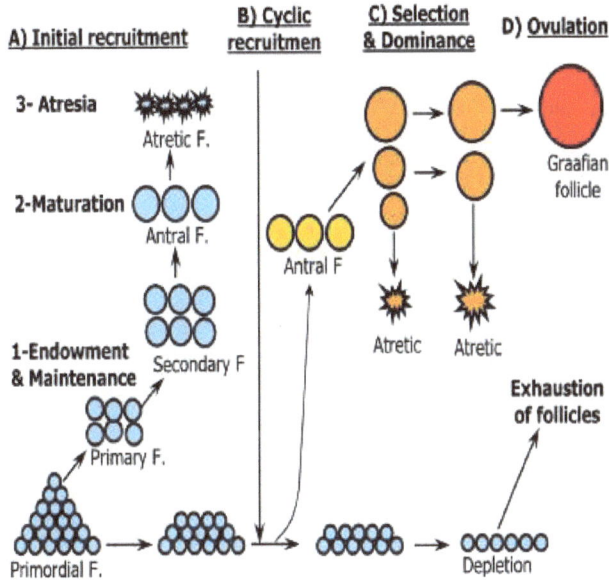

Fig. (10). Life history of the ovarian follicles

Follicles develop through primordial, primary, and secondary stages before acquiring an antral cavity. At the antral stage, most follicles undergo atresia; under optimal gonadotropin stimulation that occurs after puberty, a few of them are rescued to reach the preovulatory stage.

Eventually, depletion of the pool of resting follicles leads to ovarian follicle exhaustion and senescence.

Events of Folliculogenesis

Recruitment of Follicles

Definition: It is the process whereby a group of follicles begins to mature in the presence of sufficient pituitary gonadotropic stimulation to permit progress towards ovulation.

Types: There are two types of follicle recruitment: (a) initial recruitment and (b) cyclic recruitment. The main differences between the two types are presented in Table **1**.

Table 1. Differences between initial and cyclic recruitment of ovarian follicles.

-	Initial Recruitment	Cyclic Recruitment
Stages	Primordial	Antral of certain diameter
Its control	Ovarian Growth factors	FSH
Timing	Continuous throughout life, begins after follicle formation	Cyclic after puberty onset
Oocyte status	Start to grow, not capable of undergoing germinal vesicle breakdown	Completed growth, competent to undergo germinal vesicle breakdown
Default pathway	Atresia	Atresia or ovulation

Deviation, Selection, and Dominance

Deviation: It is the abrupt change in the diameter between the two largest follicles in the follicular wave.

Selection: It is the process whereby a follicle (s) avoids atresia and undergoes further development and becomes competent to achieve ovulation. In monovular species (*e.g.* bovine, equine, and humans), follicle selection is the process where one follicle develops from a wave (7-10 follicles) of growing follicles and becomes the only follicle with ovulatory capacity.

Dominance: It is the process where a single follicle achieves and maintains its distinction over the other recruited follicles, which undergo atresia. Follicle deviation occurs at the end of the common growth phase; the dominant follicle grows at a continuing rate, and the remaining or subordinate follicles regress or temporarily grow at a reduced rate and then regress.

Thus, in monovular species, the follicles of a wave are initially in a common growth phase that ends with a phenomenon termed deviation. The continuous growth rate of a single follicle (dominant) and reduced growth rate of the remaining follicles (subordinate follicles) characterize what is called diameter-defined deviation. Deviation begins when the largest follicle reaches an average diameter of 8.5 mm in cattle and 22.5 mm in mare (Fig. **11**). Mechanism of Follicle Deviation, Selection, and Dominance (FSH-Follicle Coupling Hypothesis): See Fig. (**12**).

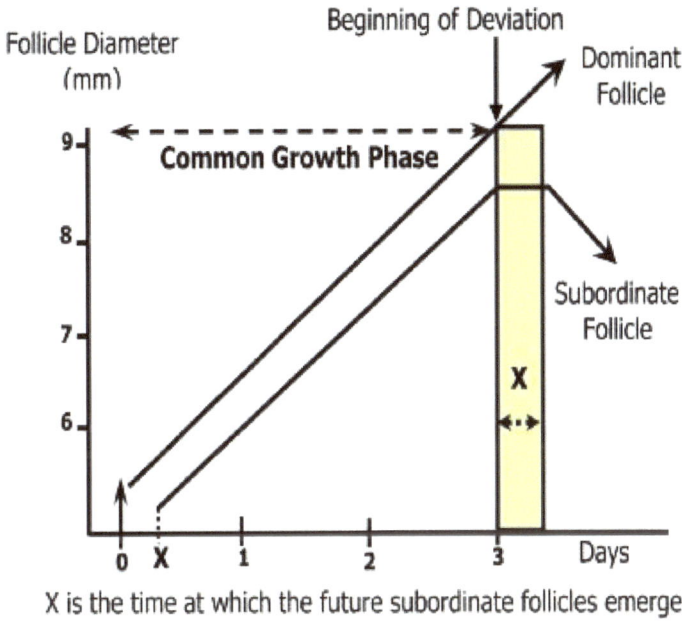

X is the time at which the future subordinate follicles emerge

Fig. (11). Schematic diagram representing the beginning of emergence (0 h) at which the follicle diameter is 4 mm in the cow; and beginning of deviation of ovarian follicles when the follicle diameter reaches 8.5 mm in the cow.

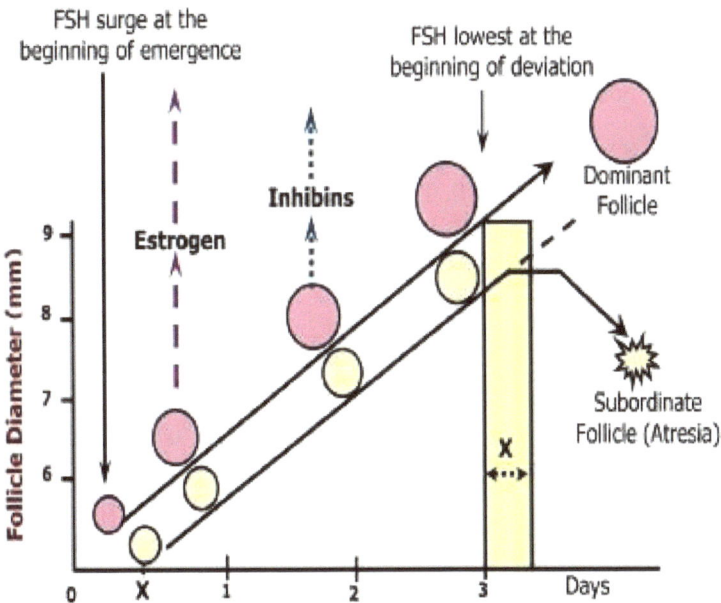

Fig. (12). FSH-Follicle coupling hypothesis. Dashed arrows mean inhibition.

1. The final stages of follicular development from 4 mm diameter to the preovulatory size are distinguished by episodes of growth termed follicular waves. In cows, there are two or three follicular waves develop during each estrus cycle.

2. These two or three follicular waves emerge approximately at 0, 9, 16 days after ovulation and are termed F1, F2, and F3 respectively as shown in Fig. (**13**).

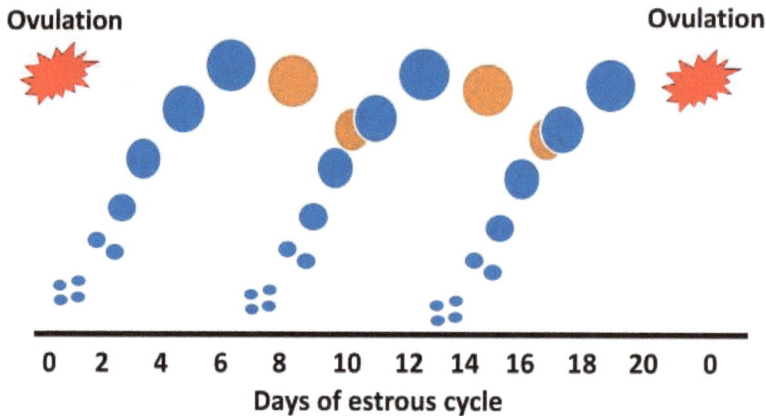

Fig. (13). Follicle development during estrus cycle (3 follicular waves).

3. Each follicular wave undergoes a common growth phase for 3 days after emergence. At the end of this phase, deviation begins and becomes established in about 8-h in cows and 1 day in mare.

4. Each follicular wave is stimulated by a surge of FSH that reaches a peak when the emerging follicles are 4 mm in the cow. The FSH levels decline during the common growth phase and the lowest level is reached 8-h in cow and 1-day after the beginning of deviation.

5. At the beginning of deviation, the more developed largest follicle produces large amounts of estrogens and inhibins, both suppressing the pituitary FSH. Thus the largest follicle can utilize the low FSH level and becomes the only follicle involved in FSH-follicle coupling. On the other hand, the smaller follicles have not reached a similar developmental stage at the beginning of deviation, but because of their continued dependency on FSH, they eventually become susceptible to the low FSH resulting in their atresia.

6. A small transient elevation in LH begins before deviation and decreases after deviation. LH receptors increase in the granulosa cells of the future dominant follicle about 8-h before the beginning of deviation. The LH stimulates the

production of estrogen and IGF-1. These intraovarian factors account for the responsiveness of the largest follicle to the low FSH.

The Fate Of The Ovarian Follicles

As previously mentioned, the ovaries contain many follicles in various stages of development. Most follicles degenerate before reaching the mature stage, but multiple Graafian follicles develop during each estrus cycle. Once a single mature follicle ruptures (in monovular species) and releases its secondary oocyte and associated cells, the remaining maturing follicles undergo atresia, the atretic follicles are eventually phagocytized by macrophages. Occasionally, in monovular species, two separate follicles develop to maturity and ovulate, leading to fraternal twins if both oocytes are fertilized [7].

Ovulation

The ovulatory surge in LH is responsible for:

a. Release of a local factor known as the meiosis-inducing factor. Under the influence of this factor, the primary oocyte of the Graafian follicle resumes and completes its first meiotic division, resulting in the formation of two daughter cells, the secondary oocyte, and the polar body. The newly formed secondary oocyte enters the second meiotic division and is arrested in the metaphase stage.
b. It Initiates the formation of many vasoactive agents collectively causing dilatation of the capillaries in the theca interna of the follicle wall. This significant change in the capillaries of the ovulatory follicle results in a 4 fold increase in the volume of the ovarian vascular compartment.
c. Concomitant with this hyperemic response, the permeability of the thecal capillaries increases to the extent that serum proteins are filtered into the interstitial spaces of the follicle.
d. The exuded serum proteins stimulate the quiescent fibroblasts in the theca externa and tunica albuginea of an ovulatory follicle and cause them to transform into proliferating cells and begin secreting a metalloproteinase that degrades the extracellular matrix of collagenous connective tissue in the follicular wall (Fig. **14**).
e. Granulosa cells continue to synthesize proteoglycans and hyaluronic acid, which attract water, thus causing an even greater increase not only in the size of the Graafian follicle but also in the loosening of the granulosa layer (Cumulus Expansion).
f. Just before ovulation, the surface of the ovary, where the Graafian follicle is pressing against the tunica albuginea, loses its blood supply. This thinned

avascular region becomes lightened and is known as Stigma. The connective tissue at the stigma degenerate, as does the follicular wall in contact with the stigma, loses its tensile strength, and eventually rupture under the force of increased intrafollicular pressure.

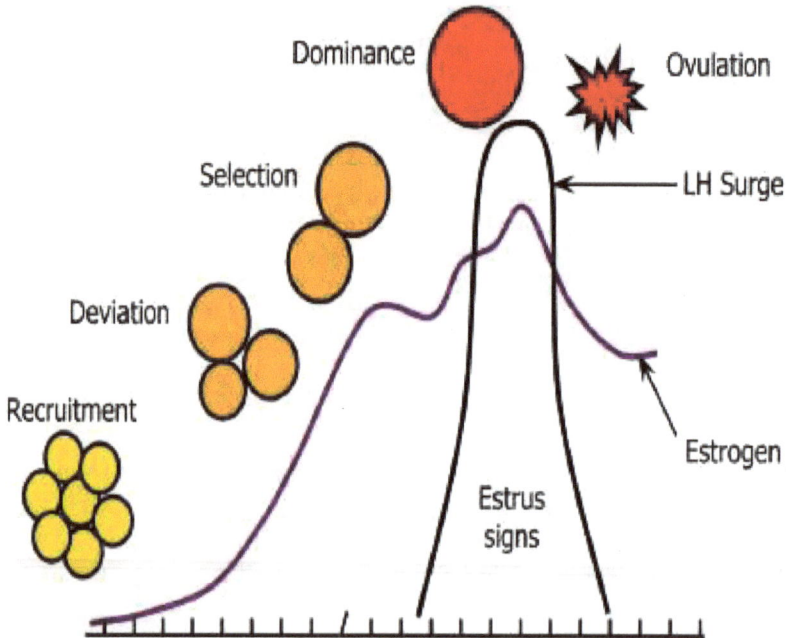

Fig. (14). Follicular Development and ovulation. LH stimulates follicle growth and a growing follicle produces high levels of estrogen. This estrogen in turn causes estrus signs and a surge of LH that triggers ovulation.

The Pattern Of Ovulation In Animals

Ovulation in mammals follows one of two basic patterns, depending on the species:

a. **Induced or reflex ovulatory animals:** This type of ovulation is stimulated by the act of copulation. Physical stimulation of the cervix triggers the release of gonadotropins from the pituitary. These gonadotropins signal the egg to resume meiosis and initiate the events that will expel it from the ovary. However, ovulation occurs after about 24 h from LH surge. The animals that utilize this type of ovulation are cat, rabbit, ferret, mink, llama, and she-camel

b. **Spontaneous ovulatory animals:** Most mammals have a **periodic ovulation** pattern independent of mating, in which female ovulates at a specific time of the year. **Environmental cues**, most notably the amount of light during the day play important role in determining the time of breeding in the domestic animals, especially the seasonal breeding animals. In sheep and goats, the decrease in daylight is the stimulus for breeding, therefore, the breeding season for these animals begins in autumn. While the horse and cat are long-day breeders that reproductive activities begin during spring and summer.

Role Of Photoperiod

a. Change in the amount of light falling on the retina produces an action potential that goes to the suprachiasmatic nucleus in the hypothalamus.
b. Then, a descending efferent fiber goes from the suprachiasmatic nucleus, reaches the spinal cord, and emerges to the superior cervical ganglia.
c. Adrenergic post-ganglionic fibers from the superior cervical ganglia supplying the pineal gland stimulate melatonin secretion. Melatonin stimulates the hypothalamus to release gonadotropin-releasing hormone (GnRH). The pattern of melatonin pulses during the day is the determining of the melatonin effect.
d. The GnRH stimulates the pituitary gland to secrete gonadotropin (FSH, LH) which causes the follicle cells to proliferate (folliculogenesis) and secrete estrogen. The estrogen has a dual action: enters certain neurons in the midbrain and evokes the pattern of mating behavior characteristic of the species (signs of estrus). Also, estrogen stimulates the LH surge from the pituitary gland, which is essential for ovulation (Fig. **15**). Ovulation occurs 24-45 h after LH surge except in mare in which ovulation occurs before the peak of LH surge.

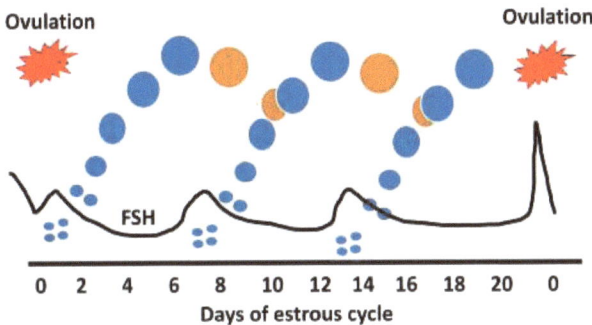

Fig. (15). FSH changes during the estrus cycle.

Luteal Phase

Luteinization

After the secondary oocyte and its associated cells are ovulated, the remainder of the Graafian follicle collapses and becomes folded, some of the ruptured blood vessels leak blood into the follicular cavity, forming a central clot, the resulting structure is called corpus hemorrhagicum. As the clot is removed, by phagocytes, continued high levels of LH convert the corpus hemorrhagicum into a temporary structure known as the corpus luteum (CL), which functions as an endocrine gland. This modified vascularized structure is composed of:

a. Granulosa-lutein cells a modified granulosa cell, which represents about 80% of the cell population of the corpus luteum. It produces progesterone and converts androgens produced by the theca-lutein cells into estrogens.
b. Theca-lutein cells modified theca cells, representing about 20% of the cell population of the corpus luteum. It is specialize in the production of progesterone, some estrogen, and androgens.

Luteolysis

Definition: It is the process by which the CL is regressed, becomes fibrotic, ceases to produce progesterone and consequently a new cycle begins all these are due to fertilization failure. The fibrous connective tissue that forms in place of CL is known as the corpus albicans, which persists for some time before being reabsorbed.

In all domestic animal except dogs and primates, the mechanisms controlling the development, secretion, and regression of the CL involves factors produced both inside and outside it.

1. CL can synthesize **oxytocin** in the same way as it is done in hypophysis. The oxytocin plays a dual role depending on the hormonal milieu:
   i. In the early stages of CL development, oxytocin supports **progesterone** secretion since there is a positive feedback loop between oxytocin and progesterone production in fully developed CL as shown in Fig. **16**.
   ii. During luteolysis, oxytocin can amplify this process by increasing the amount of prostaglandin F2α (PGF2α) secreted by the endometrium of the non-gravid uterus.
2. In addition to oxytocin, another factor known as tumor necrosis factor- α (TNFα), which affects both $PGF_{2\alpha}$ (luteolytic factor) and PGE_2 (luteotropic factor) output during the mid-and late luteal phase.
   A. If fertilization occurs, a relatively high proportion of luteotropic PGE_2 is produced which is needed for proper embryonic development, recognition, and establishment of pregnancy.
   B. If fertilization does not occur, $PGF_{2\alpha}$ output becomes greater than PGE_2, resulting in luteolysis. Therefore, TNFα has also a dual function: luteotropic and luteolytic effects.

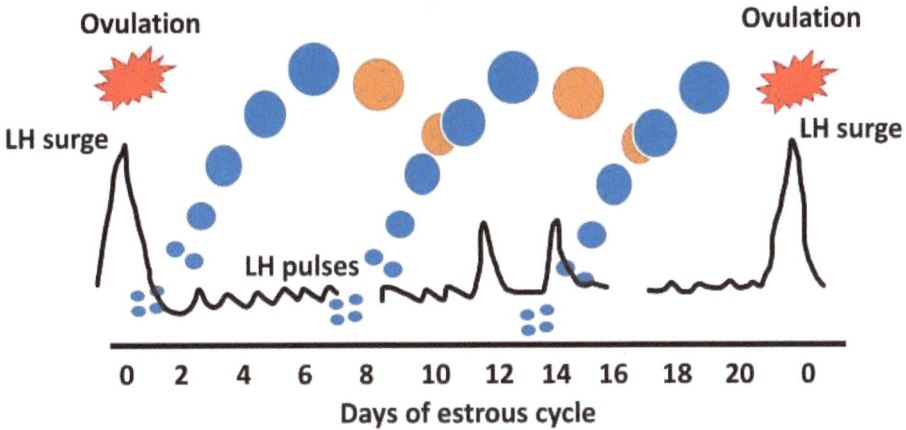

Fig. (16). LH changes during the estrus cycle.

The pathway of $PGF_{2\alpha}$ in reaching the ovaries in the mares differs from cow and mare. In mares, there is no direct unilateral effect of the $PGF_{2\alpha}$ from the uterus to the ovaries, since the uterine artery is relatively straight and caudal to the ovarian vein as it passes obliquely across the ovarian vein. Thus, in mares, the regression of CL appears to be a systemic pathway. In comparison to the mare, the local utero-ovarian pathway of $PGF_{2\alpha}$ is the mechanism of CL regression in the cow and ewe.

In Primates

The normal luteotropic factor is LH, while the luteolytic factor is the high level of estrogen. Since late in the luteal phase, LH is low and estrogen is high. In addition, if fertilization occurs, the conceptus secretes a luteotropic factor (human chorionic gonadotropin, hCG). In bitch, the presence of conceptus is the determining factor for prolonging the life span of CL. Thus CL regresses in the absence of pregnancy, and the bitch enters in a state of physiological anestrus till the second breeding season. In rabbits, the luteotropic factor is the prolactin and estrogen, luteolysis occurs as a result of a decrease in prolactin levels [8].

Reproductive Cycle

(Estrus Cycle and Menstrual Cycle)

Definition: These are cycles in which progressive alterations in the female genital tract occur as a result of ovarian changes. In primates and women, the reproductive cycle is known as the menstrual cycle due to the presence of a phase of bleeding.

In animals, the reproductive cycle is called the **estrus cycle** due to the period of heat (**estrus**) at the time of ovulation, normally the only time during which the sexual interest of the female is aroused. In spontaneously ovulating species, there is no episodic vaginal bleeding but the underlying endocrine events are essentially the same as those in the menstrual cycle.

Estrus Cycle

Phases of the Estrus Cycle

Proestrus: it is characterized by the following:

i. It coincides with the ovarian follicular phase, the ovarian follicles grows rapidly under the effect of pituitary FSH.
ii. It begins at days 17-21 of the previous cycle in cattle.
iii. Increase in estrogen secretion by the growing follicles.
iv. Increase thickness of the mucosal layers of the uterus and vagina.
v. Increased uterine vascularity and contractility with serous endometrial secretions.

Estrus (heat)

i. High levels of estrogen are secreted which stimulates (1) the preovulatory surge of LH that leads to ovulation; and (2) stimulates behavioral estrus (receptive to male).
ii. All domestic animals ovulate during this phase except cattle.

Metestrus

i. It coincides with the ovarian luteal phase, only cows ovulate at the beginning of this phase.
ii. This phase is considered a short transition phase after ovulation with decreasing estrogen and increasing progesterone levels. These high progesterone levels inhibit FSH/LH secretion to block further ovulations in monovular species; inhibit uterine contractility; stimulate endometrial glands to produce uterine milk.
iii. Recently ruptured Graafian follicle is transformed into CL.

<u>Diestrus</u>

i) Is a continuation of the luteal phase. CL becomes fully developed and progesterone exerts a dominant influence on the oviducts, uterus, vagina, pituitary, and hypothalamus.

i. CL remains functional for about 13 days in ewe and 15-17 days in sow, cow, and mare. If fertilization does not occur, CL will regress at the end of this phase.

<u>Anestrus (Anestrum)</u>

It is not a normal phase of the estrus cycle, but it is a period between the diestrus and proestrus in some domestic animals. It is characterized by quiescence of the reproductive tract. Its length varies according to various factors including the physiological state of the animal, nutrition, and species of the animal. Anestrus may be physiological or pathological:

A. Physiological anestrus is found during pregnancy in spontaneous ovulators; in unmated induced ovulators like a cat, ferret, and she-camel; in seasonal monoesters animals like the bitch.
B. Pathological one as that found in case of persistence of CL without pregnancy, that is considered a form of infertility.

Menstrual Cycle

It is the structural and functional changes that occur in the endometrium of primates and humans every month from puberty to menopause. It occurs at the same time as the ovarian cycle and is a result of it. The menstrual cycle is about 28 days in women and 25-35 days in primates, it stops during pregnancy or diseases. It is divided into three phases:

<u>Destructive Phase</u>

This phase is characterized by bleeding lasting for 3-6 days. The volume of lost blood is 50-150 ml. As a result of CL regression, a drop of progesterone levels leads to vasoconstriction of spiral arterioles supplying the endometrium, which suffers from ischemia and then necrosis. Later on, the arterioles dilate and blood rushes through them and leaks from their ischemic walls to collect under the necrosed endometrium, then separated into the uterine cavity.

Proliferative Phase (Follicular Phase)

Due to increasing estrogen levels from the growing follicles, this phase is under the control of estrogen. The surface epithelium is regenerated from the epithelium of the uterine glands that elongate and become tortuous and the blood vessels become more spiral. The endometrial stroma proliferates and the endometrial thickness increases. This phase takes about 10 days.

Secretory Phase (Luteal Phase)

The changes that occur during this phase are produced under the effect of progesterone secreted by CL. In addition, these changes provide a suitable environment for the implantation of the fertilized ovum. The uterine glands, blood vessels, endometrial thickness become more prominent and the uterine glands become secretory. The duration of this phase is about 14 days. In primates, ovulation occurs during the mid-cycle. The best time of insemination is 2-3 days before the expected time of ovulation [1].

Control of Reproductive Cycle

Before Discussing the interplay between the different neuroendocrine factors during the ovarian and reproductive cycles, you should understand the following:

1. The synthesis and release of gonadotropins (FSH, LH) are dependent on the hypothalamic factor (gonadotropin-releasing hormone, GnRH). Both gonadotropin and GnRH are secreted in a pulsatile manner.
2. The ovarian hormones regulate gonadotropin secretion through:
   a. They have a depressant effect on gonadotropin output induced by higher estrogen, progestagens, and inhibin
   b. An increase of LH secretion induced primarily by estrogens

The Estrogen Action is Characterized by:

a. Very low estrogen levels as it occurs during the proestrus phase leads to FSH secretion. There is a negative feedback between estrogen and FSH, rising estrogen level begins to suppress FSH secretion.
b. If plasma levels of estradiol increase greatly, LH secretion is enhanced leading to a **surge** of LH. The net result is positive feedback.

Thus estradiol has a dual function: low levels lead to negative feedback control on FSH; while moderate-high levels lead to positive feedback leading to stimulation of LH secretion.

Progesterone also Regulates FSH and LH Secretion Through

(a) It enhances the negative feedback effect of estrogen on FSH and LH secretion.

(b) It blocks the positive feedback effect of estradiol on FSH and LH secretion.

We mentioned previously that the phases of the ovarian cycle are the follicular phase (folliculogenesis); ovulation; and luteal phase (secretory phase). Also, the phases of the estrus cycle are proestrus, estrus, metestrus, and diestrus. You should remember that the first day of behavioral signs of estrus is considered the first day of the estrus cycle.

I. During Follicular Phase: At the end of the luteal phase (beginning of proestrus): (see Figs. **15 - 19**).

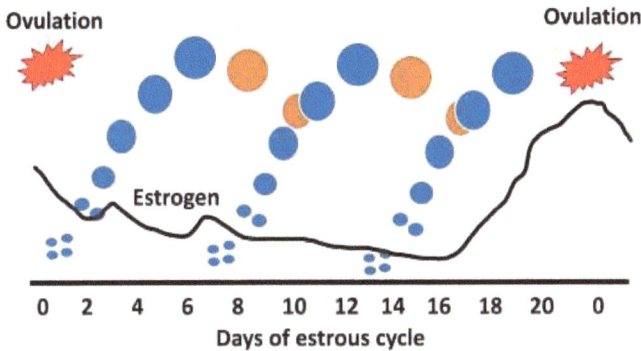

Fig. (17). Changes in estrogen levels during the estrus cycle.

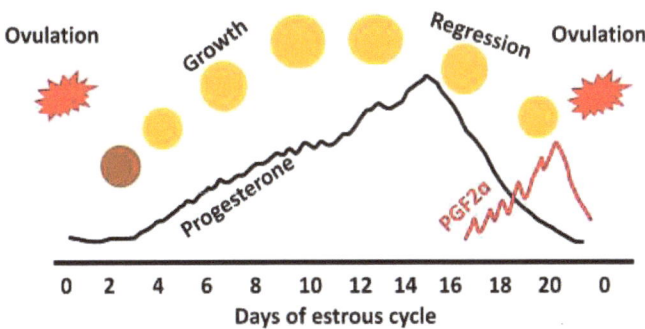

Fig. (18). Changes in progesterone level during the estrus cycle.

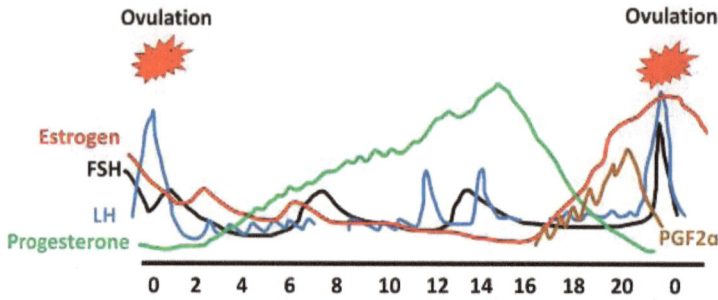

Fig. (19). Summary of Hormonal changes during the estrus cycle.

1. The levels of luteal progesterone, estrogen, and inhibin fall greatly.

2. FSH levels rise followed by a rise in LH, both permit antral growth of the follicle to proceed.

3. This leads to increase the output of estrogens, androgens, and inhibins; and these changes consequently lead to suppression of FSH while LH reaches a plateau.

4. The decrease in FSH and increase in LH leads to follicle selection that results in a further rise in estrogen (estrogen surge).

5. The estrogen surge triggers:

  i- The behavioral signs of estrus (day 1 of the estrus cycle).
  ii- An LH surge through positive feedback with estrogen that leads to ovulation.

II- Ovulation: At ovulation inhibin, androgen, and estrogen output fall, while progesterone levels rise. These hormonal changes lead to falling in LH and FSH secretion.

III- Luteal phase: It is characterized by, Rise in plasma progesterone that reaches the highest level about 8 days after LH surge. The growth of antral follicles is suppressed. At the end of the luteal phase, if conception has not occurred estrogens, progesterone and inhibins decline at luteolysis; the negative feedback is relaxed, and consequently, FSH and LH levels rise and another cycle starts.

Hormonal Control of Ovarian and Reproductive Cycle

These include:

(A) **Primary (specific) hormones,** which affect the gonads and the genital system. They include hypothalamic hormones (GnRH); pituitary hormones (FSH,

LH, prolactin); ovarian hormones (estrogens and progesterone, relaxin, inhibin, and activin).

(B) **General metabolic hormones** that affect the general metabolism and consequently the general health of the animal. They include GH, thyroxin, and cortisol.

Hypothalamic Hormones

1. GnRH: In mammals, GnRH is synthesized by hypothalamic neurons, transported by axonal flow, and stored in the median eminence. GnRH is released in pulses (about every 2 hours) and causes the liberation of two pituitary gonadotropins (FSH and LH). These two gonadotropins participate in the control of reproduction in male and female mammals. Secretion of GnRH is increased during coitus in the cat and is followed by an LH surge after 15 minutes. Pheromones (airborne or water-borne) and light cycle play important role in the control of GnRH secretion.

Pheromones

These are substances that are secreted by one sex and when detected by the other sex it induces activation of its sexual derives. In some males, special glands are located in the head region near the horns called pole glands. These glands secrete volatile substances that can be smelt by females and increase their sex drive. Also, the sloughed keratinized epithelial lining the vagina during the heat period become decomposed by the action of flora inhabiting the vagina and the result is a wide variety of volatile fatty acids specific for that species. These volatile fatty acids are a signal for the male to copulate with that female. In bitch a special substance called methylparahydroxybenzoate is produced that specifically attract the males to the females [9].

Oxytocin

Source: It is synthesized by the supraoptic and paraventricular nuclei in the hypothalamus. It is also produced in the corpus luteum of ruminants (cow and ewe), primates, and women.

Release: oxytocin is completely under reflex control of the hypothalamus. Afferent impulses reach the hypothalamus from:

a. Mechanical receptors in teats during suckling. The suckling reflex is the most powerful stimulant of oxytocin secretion.

b. Receptors in the vulva during coitus.

c. Receptors in the wall of the distended gravid uterus at the end of pregnancy and beginning of parturition.

The efferent of these reflexes is the hypothalamo-hypophyseal tract through which oxytocin is transported to the posterior pituitary and stored there till released at need.

Anterior Pituitary Hormones

Follicle-stimulating Hormone (FSH)

It plays the following functions during folliculogenesis:

1. Although the initial follicular growth (recruitment) is independent of FSH, it is required to support and prevent their apoptosis.

2. It is responsible for cyclic recruitment *i.e.* growth of follicles towards the preantral stages.

3. It initiates steroidogenesis (aromatization of granulosa androgen to estrogen)

4. FSH stimulation along with the estrogen environment allows LH receptor formation on primed granulosa cells in the dominant follicle in preparation for LH surge. (N.B. the dominant follicle has more FSH receptors to respond to the decreasing level of FSH).

5. It stimulates the synthesis of inhibin B (in primates) during the early follicular phase.

6. It up/down-regulates its receptors.

Luteinizing Hormone (LH)

1. It stimulates theca cells to continue to make androgen for granulosa conversion to estrogen/progesterone

2. It stimulates the oocyte to resume meiosis (but not complete until sperm enters and the 2nd polar body is released).

3. It plays an important role in follicle selection and dominance.

4. LH surge is responsible for ovulation.

5. It controls inhibin A (in all animals) during the luteal phase.

6. It stimulates corpus luteum formation and initiation of progesterone production.

Ovarian Hormones

A. Protein hormones: They include Inhibins, activins, and relaxin.

Inhibins: as mentioned before (see male reproduction) that there are different types of inhibins and activins. Both inhibin A and inhibin B are present in the ovarian follicles of primates and women. In ruminant species, inhibin B is not present but only inhibin A is present.

In primates, Inhibin B is synthesized by the growing follicles and reaches its peak at the mid-follicular phase. Inhibin A is a product of the dominant follicle, therefore, it increases in the late follicular phase, and reaches its peaks at the mid-cycle and subsequently in the mid-luteal phase.

Therefore, you can conclude the following:

a. In small antral follicles, inhibin B is an immediate product of FSH stimulation, and its increase results indirectly from FSH-stimulated follicle growth. Inhibin B secretion from small antral follicles is enhanced by estrogen.
b. Inhibin A is stimulated at a later stage of follicular growth and becomes the predominant form of inhibin secreted in response to both FSH and LH from the preovulatory follicle.

Thus, the increase in both FSH and estrogen in the early follicular phase results in an increase in inhibin B whereas, FSH with the selective LH rise in the late follicular phase favors inhibin A secretion [10].

Activins: Activin A is a granulosa cell-derived factor and functions as an intrafollicular modulator of both granulosa cells (autocrine and paracrine actions) and theca cells (paracrine action). On the other hand, activin B is present in the ZP layer, being synthesized by the oocyte itself, and the pituitary, being having a paracrine role in regulating FSH secretion. The main functions of activins include the following:

a. Promote cell proliferation
b. Up-regulate FSH receptors
c. Activin B, which is present in ZP until fertilization, is proteolytically cleaved by the lytic enzymes released from the cortical granules. The specific localization of activin B in ZP of immature or mature oocytes and its absence

after fertilization could imply a regulatory role for activin B in sperm binding or fertilization [10].

d. Activin A stimulates oocyte development and maturation; this action is opposed by follistatin.

e. Activin B-follistatin network in pituitary regulates FSH secretion during the estrus cycle. For example, estrogen may directly influence FSH secretion through local modulators such as activin B and follistatin.

f. Enhance FSH-dependent processes by granulosa cells such as inhibin production and steroidogenesis; while it suppresses LH-induced androgen production by theca cells.

Estrogens and Progesterone

Functions of Estrogen

Actions on the Sex organs

i. It stimulates the growth of the secondary sex organs *e.g.* uterus, fallopian tubes, vagina, external genitalia except for the clitoris, which is developed by androgens.

ii. It increases the number of contractile proteins and oxytocin receptors in uterine myometrium *i.e.* It sensitizes the uterine muscle to the action of oxytocin at the end of pregnancy.

iii. It increases the vascularization and the rhythmic contraction of fallopian tubes.

iv. It relaxes (opens) the cervix that facilitates the passage of sperms during estrus and facilitates delivery of the fetus during parturition. It stimulates the cervix to secrete thin watery secretions

v. It increases the growth (increases the thickness) of the vagina. It stimulates glycogen deposition in the mucosal cells of the vagina, and the glycogenolysis yields lactic acid, which makes acidic vaginal secretion.

vi. It is responsible for the growth of the duct system, stroma, and blood vessels of the mammary gland.

Action on the Female Sex Characters

i. It is responsible for the appearance of estrus signs in animals.

ii. It is responsible for the characteristic distribution of the stored fat that gives the feminine character.

iii. It does not stimulate the growth of vocal cords, so the female keeps the high-pitched sharp voice.

iv. It is responsible for the development of the broad pelvis.

General Metabolic Actions

i. It has anabolic action through stimulation of BMR and enhancing growth; and produces positive nitrogen balance.
ii. It stimulates Na^+ and water in the body during pregnancy.
iii. It has a potent effect on bone growth. It increased Ca^{2+} deposition in bone by increasing the activity of osteoblasts (bone-forming cells). It hastens the closure of the epiphyseal plate of the long bone.
iv. It has a potent stimulant for prolactin production.

Functions of Progesterone

1. It is responsible for uterine changes that occur during the secretory phase. These changes occur after the uterus has been prepared by estrogen. These changes include: the uterine glands become larger, more tortuous, and full secretory; the spiral arterioles become more coiled and develop to reach endometrium; the endometrium increases in thickness

2. It helps implantation of the fertilized ova by providing a suitable environment, maintaining the secretory phase of the endometrium (production of uterine milk, for zygote nutrition), and preventing further ovulation and estrus cycle during pregnancy.

3. It stimulates the growth of glandular parts of the mammary gland.

4. It depresses the myometrial activity and provides a quiescence state in the uterus *i.e.* maintains pregnancy.

5. It is responsible for a tight closure of the cervix during pregnancy

6. It stimulates the vagina to secrete thick mucous.

7. It has thermogenic action through increasing resting metabolic rate.

Uterine Prostaglandin

a. Both PGE and $PGF_{2\alpha}$ play important roles in the ovulation process. PGE induces hyperemic response whereas $PGF_{2\alpha}$ helps the contraction of smooth muscles of the Graafian follicle resulting in its rupture.
b. PG has an important role in the motility of the oviduct during ovulation to capture the ovum.
c. PGs play a role in fertilization by liquefying the cervical mucous making it more receptive to the sperm movement; and by enhancing the antiperistaltic

contractions of the uterus and fallopian tubes, helping the backward movement of sperms toward the fallopian tube where fertilization occurs.

d. $PGF_{2\alpha}$ has an important role in the regulation of the estrus cycle through CL regression (luteolysis). Clinically it is used in synchronization of the estrus cycle, *i.e.* making all the females in a herd come into the estrus phase at one time.

e. $PGF_{2\alpha}$ plays an important role in parturition being magnifying the oxytocin effect on uterine contractions (Ferguson reflex, see parturition) and stimulating the abdominal and the uterine muscle to vigorously contract to expulse the fetus.

Pregnancy

The main events of pregnancy include the following:

1. Fertilization of the ova

2. Maternal recognition of pregnancy

3. Implantation of the fertilized ova in the uterine wall.

4. Embryonic development and differentiation.

5. Placentation.

Before discussing these events, you should know about the ovulated oocyte (its structure), this is because each component of its structure has an important role in successful fertilization.

Structure of the Ovulated Oocyte

Cytoplasm: All the material necessary for the beginning of the growth and development must be stored in the cytoplasm of the mature ovum. So the egg has a remarkable cytoplasmic storehouse that is has accumulated during its maturation. The cortical granules are structures, homologs to acrosomal vesicles of the sperm, are Golgi-derived organelles containing proteolytic enzymes; mucopolysaccharides; adhesive glycoproteins; and hyalin protein. The enzymes and mucopolysaccharides are active in preventing other sperms from entering the egg after the first sperm has entered. The hyaline and adhesive glycoproteins surround the early embryo and provide support for the cleavage-stage blastomeres.

Zona Pellucida (ZP): It is an amorphous substance separating the egg from the surrounding follicular cells. The ZP is composed of glycoproteins, ZP_1, ZP_2, and

ZP_3 secreted by the egg. There is a gap junction between the egg and ZP; and between follicular cells and ZP. Therefore, there is a communication between the egg and the follicular cells (Fig. **20**).

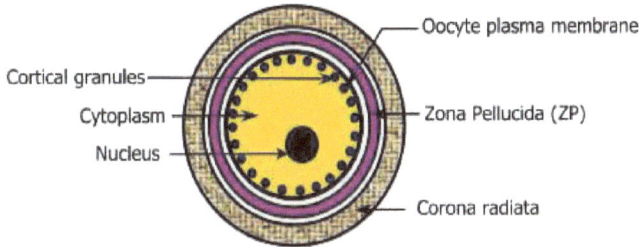

Fig. (20). Structure of the ovulated mature oocyte.

Cumulus: The egg is surrounded by a layer of cells called the cumulus, which is made of the ovarian follicular cells that were nurturing the egg at the time of its release from the ovary. The innermost layer of cumulus cells, immediately adjacent to the ZP, is called the corona radiata.

Fertilization

The events of fertilization usually include: (1) contact and recognition between sperm and egg; (2) regulation of sperm entry into the egg; (3) fusion of genetic material from the two gametes; (4) cortical reaction to prevent polyspermy and activation of egg metabolism to start development (Fig. **21**).

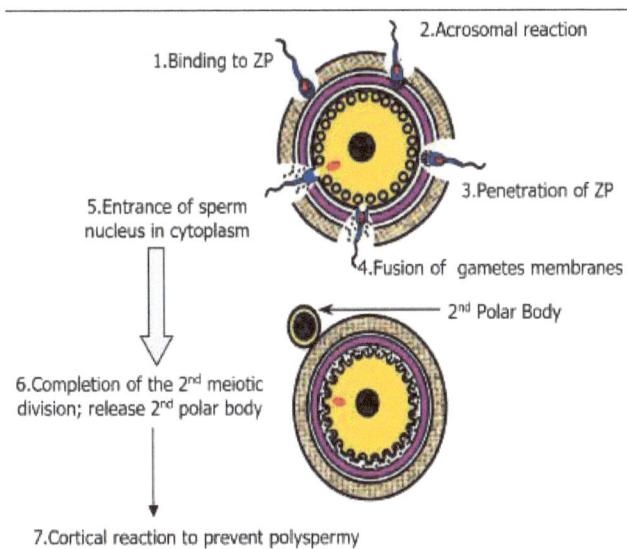

Fig. (21). Mechanism of ovulation.

1. **Contact and recognition between sperm and egg:** The ZP3 molecules of the zona pellucida have two regions: (a) the sperm receptor that recognizes integral proteins of the sperm plasmalemma; and (b) the other region binds to receptor proteins located in the head of the sperm, triggering the acrosomal reaction. Thus, the recognition of the sperm and acrosomal reaction occurs only after the sperm has bound to the ZP. This reaction results in the release of the acrosomal enzymes into the ZP. The liberated enzymes, acrosin, digest the ZP permitting the flagellar movement of the sperm to propel the sperm toward the oocyte. Once the sperm penetrates the width of ZP, it enters the perivitelline space, located between the ZP and the oocyte cell membrane, and can reach the oocyte (Fig. **21**).

2. **Regulation of sperm entry into the egg:** Fusion between sperm and egg is mediated by protein molecules whose hydrophobic group merges the sperm and egg plasma membranes. In mammals, fertilin proteins in the sperm bind to integrins in the egg and allow the membranes to fuse.

3. **Fusion of genetic material from the two gametes.** The entry of the sperm nucleus in the cytoplasm of the secondary oocyte triggers it to resume and complete its second meiotic division. This results in an unequal division of the cytoplasm, forming two haploid cells, the ovum, and the second polar body. The ovum nucleus (female pronucleus) and the sperm nucleus (male pronucleus) migrate toward each other, and finally fuse forming a zygote with the diploid number of chromosomes and thus completing the events of fertilization.

1. **Cortical reaction** to prevent polyspermy and activation of egg metabolism to start development. The contact between the sperm and the oocyte is responsible for the cortical reaction, which prevents polyspermy, the process of more than a single sperm fusing the egg. The cortical reaction has two mechanisms in preventing polyspermy:
   a. Fast block to polyspermy: It involves a change in the resting membrane potential of the oocyte plasma membrane that prevents contact between the oocyte and another sperm.
   b. Slow block to polyspermy. It involves the release of the contents of numerous cortical granules located in the oocyte's cytoplasm into the perivitelline space. Enzymes within the cortical granules act to hydrolyze ZP_3 molecules, the sperm receptors, in the ZP, thus preventing additional sperm from reaching the oocyte. The mechanism of the cortical granule reaction is similar to that of the acrosomal reaction. Upon fertilization, the intracellular calcium ion concentration of the egg increases greatly. In this high calcium ion environment, the cortical granule membrane fuse with the egg plasma membrane, releasing their contents.

Maternal Recognition of Pregnancy

For most species, a critical need arises early in gestation for the mother to recognize that she is pregnant. Progesterone levels in maternal blood must remain at a high level to keep endometrial maintenance in a state favorable to embryonic survival. The conceptus must signal its presence to the mother to prevent luteolysis of the corpus luteum. The need to prolong luteal life is accomplished by:

a. Production of positive luteotropic factors (*i.e.* hormones that maintain the corpus luteum and act like LH).

b. Prevention of normal luteolytic factors (*e.g.* prevent the production of uterine $PGF_{2\alpha}$)

    I. **In primates and humans:** The blastocyst prolongs the life of the corpus luteum by producing a luteotropic factor known as human chorionic gonadotropin (hCG). hCG is synthesized by the syncytiotrophoblast of the implanting blastocyst and can be released into maternal circulation as early as 6-7 days post-fertilization. This hormone binds to LH receptors on the luteal cells and exerts a luteotropic action and prevents luteolysis.

    II. **In large domestic animals:** Luteolysis is normally induced by uterine $PGF_{2\alpha}$. In addition, chorionic gonadotropins are not detected in the blood of pigs, cows, and sheep bearing pre-implantation conceptus. However, in pigs, the conceptus produces estrogens, which help prolong luteal life. In cattle and sheep, the conceptus secretes a soluble chemical known as interferon-τ (IFNτ), which modifies uterine secretion of PGF2α. The equine has an unusual mode of controlling pregnancy: (a) Like cows, the conceptus secretes interferon-τ (IFNτ); (b) Equine fetal gonads also produces estrogen in mid-gestation; and (c) Equine is the only domestic species to have a chorionic gonadotropin, named eCG (the old name is pregnant mare serum gonadotropin, PMSG). eCG is secreted by transient structures called endometrial cups and acts to maintain the primary corpus luteum, and induce the formation of several secondary corpus luteum.

    III. **In cat and bitch** In cat. The corpus luteum lasts for 35-45 days after ovulation, whether pregnant or not. The placenta extends luteal function for the rest of pregnancy (about 63 days). In bitch, the presence of conceptus prolongs the luteal life [9].

Embryonic Development and Differentiation

As the zygote (fertilized ova) continues its journey through the oviduct on its way to the uterus, it undergoes numerous mitotic divisions, becoming the spherical clump cells known as the morula. With further divisions and modifications, the morula is transformed into the blastocyst, composed of a hollow ball of cells. The lumen of the blastocyst contains fluid and a few cells at one pole. The peripheral cells are known as trophoblasts, and the cells trapped inside the blastocyst are the embryoblasts. The blastocyst enters the uterine cavity about 4 days after fertilization and begins to embed itself into the uterine wall, a process known as implantation. In bovine, the cleavage rate of the zygote from one cell to 2, 4, 8, 16 cells is approximately every 20-h. From fertilization through blastocyst the conceptus is within the ZP, which ruptured (hatch) during the implantation process (Table **2**).

Table 2. The time implantation, luteal regression, and gestation period in different animal species.

Species	Embryo Attachment	Luteal Regression	Duration of Pregnancy
Cow	30-45 days	16-18 days	282 days
Ewe	16	12-13	150-155
Mare	35-105	14-15	330
Women	6-8	12-14	270
Bitch	12-14	61	61
Pig	18	14-16	114
Rabbit	7-8	12*	29
Rat	4-5	10-12*	22

*=pseudopregnancy.

Implantation

There are two types of implantation:

a. **Invasive implantation:** It means that the blastocyst invades and erodes maternal tissues *i.e.* the blastocyst burrows into the endometrium. It occurs in humans, most primates, bitch, cats, rabbit, and mouse
b. **Non-Invasive implantation:** It means that all tissue layers remain intact and the blastocyst essentially attaches to the surface of the endometrium. It occurs in pigs, ewe, cows, and mare.

Role of Steroid Hormones in Implantation

Implantation depends on ovarian steroid support, progesterone and estrogen.

a. Progesterone domination is required if the uterus and implanting blastocyst are to engage. Progesterone maintains the secretory activity of the endometrial glands (uterine milk production) while suppressing myometrial contractions.

b. Estrogen affects the uterine epithelium in two ways: It stimulates the release of glandular secretions, which stimulate the growth of the blastocyst, and increases the responsiveness of the epithelial cells to the blastocyst. In most species, estrogen comes from the ovary, but in some species, it comes from the conceptus (e.g pig).

Nutrition During Embryonic Development

a. The early embryonic stage (from zygote till morula) depends totally on its maternal cytoplasmic inheritance.
b. Blastocyst stage: The blastocyst derives its nutrients from uterine secretions (uterine milk) that bathe it; this is called histotrophic nutrition. As the blastocyst develops further into a growing fetus, need to switch to a more efficient mode of nutrition-placental nutrition (hematotropic nutrition).

Placentation

The placenta develops from the chorion (fetal origin) and decidua basalis (maternal origin).

Placental Classification

According to the Degree of Separation of Maternal and Fetal Tissues

The classification of the placenta depends upon the number of cell layers that form the fetoplacental unit. Normally 6 layers separate the blood of the mother from that of the fetus. These layers include (a) maternal layers (endometrial epithelium, endometrial interstitium, and endometrial endothelium); (b) fetal layers (chorionic epithelium, chorionic interstitium, and chorionic endothelium).

a. Epitheliochorial placenta in which, all the six layers are present; this type present is in mare, sow, and ruminant.
b. Endotheliochorial placenta, in which the 3 fetal layers and only 2 maternal layers are present like bitch and cat.
c. Hemochorial placenta, in which the 3 fetal layers are present with no maternal layers. The fetal membranes swim in the maternal blood like primates.
d. Hemoendothelial, in which there are no maternal layers but there is only one fetal layer as in rabbit.

According to the Mode of Attachment Between the Placenta and the Uterus

a. Diffuse placenta, forms over a broad area as in equine, camel, and sow.

b. Cotyledonary: placenta forms in focal areas "cotyledons" as in cow, buffalo, sheep, and goat.

c. Zonary: placenta forms in a band as in bitch and cat.

d. Discoidal: the shape of the placenta is discoid as in primates, rabbits, guinea pigs, rats, and mice.

Functions of Placenta

A. **Respiratory function:** O_2 and CO_2 pass across the placenta by simple diffusion. The fetal hemoglobin has more affinity and carrying capacity for oxygen than adult hemoglobin.

B. **Nutritive function:** The transfer of nutrients from the mother to the fetus is achieved by: (a) simple diffusion, *e.g.* water and electrolytes; (b) facilitated diffusion, *e.g.* glucose; (c) active transport, *e.g.* amino acids; and (d) pinocytosis, *e.g.* large protein molecules.

C. **Excretory function:** Waste products of the fetus as urea are passed to maternal blood by simple diffusion through the placenta.

D. **Barrier function:** Placental barrier is incomplete. It allows the passage of antibodies (IgG only), hormones, antibodies, sedatives, some viruses. Substances of large molecular size as heparin and insulin can not pass. In herbivores, the placenta does not transfer antibodies during fetal life. So the new borne is free from maternal antibodies and should receive the colostrum on the first day after delivery.

E. **Endocrine function:** The placenta acts as an endocrine gland, which secretes several hormones essential for the maintenance and termination of pregnancy. These hormones are:

Chorionic Gonadotropins

The main chorionic gonadotropins secreted during pregnancy are hCG in primates and women; and eCG in equines. The differences between the two hormones are summarized in Table **3**.

Table 3. The main differences between hCG and eCG.

hCG (human and primates)	eCG, PMSG (equines only)
Produced from the trophoblast layer of the embedded fertilized ova.	Produced from the cuplic structures in the endometrium
Produced 10 days after fertilization; reaches maximum levels after 10 weeks; then decline and remain low	Produced 30 days after implantation; reaches a maximum at mid-gestation
Can be detected in both urine and serum samples	Can be detected only in serum samples

hCG (human and primates)	eCG, PMSG (equines only)
It is LH in nature and weak FSH effect	It is FSH in nature
Actions: It prevents luteolysis Stimulates corpus luteum growth Stimulates corpus luteum to secrete estrogen and progesterone till the 10th week of gestation for continual endometrial growth. It acts on the placenta itself from the 4th month to secrete estrogen and progesterone. Suppresses maternal immune function i.e reduces the possibility of fetus immunorejection	**Actions:** It prevents luteolysis Since it is FSH in nature, it may stimulate follicular growth during pregnancy, which leads to an increase in estrogen secretion and the appearance of estrus during pregnancy. Secondary corpus luteum will be formed when these follicles ovulate and more progesterone is produced for further pregnancy maintenance.

Placental Lactogen (PL)

It is a protein hormone secreted from the syncytiotrophoblast in most domestic animals; primates and humans. It is also named chorionic somatomammotropin (CS). It is lactogenic and has a smaller growth-stimulating activity. Structurally, it is a GH-like protein in humans and primates; while it is a prolactin-like protein in rodents and ruminants.

It can be detected in early pregnancy and large amounts are found in maternal blood but do not reach fetal blood. PL is considered maternal GH of pregnancy because maternal pituitary GH is not increased during pregnancy. The main actions include: 1) stimulating lipolysis to increase free fatty acids, which provide a source of energy for dam and fetal nutrition; 2) stimulating fetal and placental growth since its level is proportional to the placental size; 3) stimulating nitrogen, K^+, and Ca^{2+} retention.

Steroid Hormones

Role of Estrogen During Pregnancy

Estrogen is secreted during pregnancy either from the corpus luteum or from the placenta. It has a synergistic effect with progesterone in increasing the vascularity and thickness of the endometrium as well as increasing the metabolic rate of the dam to face the burden of pregnancy. High levels of estrogen in late gestation induce myometrial oxytocin receptors, thus preparing the uterus for parturition. Induces gap junctions in myometrium so it acts as syncytium; increasing communication and coordinating myometrial contractions during parturition. It stimulates the development of the mammary gland during pregnancy since it is considered one of the many hormones necessary for both ductal and alveolar growth in the mammary gland.

Role of Progesterone During Pregnancy

Pregnancy needs a quiescence state in the uterus; closed cervix; and high metabolic rate in the mother. Progesterone is the hormone responsible for establishing these functions: (1) it promotes a tightly contracted cervix; (2) it provides the proper environment for fetal survival (*i.e.* increase the vascularity of the endometrium, stimulates uterine milk secretion in early pregnancy); (3) it suppresses the myometrial contraction induced by estrogen. This suppression effect is called progesterone block. Around the end of gestation, the block is antagonized by rising levels of estrogen to become dominant over progesterone facilitating parturition.

Progesterone secretion is from the placental trophoblast since it uses cholesterol as the substrate derived from the maternal circulation. The placenta is capable of synthesizing adequate and supportive levels of progesterone at a specific time in the different animal species (Table **4**). In some species such as goats, pigs, rabbits and bitch, the placenta never secretes enough progesterone, therefore, they depend upon corpus luteum throughout pregnancy.

Table 4. The time at which the placenta take-over the progesterone support.

-	Time of placental take-over	Gestation Length (days)
Primates	5-8 weeks	Differs
Cow	150 days	270 days
Ewe	50 days	147-150 days
Mare	70 days	340 days
Cat	45 days	63 days
Goat	150 days	150 days
Sow	115 days	115 days
Rabbit	30 days	30 days

Relaxin

It is a polypeptide hormone that is produced by the corpus luteum, uterus, placenta, and mammary gland in primates, women, and rats. During pregnancy, it relaxes the pubic symphysis and other pelvic joints to overcome the increase in uterine volume and softens and dilates the uterine cervix. Thus, during pregnancy, it inhibits uterine contractions and may play a role in the development of the mammary gland; and at the end of pregnancy, it facilitates delivery.

Other placental hormones: The placenta produces hypothalamic and pituitary-like peptides, including GnRH, TRH, CRH, GHRH, somatostatin, ACTH, and TSH. CRH levels become very high in maternal plasma, and peak during labor. In addition, placental ACTH and TSH augment maternal adrenal and thyroid gland activity. The placenta also synthesizes 1,25-hydroxyvitamin D, which helps regulate calcium homeostasis and skeletal formation in the fetus.

Hormonal Control of Pregnancy

Pituitary Hormones

a. **Gonadotropins:** The secretion of pituitary gonadotropins is stopped during pregnancy due to the high levels of estrogens and progesterone (negative feedback).
b. **Prolactin:** It is secreted in increasing amount during pregnancy under the effect of estrogen increase. It is essential for the development of the mammary gland (during pregnancy) and lactogenesis process (after parturition).
c. **ACTH:** The increase of ACTH during pregnancy results in an increase of glucocorticoids (cortisol) from the adrenal cortex. Glucocorticoids help in the mobilization of amino acids from the mother to the fetus to meet the greater demand for protein synthesis and fetal growth.
d. **Growth hormone:** The maternal growth hormone secretion is decreased during pregnancy through negative feedback with PL.

Ovarian Hormones:

See above (role of corpus luteum in the maintenance of pregnancy).

Placental Hormones:

Mentioned before, See above

Other Hormones

a. **Aldosterone** secretion increases throughout pregnancy because the high levels of estrogen induce augmentation of renin and angiotensinogen levels. The higher levels of angiotensin II stimulate aldosterone secretion. This induces a positive sodium balance, which is needed to support a high maternal plasma volume and to build up the extracellular fluid of the fetus.
b. **Thyroxin and Cortisol:** The levels of these hormones increase during pregnancy under the effect of estrogen, which increases their binding globulin. Thyroxin is essential for increasing the metabolic rate of the dam. Cortisol is required for mammary gland development, in addition to its higher metabolic effect.

c. **Parathormone:** Its secretion also increases as this hormone is essential to maintain normal calcium levels as the growing fetus consumes a considerable amount of calcium for bone formation. Estrogen is essential for calcium deposition in the bone while parathormone facilitates calcium withdrawal from the maternal bone to their blood.

CONCLUSION

In this introductory chapter, the physiological concepts of male and female reproductive systems development and functions were illustrated. These important physiological aspects are the basics for the laboratory and clinical applications for assisted reproductive technologies that are fully described in the coming chapters of this book.

CONSENT FOR PUBLICATION

Not applicable

CONFLICT OF INTEREST

The authors declare no conflict of interest, financial or otherwise.

ACKNOWLEDGEMENT

Declared none.

REFERENCES

[1]   Hall J. Guyton and Hall Textbook of Medical Physiology. 2015. 13th Edition. Eds: John Hall.

[2]   Heindel JJ, Treinen KA. Physiology of the male reproductive system: endocrine, paracrine and autocrine regulation. Toxicol Pathol 1989; 17(2): 411-45.
        [http://dx.doi.org/10.1177/019262338901700219] [PMID: 2675292]

[3]   Mawhinney M, Mariotti A. Physiology, pathology and pharmacology of the male reproductive system. Periodontol 2000 2013; 61(1): 232-51.
        [http://dx.doi.org/10.1111/j.1600-0757.2011.00408.x] [PMID: 23240952]

[4]   Barrett KE, Barman SM, Brooks HL, Yuan JX-J. Ganong's Review of Medical Physiology, 26th Edithion. McGraw-Hill Education 2019.

[5]   Haschek WM, Rousseaux CG, Wallig MA. Male reproductive system. 2010; 553-97.
        [http://dx.doi.org/10.1016/B978-0-12-370469-6.00018-0]

[6]   Roberts KP, Pryor JL. Anatomy and Physiology of the Male Reproductive System 1997; 1-21.
        [http://dx.doi.org/10.1007/978-1-4612-1848-7_1]

[7]   Bates GW, Bowling M. Physiology of the female reproductive axis. Periodontol 2000 2013; 61(1): 89-102.
        [http://dx.doi.org/10.1111/j.1600-0757.2011.00409.x] [PMID: 23240945]

[8]   Vogazianou A. Anatomy and physiology of the female reproductive system. 2019; 739-52.
        [http://dx.doi.org/10.1007/978-3-319-99817-6_38]

[9]   Klein B. Cunningham's Textbook of Veterinary Physiology. 6th ed., Saunders publishing 2019.

[10]   Appiah Adu-Gyamfi E, Tanam Djankpa F, Nelson W, *et al.* Activin and inhibin signaling: From regulation of physiology to involvement in the pathology of the female reproductive system. Cytokine 2020; 133: 155105.
[http://dx.doi.org/10.1016/j.cyto.2020.155105] [PMID: 32438278]

CHAPTER 2

Sperm Assessment and Processing

Krishnavignesh Lakshmanan[1], Deepa Onnarakatt[1,*], V. M. Thomas[1], Smisha Sridev[1], Anuja Gopalakrishnan[1] and **Poongothai Muthuswamy[2]**

[1] *Department of Reproductive Medicine, Chennai Fertility Center and Research Institute, Chennai, Tamil Nadu, India*

[2] *Department of Biotechnology, Dr. NGP Arts and Science College, Coimbatore, Tamil Nadu, India*

Abstract: Semen analysis has become a definitive tool for understanding the male fertility factor. The volume of the ejaculation indicates the functioning and patency of testicular organs, whereas the quantity and quality of sperm define one's fertility profile. Thus, semen analysis is the cornerstone for diagnosis as well as for medical management. Semen comes with both cellular and non-cellular components. For the management of male infertility, it is necessary to have better quality sperm. As a result, semen processing has become an integral part of any medically assisted reproduction. There are different standardized preparation methods available that include sperm wash, swim-up technique, pelleting method, density gradient, *etc.* The aim of this chapter is to discuss semen analysis, its relevance in understanding male fertility factors, and the available sperm preparation methods. Based on the literature, it can be concluded that a proper semen assessment, according to the latest WHO guidelines, is mandatory for diagnosing male fertility issues. The appropriate sperm preparation method and good quality of sperms combined with the skill set of embryologists and environmental conditions will be the decisive factor in any IUI/IVF/ART success.

Keywords: Density gradient, Pelleting method, Semen analysis, Semen processing, Sperm wash, Swim-up technique.

INTRODUCTION

The creation begins with conception. Conception is achieved by two germ cells called sperms and oocytes. All conceptions are not achieved naturally, thus resulting in infertility. It turns out to be a global problem since it affects about 20% of couples worldwide. Male and female factors equally contribute to this condition. For the male fertility evaluation, a semen analysis (SA) is the primary screening procedure. It is advised to perform a minimum of two standardized

* **Corresponding author Deepa Onnarakatt:** Department of Embryology, Chennai Fertility Center and Research Institute, 79/129, Nelson Manickam Road, Aminjikarai, Chennai 600029, Tamil Nadu, India; Tel: +91 9840588473; E-mail: deeps688@gmail.com

semen analyses at an interval of 4 weeks apart. SA is repeated since semen parameters vary from time to time in one person as like other fluid parameters. It is also done to neglect technical or manual errors. World health organization (WHO, 2021) instructed to perform at least one diagnostic semen analysis before initiation of a treatment cycle. It has provided guidelines, reference values, and protocols to perform a semen analysis (Table **1**) [1].

Table 1. WHO 2021 guidelines for seminal parameters and reference values [1].

Parameter	Lower reference value
Semen volume	1.4 ml
Total sperm number (10^6 per ejaculate)	39
Sperm concentration	16 M/mL
Total motility (PR + NP, %)	42%
Progressive motility (PR, %)	30%
Vitality (live spermatozoa, %)	54%
Sperm morphology (normal forms, %)	4%
pH	≥ 7.2
Peroxidase-positive leukocytes (10^6 per ml)	<1.0
Immunobead test (motile spermatozoa with bound beads, %)	≥ 50
Seminal zinc (mol/ejaculate)	≥ 2.4
Seminal fructose (mol/ejaculate)	≥ 13
Seminal neutral glucosidase (mU/ejaculate)	≥ 20

Semen processing is a vital part of any IVF (*in vitro* Fertilization) treatment. It can be done for ongoing procedures as well as for cryopreservation for future use. It is directly related to the success of the procedure. A variety of semen preparation methods are followed in IVF laboratories. In this chapter, we are going to discuss various insights into sperm assessment and processing in detail.

WHAT IS SPERM?

Sperm is the fundamental unit of the male reproductive system. Sperm, illustrated in Fig. (**1**) is a haploid microscopic male reproductive cell. It is produced in the seminiferous tubules of testes, matures and stored in the epididymis, and transported through the ejaculatory duct along with the fluids secreted from the accessory glands [2].

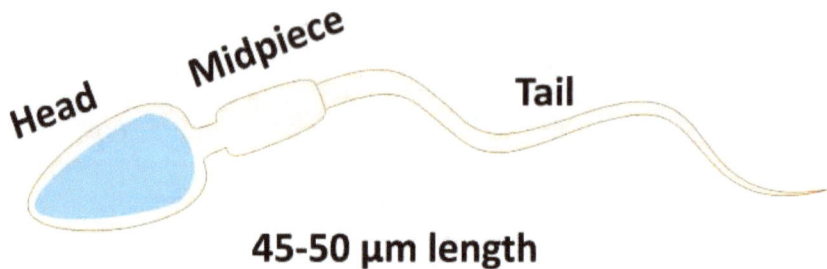

Fig. (1). Normal human sperm.

The head of mature sperm is made up of a well-defined nucleus and acrosome region. The nucleus contains the 23 chromosomes of parental DNA ranging from 30-35% while the acrosome contains proteolytic enzymes that constitute around 65-70%. Unlike most somatic cells, it lacks a large cytoplasm. The head is followed by a mid-piece that contains mitochondria, and a tail facilitates its mobility [3].

WHAT IS THE DIFFERENCE BETWEEN SPERM AND SEMEN?

Semen is a term that is often used interchangeably with sperm, but both are different, where semen refers to the seminal fluid that contains millions of sperm cells along with other plasma fluid. It can look white, yellow, or slightly greyish in color and viscous. The composition of the semen is of mature and immature sperm cells, white blood cells, and other non-cellular components like spermine, prostaglandins, electrolytes, *etc* [4].

Semen Collection

Semen collection is a vital step. Clear instructions should be given to the patients for collecting the entire sample in a wide-mouth non-toxic sterile container (Fig. **2**). The patient can be asked to collect the sample through masturbation in a room near the andrology lab. If the patient faces difficulties in collecting samples, it may be advisable for taking phosphodiesterase (PDE) inhibitors (prescribed based on clinician's advice), home collection in a sterile container or a non-spermicidal condom, or through electroejaculation. After home collection, the sample should be transported to the examining laboratory within 45 minutes (avoid extreme temperature), since the semen analysis should be completed within 1-hour post ejaculation. The semen container can be kept at 37°C or room temperature (RT). The detailed reports should be maintained for every sample (Fig. **3**). All semenology procedures should be carried out in an aseptic condition (Fig. **4**) [1].

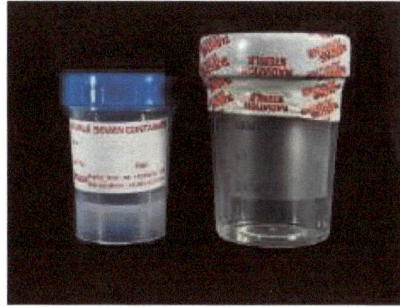

Fig. (2). Semen collection container.

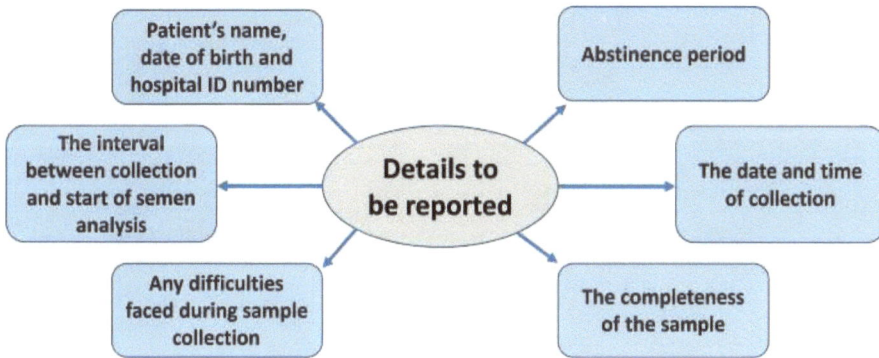

Fig. (3). Details to be reported.

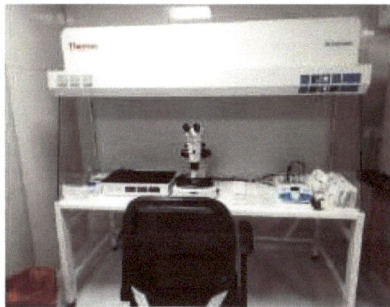

Fig. (4). Semen processing unit.

Semen Analysis

SA is a routine screening procedure performed to examine male infertility through macroscopic and microscopic procedures. It remains the primary tool for the analysis of male infertility. Seminogram plays a vital role in the investigation of male infertility to determine the function of sperm and to provide a reference point for the use of medically assisted reproduction (MAR) [5]. In many cases, there is a chance for a patient to get diagnosed with multiple male factors like Oligoasthenozoospermia, Oligoteratozoospermia, Oligoasthenoteratozoospermia (OAT), Asthenoteretazoospermia, Severe Oligoasthenoteratozoospermia (SOAT), *etc* (Table **2**), so it is important to check all the parameters as prescribed by WHO and report accordingly. A detailed description of the seminogram is depicted in Flowchart 1.

Flowchart 1: Seminogram.

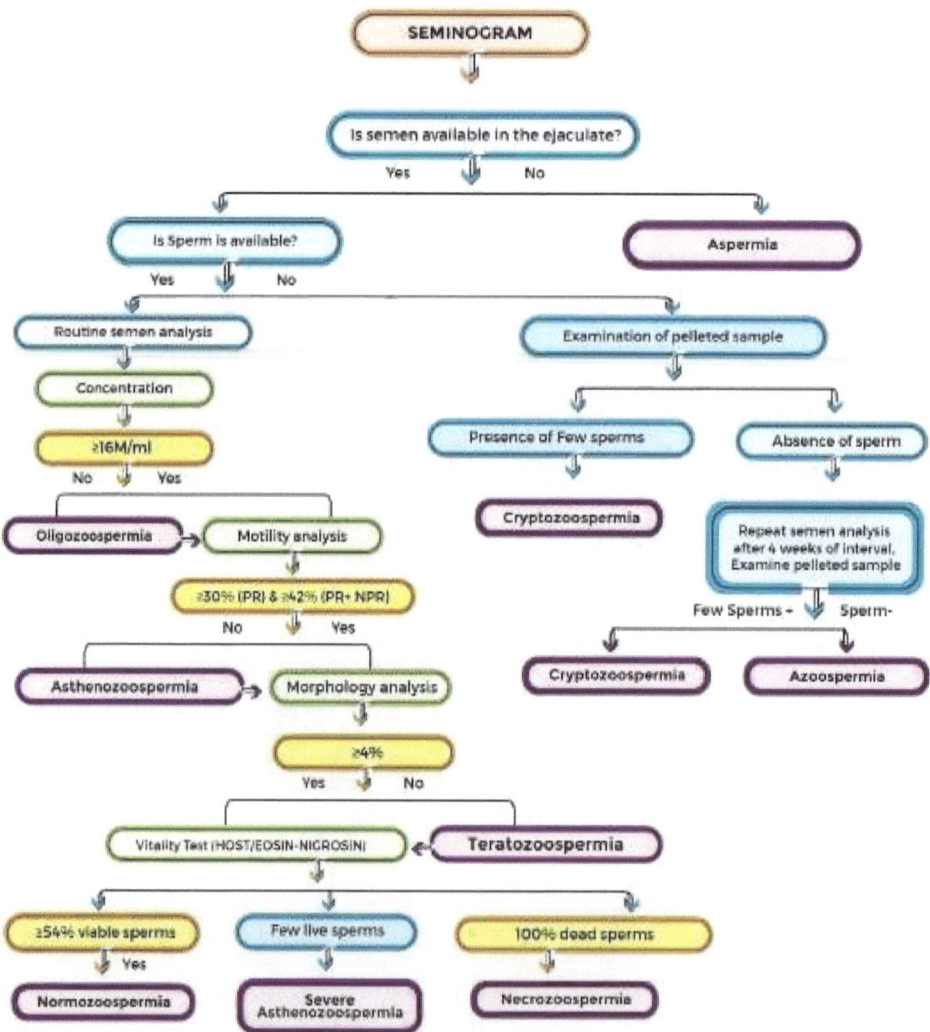

Foot Note: PR- Progressive motile sperms; NPR-Non- Progressive motile sperms; HOST- Hypo osmotic swelling test

Table 2. Terminologies associated with male factor fertility [1, 17].

TERM	DEFINITION
Normozoospermia	Satisfies all the seminal parameters prescribed by WHO 2021 criteria (Concentration, motility, morphology, vitality)
Oligozoospermia	Decreased sperm concentration (<16 million sperm/mL)
Asthenozoospermia	Decreased percentage of motility (<30% (PR) ; <42% (PR+NPR))
Teratozoospermia	Decreased percentage of morphology (normal forms <4%)
Asthenoteratozoospermia	Decreased percentage motility and morphology
Oligoasthenozoospermia	Decreased concentration and motility
Oligoasthenoteratozoospermia (OAT)	Decreased concentration, motility, and morphology
Cryptozoospermia	No sperm seen in the ejaculate, but sperm found in a centrifuged pellet
Hematospermia	Red blood cells observed in the ejaculate
Leukocytospermia, pyospermia	>1×10^6 WBC/ml observed in the ejaculate
Necrozoospermia	All sperms are dead confirmed by vitality test
Globozoospermia	Round spermatic heads, absence of acrosome
Polyzoospermia	Very high concentration of sperms in the semen (>200M/mL)
Hyperspermia	High semen volume (>7.6 mL)
Hypospermia	Low semen volume (<1.5 mL)
Azoospermia	Complete absence of sperm in two centrifuged pellets of semen samples
Aspermia	No semen in the ejaculate

Macroscopic Investigation

The macroscopic investigation provides an overview of the sample. This includes:

Appearance of the Ejaculate

The appearance of the ejaculate is normally opalescent or translucent in color. The sample can appear red-brown it may due to the presence of red blood cells, or yellow in a man with jaundice or with a high number of WBC's or taking certain vitamins or medications. If the sperm concentration is minimal it may appear less opaque [1].

Volume of the Ejaculate

The ejaculated semen volume is measured by the pre-weighed, clean, disposable container. The semen volume is primarily contributed by the prostate gland and the seminal vesicles with a low amount of fluids from the epididymis and bulbourethral glands. Exact volume is important for the semen examination, as it includes the total number of sperms and non-sperm cells.

The volume of the semen sample can be calculated as

Mass = Weight of the collection container with semen (sample) **(B)** –

A pre-weighed, clean, semen container **(A)**

The density of the semen can be assumed to be 1 g/ml (range 1.04-1.10g/ml) [6].

Formula for calculating **Semen volume = Mass/Density**

It also can be measured using a wide-mouth modified graduated glass measuring cylinder. As per recent WHO guidelines the normal range of semen volume should be ≥1.4 ml. If the ejaculated sample is lesser than the above-mentioned volume it is described as "Hypospermia", whereas when the sample exceeds the upper limit it is termed as "Hyperspermia".

Liquefaction

Semen is a thick gel at the time of ejaculation. The complete human semen usually liquefies conversion of ejaculated semisolid coagulated semen sample into completely homogeneous watery liquid this was done by proteolytic enzymes, within 15 minutes at RT, although in some cases it may take up to 1 hour [6]. If it is not liquefied within 1 hour then enzymatically (using bromelain) or mechanically (aspiration of the sample in and out repeatedly using 18 or 19 gauge needle attached to a syringe) it can be liquefied. Delayed liquefaction may indicate an infection or problem with the male accessory glands.

Viscosity

Semen viscosity measures the seminal fluid's resistance to flow. High viscosity is determined by the elastic property of the semen sample when it is free dropped from a pipette. A thread length greater than 2 cm is considered viscous (Fig. **5**). High viscosity after complete liquefaction can affect motility and concentration. In cases of moderate or severe viscosity, a preweighed vial of trypsin powder (5 mg) can be used to break the viscosity.

Semen pH

The semen sample is primarily made up of the alkaline fluid from the seminal vesicles and the acidic fluid from the prostate. The pH of the two fluids combined should be in the range of 7.2–8.2. The normal pH of the semen sample ranges from 7.2 to 8.0. A pH of <7.0 may indicate an obstruction in the ejaculatory duct. This results in an absence of alkaline secretions from the seminal vesicle fluid. A higher pH could indicate the presence of an underlying infection [1].

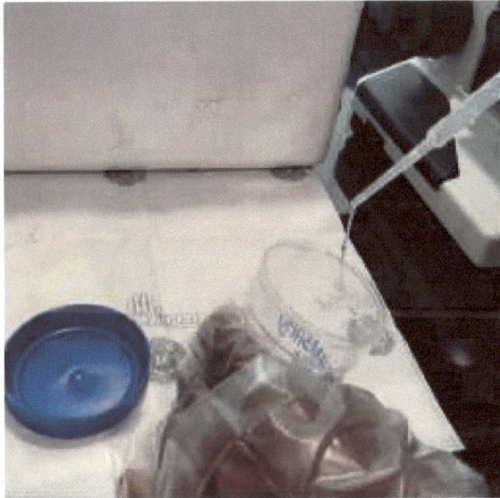

Fig. (5). Semen viscosity.

Microscopic Investigation

Microscopic investigations are an integral part of semen analysis. The microscope should include the following accessories; objectives with ×10 (yellow), ×20 (green), ×40 (blue) are required for evaluation of motility, vitality, and counting of spermatozoa and non-sperm cells [6]. Morphology and vitality evaluation should be done at × 100 (white) objective under oil immersion in the bright field.

Concentration

The concentration of sperm is calculated by the number of sperm present in one ml of the semen sample. The total sperm number or count is the total number of sperm present in the entire semen ejaculate. The semen concentration is calculated by properly mixing the semen through pipetting or vortex of the ejaculate. The sample is placed in a counting chamber, which is composed of 10 X 10 grids that contain a total of 100 squares in 10 rows and 10 columns (Fig. **6**). It was placed under a phase-contrast microscope under 20 X objective for counting. The count is given in millions. The lower reference value of the normal sample is $\geq 16 \times 10^6$/ml.

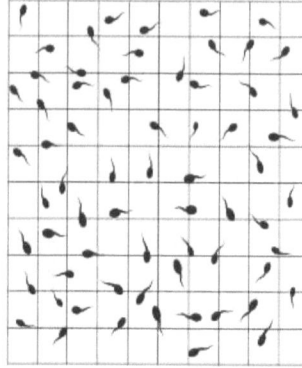

Fig. (6). Sperm assessment in Makler chamber.

Total sperm count = concentration of the sample **X** total volume of the sample.

Motility

Assessment of sperm motility is important in the seminogram as it is directly related to pregnancy. WHO recommends the use of chambers 20 microns deep that will enable the free movement of sperms. Sperm motility has been classified into three categories namely progressive (sperms that will move in a linear fashion or making large circles with $\geq 25\mu m/s$ speed is considered as rapid progressive and the sperms with the speed of 5 to < 25 $\mu m/s$ are noted as slow progressive), non-progressive (all other patterns of motility with an absence of progression, weak flagellar and twitching movement with the speed < 5 $\mu m/s$), and immotile (no movement). The standard reference values were mentioned in Table **1**. If the majority of sperms are non-motile, the chance of fertilization of the oocyte will be minimal.

$$\text{Total motility count (TMC) \%} = \frac{\text{Total no. of motile sperms}}{\text{Sperm concentration}} \times 100$$

$$\text{Progressive sperms (PR) \%} = \frac{\text{Total no. of Progressive sperms}}{\text{Sperm concentration}} \times 100$$

$$\text{Non- progressive sperms (NPR) \%} = \frac{\text{Total no. of non- progressive sperms}}{\text{Sperm concentration}} \times 100$$

$$\text{Immotile sperm (IM) \%} = \frac{\text{Total no. of immotile sperms}}{\text{Sperm concentration}} \times 100$$

Sperm Morphology

The next step in SA is to identify the percentage of sperms with normal forms. The morphological classification is examined based on Kruger's strict criteria and the latest WHO guidelines. The sperm head must appear as a smooth oval shape with a well-defined acrosome that covers 40%–70%. There should be no abnormalities in the neck, midpiece, and tail. The tail should be thinner than the midpiece (which should be < 1 μm in width and length), uncoiled, and ~ 45 μm. Finally, there cannot be any extra residual cytoplasm > 20% of the area of the sperm head [7]. The WHO reference values were mentioned in Table **1**. The sperms with different morphology were represented in Fig. (**7**).

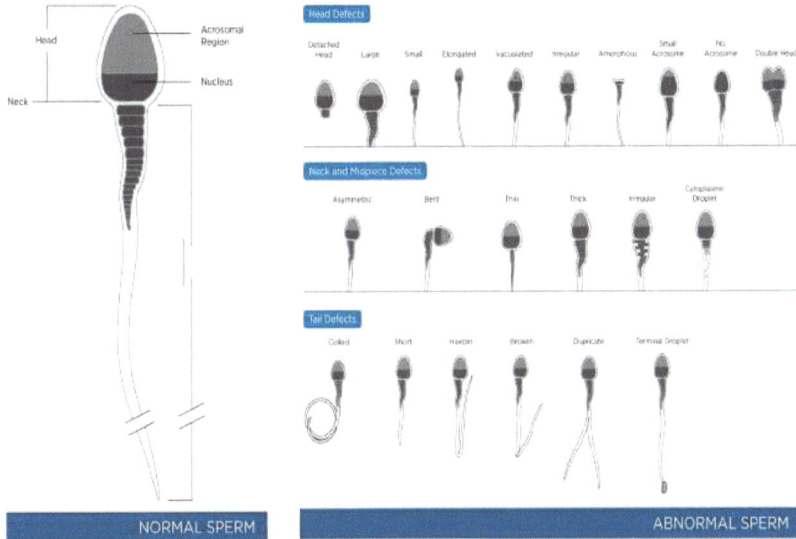

Fig. (**7**). Sperm morphology assessment.

Round Cells

In the semen sample, two forms of the round cells can be observed, it can be either immature germ cells or leukocytes, or both. The presence of excessive immature germ cells in the ejaculate may indicate an environment of oxidative stress (OS) during spermatogenesis. Therefore, it is important to differentiate between these two round cells. To distinguish between the two, round cells are counted in a high-power field. Leukocytospermia (Table **2**) and round cells can be identified by immunohistochemical staining using flow cytometry, the peroxidase test, the Endtz test, and monoclonal antibodies. The presence of more than 5 round cells or >1 × 10^6 round cells on the wet preparation indicates testing for the presence of white blood cells according to WHO guidelines. The presence of >1.0 × 10^6 round cells may contain activated leukocytes, which can produce ROS that is harmful to the spermatozoa. It is a well-known indicator of infection or inflammation in the male sex glands and the urogenital tract [8].

Sperm Agglutination

Sperm agglutination is described as the motile spermatozoa sticking to one another, the attachment can be head to head agglutination, tail to tail, or in a mixed way [6]. The specimen is observed for sperm agglutination by preparing a drop (50μL) of semen into a warm microscopic slide covered by a coverslip. The presence of sperm agglutination with a shaky sperm head was suggestive of the existence of an immunological cause of infertility. Sperm to sperm agglutination (site-specific agglutination) may indicate the presence of anti-sperm antibodies while sperm to non-sperm elements (nonspecific agglutination) can indicate accessory gland infection [1, 9]. The different grades of sperm agglutination are represented in Fig. (**8**).

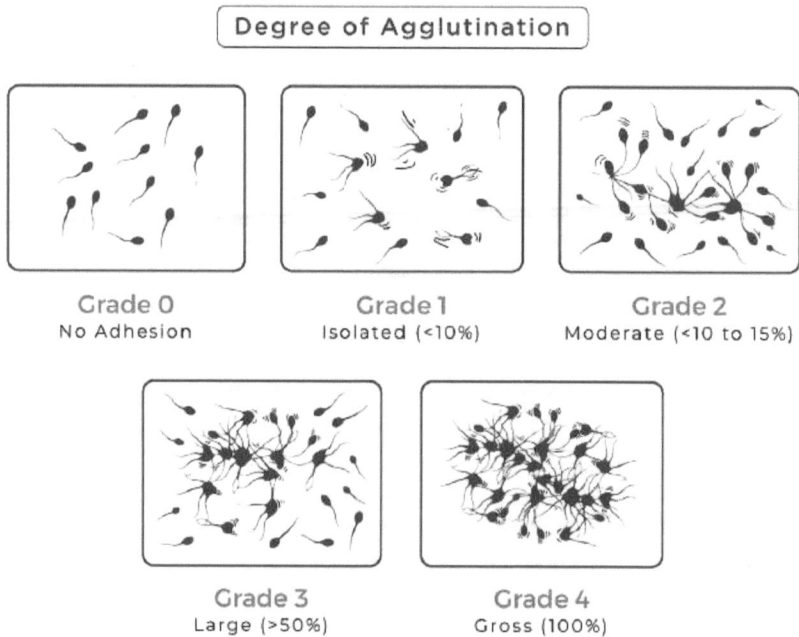

Fig. (**8**). Different grades of sperm agglutination.

Aggregation

Aggregation is defined as the attachment of immotile sperms to one other or non-sperm cells, debris, or mucus strands. A small aggregate of dead sperm and other material is commonly seen in the usual semen sample, whereas large aggregates, often containing hundreds of sperm, are considered to be abnormal [1].

Vitality

Sperm vitality is evaluated by examining cellular integrity, assessing the ability of the sperm plasma membrane to exclude extracellular substances. A basic requirement for fertilization is a live sperm. The latest WHO guidelines recommended the use of the eosin-nigrosin test and the hypo-osmotic swelling (HOS) test for the diagnostic investigation of sperm vitality. If the entire spermatozoa in the sample are found immotile it is important to select a live sperm for ICSI [10].

Eosin-Nigrosin

Eosin-Nigrosin (E/N) is a staining procedure that differentiates viable sperms from dead ones. It is recommended when the initial motility is gone below 25%. Nigrosin increases the contrast between the background and the sperm's heads and aids to visualize sperm more easily. After a motility assessment, this staining technique must be carried out immediately in the same ejaculate. The staining solution consists of Eosin–Y (1%) and Nigrosin (10%). At least 200 sperm were assessed under 100 × bright field objective. White (unstained) or light pink sperm are classified as live, whereas a dead sperm appears dark pink (Fig. **9a**). The vitality range of the normal sample should be ≥ 54% (WHO, 2021). For the sample with motility below 25%, viability should be ≥25% [11, 12].

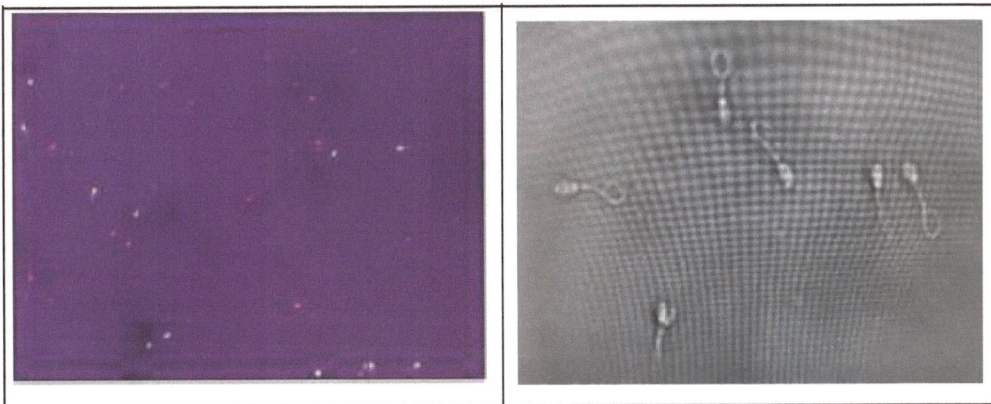

Fig. (9). a- Vitality test - Eosin-Nigrosin assay [14]. b- Hypo-osmotic swelling test [15].

Hypo-osmotic Swelling (HOS) Test

It was originally described as a test for sperm function. Now it is most appropriately considered as an ideal test for sperm vitality. The principle behind the HOS test is that viable live sperms can withstand moderate hypo-osmotic stress. Swelling and curling of tails can be seen in live sperms (Fig. **9b**). In the dead sperms, as its plasma membrane are not intact, it will not swell, while the senescent cells undergo uncontrolled swelling that leads to rupture of the plasma membrane. HOS test can be assessed at a magnification of 40x. Identify several swollen cells in about 200 spermatozoa [13].

Calculate the mean percentage of swollen and curling cells:

$$\text{Swelling Percentage} = \frac{\text{Number of sperm with swollen and curling tails} \times 100}{\text{The total number of sperms counted}}$$

Normal values (fertile): > 54% spermatozoa with swollen and curling tails

Abnormal values (infertile): < 54% spermatozoa with swollen and curling tails.

It is useful in selecting non-motile but viable sperm for assisted reproductive technologies. A HOST result of lesser than 50% is associated with increased miscarriage rates [16].

Biochemical Evaluation

Evaluation of various biochemical components present in seminal plasma can provide good inputs on the quality and quantity of the sperms as well as the process of spermatogenesis. Among several biochemical components fructose, zinc, and citric acid are directly related to sperm motility. Fructose can also provide information on the status of seminal vesicles, potential ejaculatory duct obstruction, and endocrine anomalies [18]. Zinc plays a regulatory role in the process of capacitation and acrosome reaction. Citric acid helps in maintaining seminal pH and is also connected with coagulation and liquefaction [19]. The concentration of α-glucosidase is directly correlated with sperm quality as well as it reveals the epididymal obstruction and it can be evaluated [20]. When the level of GGT is sub-optimum in the seminal plasma, it indicates accessory genital gland infections [21]. That is why it is necessary to evaluate the biochemical components of semen.

Why Semen Analysis has to be Repeated?

Due to the high degree of individual variability in spermatogenesis, demands the seminogram be repeated. Seasonal variations in morphology and concentration in the spermatozoa have also been observed. The sperm concentration with normal morphology is higher in the winter season, compared to fall, spring, and summer [22].

Sperm Counting Chambers

Sperm count is an essential part of the semen evaluation process. A device that is used to determine the concentration/motility of cells is called a counting chamber. One of the important factors influencing the count and motility of the sperm is the depth of the counting chamber. It should be a simple-to-use device for rapid and accurate sperm count, motility from the undiluted specimen. Visual estimation of sperm quality and quantity is inexpensive and simple, but the accuracy and reliability depend on the subjective assessment by the technician/technique and the tool. As technology and techniques improve, a modern device has evolved for fast and accurate analysis of undiluted samples with ease. Currently, varieties of devices are available for sperm counting that includes DROP, Standard Count, Cell Vision, MicroCell, 2X-CEL, Makler, JCD, Burker, CellVU, Leja, Macro, GoldCyto, and Geoffrey. The Neubauer hemocytometer (Fig. **10**) and Makler chambers (Fig. **11**) is the most common chambers used worldwide [23].

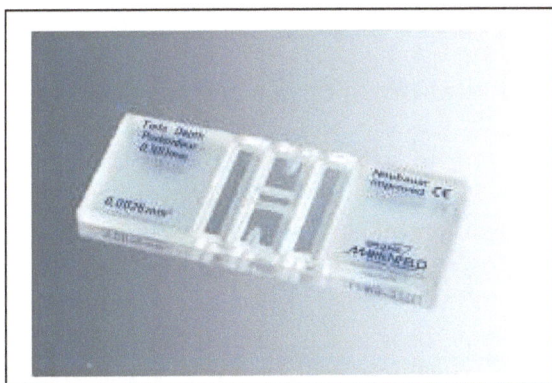

Fig. (10). Neubauer hemocytometer.

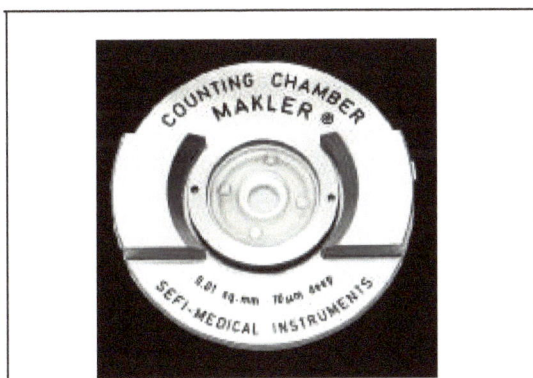

Fig. (11). Makler chamber.

Semen Analysis by CASA Method

CASA is an automated system, developed to provide reliable and precise data on the sperm concentration, viability, dynamics, or morphology of spermatozoa. It helps to carry out the statistical analysis of the sperm population, based on the development of continuous spermatozoa images, digital processing, and data analysis with the help of a video camera, video capture card, and computer. In addition, it can analyze the morphology and motion parameters of spermatozoa. The parameter includes; 1. Sperm motion velocity (curvilinear velocity (VCL, 1ms1), straight-line velocity (VSL, lms1), and average path velocity (VAP), 2. Velocity ratio (linearity, straightness (STR), wobble (WOB), and 3. Sperm wobble characteristics (amplitude of lateral head displacement (ALH), beat/cross frequency (BCF), and mean angular displacement (MAD, degrees). Literature suggests that CASA requires further improvement for a wider and routine application in clinical practice [24].

SQA

SQA-V is a simple and inexpensive device and provides a quantitative estimation of motility and morphology of sperm according to WHO 5th criteria. It is a self-testing and self-calibrating instrument and passes the self-test before analyzing samples. The SQA-V technology is based on the principle of electro-optical signal processing in combination with a series of algorithms to calculate the traditional parameters of semen assessment. It has two independent channels, one measuring sperm concentration and the other measuring motile sperm concentration, transmitting analog signals for analysis. According to the SQA-V manufacturer's instructions, a disposable testing capillary was filled with a fully liquefied, non-diluted, thoroughly mixed semen sample and run on the SQA-V at RT [25]. It can also be used as a screening tool for semen quality determination and management of male infertility [26].

Semen Processing

Semen processing/preparation is defined as the separation and harvesting of good, progressively motile spermatozoa from the rest of the seminal plasma. The procedure will enable the processing of a larger quantity of semen sample and it should eliminate the toxic substances, dead and immotile sperms, and all other cells present in the given semen sample without harming the good viable sperms. It should be easy, quick, and cost-effective. It is a crucial process in assisted reproductive treatments (ART) to select viable sperm and remove seminal plasma [27]. An array of different sperm processing techniques [28, 29] is followed that has been described below (Fig. **12**).

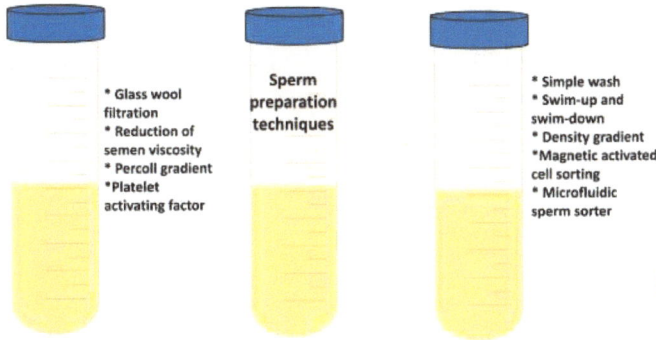

Fig. (12). Different sperm preparation techniques.

Simple Wash Method

This method is used when the semen sample is very poor and it mainly removes seminal plasma from the sperms. A simple wash of sperm essentially removes the seminal plasma without any significant decrease in total sperm count. After post-wash, the hypermotility may increase primarily due to the removal of seminal plasma. This method involves diluting the semen with twice or more the amount of culture medium and allowing the spermatozoa to settle down by applying the centrifugal force of not more than 800 g. This procedure takes 20 to 40 minutes. The procedure has explained in flowchart 2.

Flowchart 2: Simple Wash Method

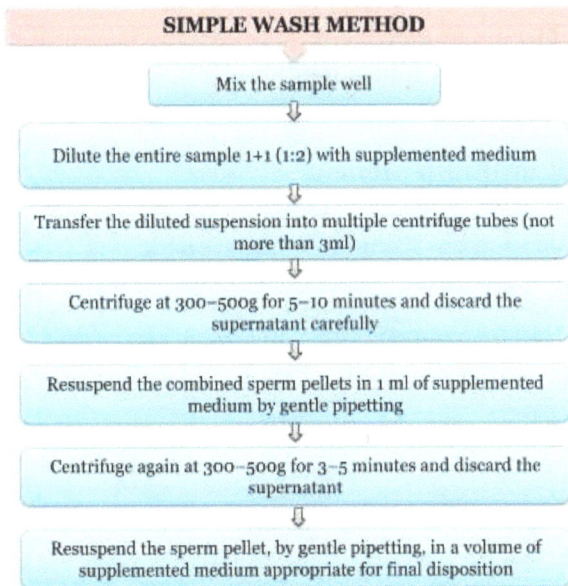

Sperm Migration Based Methods

In vivo, the spermatozoa in the ejaculate naturally migrate towards the cervix and uterus after deposition in the vagina. Similarly, spermatozoa migrate towards the layered culture medium *in vitro*. The "Swim Up" method utilizes this phenomenon where the progressively motile spermatozoa swim towards and accumulate in the culture medium phase after 15-60 minutes of incubation. This can be performed in two ways-

a. Direct swim-up – The culture medium is directly layered over or underneath (Swim down) the semen. The procedure is described in Flowchart 3.

Flowchart 3: Direct Swim Up Method

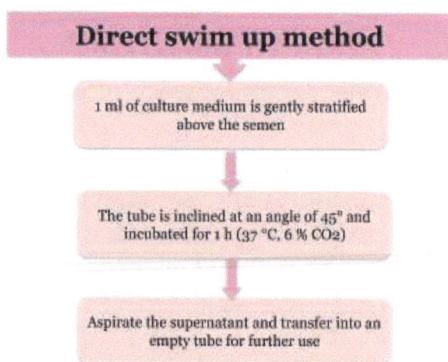

Direct swim up method

1 ml of culture medium is gently stratified above the semen

The tube is inclined at an angle of 45° and incubated for 1 h (37 °C, 6 % CO2)

Aspirate the supernatant and transfer into an empty tube for further use

b. Swim-up from a washed pellet – The semen sample can be mixed with culture medium and centrifuged. The supernatant is removed, after which the pellet is layered with a more culture medium [29]. This process has explained in a Flowchart 4.

Flowchart 4: Pellet Swim Up Method

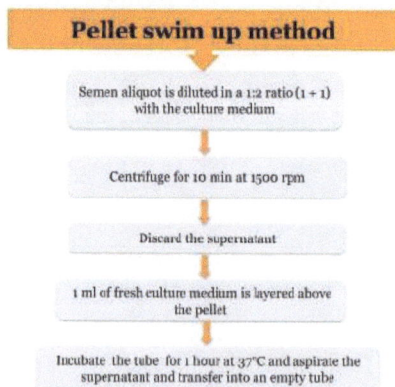

Pellet swim up method

Semen aliquot is diluted in a 1:2 ratio (1 + 1) with the culture medium

Centrifuge for 10 min at 1500 rpm

Discard the supernatant

1 ml of fresh culture medium is layered above the pellet

Incubate the tube for 1 hour at 37°C and aspirate the supernatant and transfer into an empty tube

The Method Based on the Specific Density

Based on the specific density the preparation methods have been classified into two types are Continuous-step density gradient method and discontinuous density gradient centrifugation methods [29].

Continuous-step Density Gradient

The continuous-step density gradient technique is a simple procedure performed to concentrate the progressively motile spermatozoa selectively. The semen sample was placed on 6.0 ml of percoll solution (80%), and the density gradient was formed by mixing the ejaculate and percoll with an L-shaped rod for a couple of shots. Post centrifugation at 400 x g for 30 mins, yielded a rich number of progressively motile spermatozoa in the pellet, while immotile sperm and other cell forms persisted in the upper layer of the density gradient. In the majority of the cases, it was found an increase in sperm motility by 80%, independent of the initial motility [30]. The fertility rate was therefore boosted.

Discontinuous Density Gradient Centrifugation

In cases of extreme oligozoospermia, teratozoospermia, or asthenozoospermia, this is the method favored to harvest a good quantity of motile sperms. It enables to segregate good quality sperms by a density discontinuous gradient from dead sperms, leukocytes, and the other seminal plasma components. The colloidal silica coated with silane of the gradient; facilitates the separation of cells based on their density and motility. Morphologically normal, immature, and abnormal spermatozoon have different densities of 1.10 g/ml, 1.06, and 1.09 g/ml respectively. During the centrifugation process, each spermatozoon is placed at the level of the gradient that corresponds to its density. Thus, the resulting interphases are discarded between seminal plasma and the 40% upper layer containing leukocytes, cell debris, and 40% and 80% containing morphologically defective sperm with low motility. A pellet at the bottom of the tube forms concentrated viable and morphologically good progressive sperms. At last, the sample is free from debris, immature and dead sperms, leukocytes, and bacteria [31]. Discontinuous density gradient can also be performed using three-layer density gradient (90%, 70%, and 40%) solutions and two-layer with 45% upper and 90% lower density gradient solutions. Out of these methods, double density gradient centrifugation (DGC) (Flowchart 5) was the commonly used method.

Flowchart 5: Double Density Gradient Method

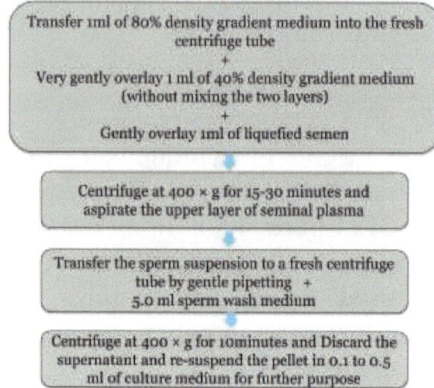

Double density gradient method

Transfer 1ml of 80% density gradient medium into the fresh centrifuge tube
+
Very gently overlay 1 ml of 40% density gradient medium (without mixing the two layers)
+
Gently overlay 1ml of liquefied semen

Centrifuge at 400 × g for 15-30 minutes and aspirate the upper layer of seminal plasma

Transfer the sperm suspension to a fresh centrifuge tube by gentle pipetting +
5.0 ml sperm wash medium

Centrifuge at 400 × g for 10minutes and Discard the supernatant and re-suspend the pellet in 0.1 to 0.5 ml of culture medium for further purpose

Magnetic Activated Cell Sorting (MACS)

This method separates apoptotic spermatozoa from non-apoptotic ones. The cells undergoing apoptosis have phosphatidylserine residues in their membrane which has a strong affinity towards Annexin V conjugated to colloidal super magnetic beads which separate apoptotic cells when applied with a magnetic field [32]. This method can be used in conjunction with DGC to maximize fertilization potential and embryo quality (Flowchart 6).

Flowchart 6: Magnetic Activated Cell Sorting (MACS)

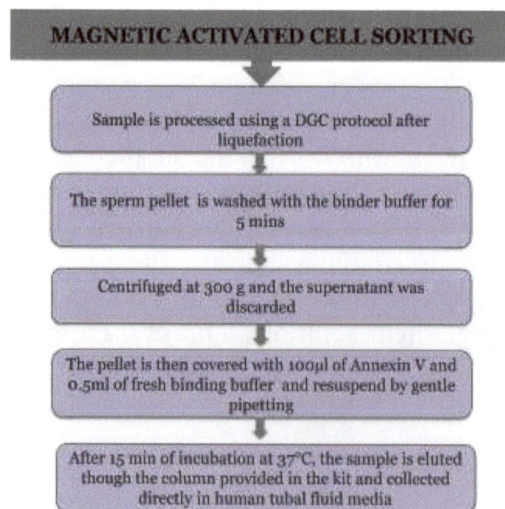

MAGNETIC ACTIVATED CELL SORTING

Sample is processed using a DGC protocol after liquefaction

The sperm pellet is washed with the binder buffer for 5 mins

Centrifuged at 300 g and the supernatant was discarded

The pellet is then covered with 100µl of Annexin V and 0.5ml of fresh binding buffer and resuspend by gentle pipetting

After 15 min of incubation at 37°C, the sample is eluted though the column provided in the kit and collected directly in human tubal fluid media

Microfluidic Sperm Sorter

This technique is employed to harvest morphologically normal sperm cells with high motility and with no debris based on fluid dynamics in a microenvironment or nano environment, relying on variables such as fluid density, viscosity, velocity, and size/geometry of the environment. Sperm selection in this method is based on the two gravity-driven laminar flows within the central microfluidic channel. The semen and the medium dispensed into inlet A and B, respectively, flow parallel to each other and exit through their respective outlets (A/D and B/C). Dependent on their ability to swim over these two outlets, the sperm cells are sorted. Only motile sperms can swim toward outlet C whereas immotile spermatozoa keep flowing to outlet D. Hence, this technique does not allow sperms to undergo any sort of physical stress from sources such as a centrifuge and thus minimize DNA damage [33].

Glass Wool Filtration

It separates motile spermatozoa from the rest of the seminal components by filtration through densely packed glass wool fibers. The debris, immotile sperm cells, and leukocytes can be separated through filtration. Glass wool filtration reported 87.5% efficiency in eliminating leukocytes in the semen sample. This is significant since leukocytes are the main causative for ROS in the ejaculate. Following filtration, the semen sample is centrifuged to eliminate seminal plasma from live sperm cells. The production of ROS can be reduced by performing centrifugation in the absence of non-viable spermatozoa and leukocytes [34]. Flowchart 7 provides a glass wool filtration process. The advantages and disadvantages of different sperm preparation techniques were given in Table **3** [35]. Based on the nature of the sample the preparation methods were recommended (Table **4**) [26].

Antisperm Antibody

Antisperm antibodies can be observed in seminal plasma, serum, or attached to sperm. This condition is reported in 3-12% of men undergoing infertility treatment. There are multiple procedures have performed to spot antisperm antibodies. It includes direct immunobead test (DIT), tray agglutination test (TAT), mixed agglutination reaction (MAR), indirect MAR or immunobead test (IBT), sperm immobilization test (SIT), and enzyme-linked immune sorbent assay (ELISA) and flow cytometry. Different classes of immunoglobulins can be distinguished through ELISA [36, 37].

Flowchart 7: Glass Wool Filtration

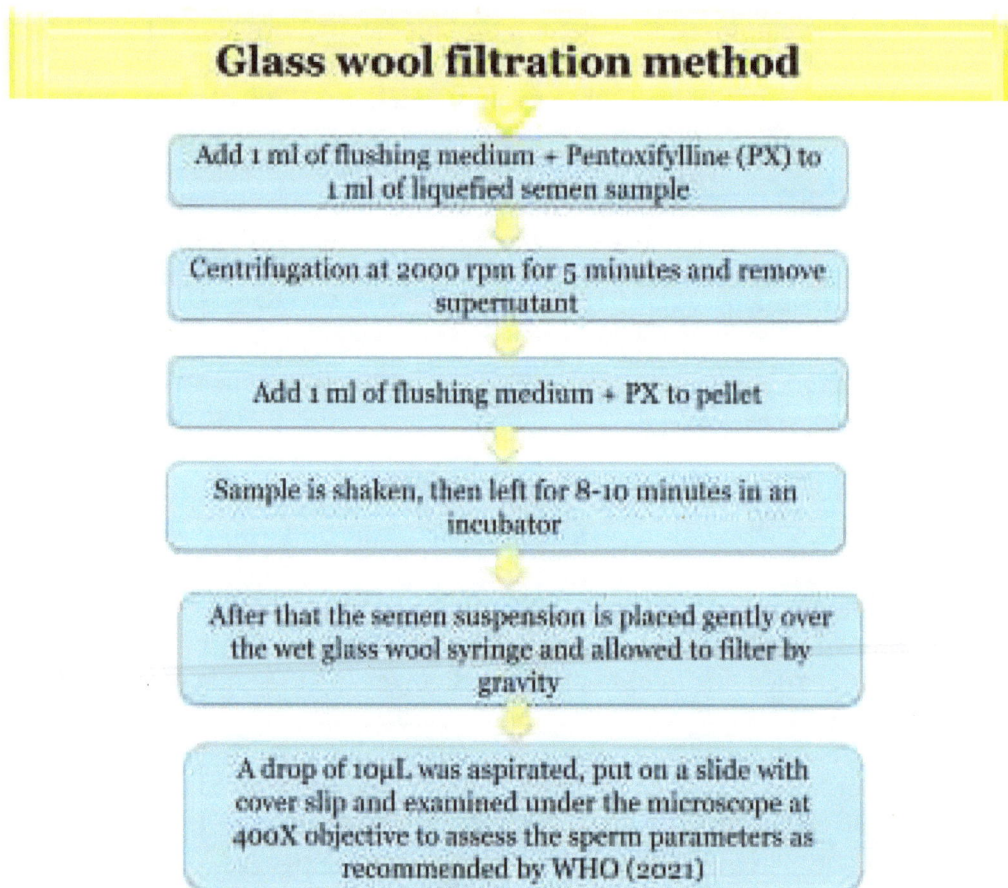

Glass wool filtration method

Add 1 ml of flushing medium + Pentoxifylline (PX) to 1 ml of liquefied semen sample

Centrifugation at 2000 rpm for 5 minutes and remove supernatant

Add 1 ml of flushing medium + PX to pellet

Sample is shaken, then left for 8-10 minutes in an incubator

After that the semen suspension is placed gently over the wet glass wool syringe and allowed to filter by gravity

A drop of 10μL was aspirated, put on a slide with cover slip and examined under the microscope at 400X objective to assess the sperm parameters as recommended by WHO (2021)

Table 3. Advantages and disadvantages of sperm preparation technique [35].

Sperm Preparation Technique	Advantages	Disadvantages
Simple wash	Simplest method, uses low amount of media and cost effective	All the components of semen that can be pelleted
Sperm migration based methods	• Retrieved comparatively clean and rich progressively motile sperms. • Minimal use of centrifugation reduces centrifugation related iatrogenic damage.	• Pellet retains the dead sperm and leukocytes and has risks related to oxidative stress. • Sperm segregation is based on their motility. • Diffusion of substances like prostatic zinc from the semen to the overlaid medium.

Density gradient method	• Segregates sperms based on the functional maturity, DNA integrity and morphology. • Used to remove bacteria, debris, leukocytes, viruses, etc. • Gives a good yield for Oligoasthenoteratozoospermic samples (OATs). • Requires less time than a swim-up method.	• Costly method • Yield is low when compared to swim up • Increased chances of iatrogenic damage due to 2-3 centrifugation cycles.
Magnetic Activated cell sorting (MACS)	• MACS acts at molecular level. • It is the only known technique that separates apoptotic sperms from non-apoptotic sperms. • Rapid, convenient and noninvasive.	• MACS which removes apoptotic sperms needs to be used in conjunction.
Glass wool filtration	• Easy to perform. • Recovers sperms with good motility and normal chromatin condensation. • Eliminates leukocytes. • Low ROS	• Relatively expensive. • Debris and immotile sperms may pass through the mesh and be present in the sample even after filtration.

Table 4. Recommended preparation methods based on semen sample [35].

Types of sample	Recommended method
Normozoospermia	DGC / MFSS
Teratozoospermia	DGC / MACS
Asthenozoospermia	DGC / MACS
Oligoasthenoteratozoospermia	Wash and Swim-up / DGC
Leukocytospermia	DGC / MACS
Hyperviscosity	Pellet swim-up
Hypospermia	Wash and Pellet
Heavy debris	DGC
Retrograde ejaculate	Wash and pellet
PESA sample	DGC / wash
TESE sample	Wash and pellet / DGC
M TESE sample	Wash and pellet / DGC
Frozen-thawed surgical sample	Wash and pellet
Frozen-thawed ejaculate	DGC
Sperm with a high DFI	DGC / MACS
Previous ICSI failure	IMSI / MACS / PICSI
Unexplained infertility	MACS / IMSI / PICSI
Previous RIF	MACS / PICSI
Patient Preference	As per discussion / counseling

Preparing Retrograde Ejaculation Samples

Retrograde ejaculation (RE) is a significant cause of anejaculation or aspermia, leaving the patient infertile. It is a type of ejaculatory dysfunction in which seminal fluid is propelled proximally towards the bladder after leaving the ejaculatory ducts. It happens due to the partial closure of the bladder sphincter muscle during ejaculation. Post ejaculation up to 15% of sperms can be seen in urine normally. If this percentage is increased, then it is an indication of RE. RE can be complete, in which case no semen is propelled in the antegrade direction, or partial, in which there is still some antegrade propulsion present. Among male infertility, RE shows up to 4-2% cases. Flowchart 8 describes the processing of RE urine samples [38].

Flowchart 8: Sample Preparation for Retrograde Ejaculation

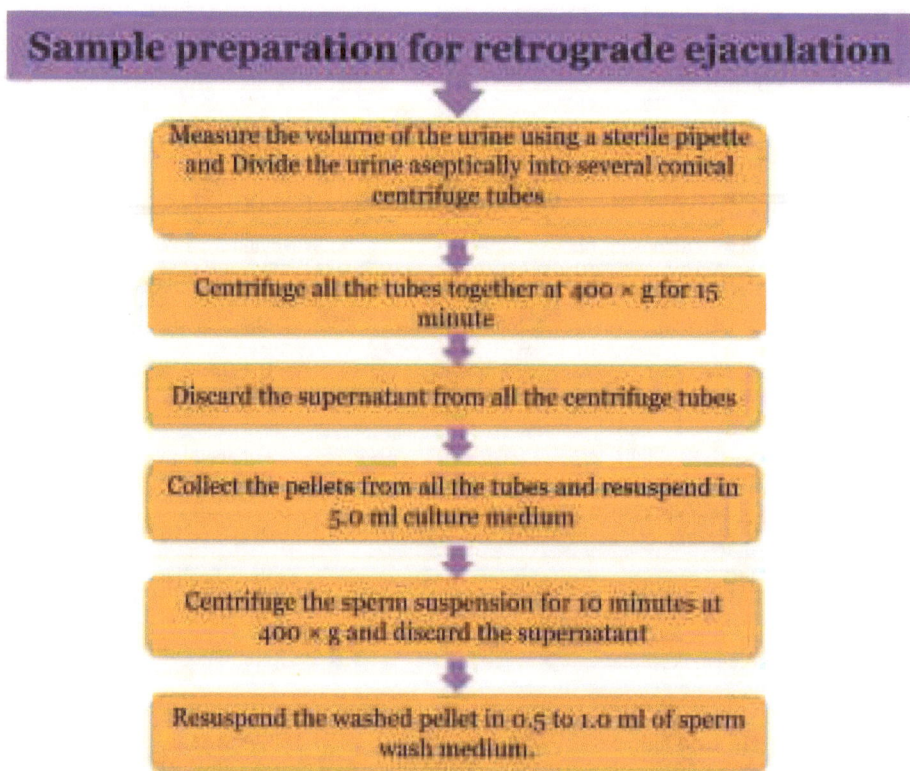

Processing of Infected Sample

The semen samples of patients with HIV, HCV, and HbsAg going for MAR need special processing to prevent uninfected female partners from acquiring the infection. Initial DGC separates virus-infected seminal plasma and non-sperm cells from virus-free sperms which form a pellet. This pellet is then processed by the swim-up method to harvest virus-free motile spermatozoa. For ART, only virus-free samples should be used. Therefore, it is mandatory to perform viral screening using reverse transcriptase- polymerase chain reaction (RT-PCR) in all the prepared samples before use. The findings are positive; there is still a chance of delivering a virus-free baby from the infected couple through sperm preparation [39]. The sperm wash procedure for the infected sample has been described in flowchart 9.

Flowchart 9: Preparation for Infected Sample.

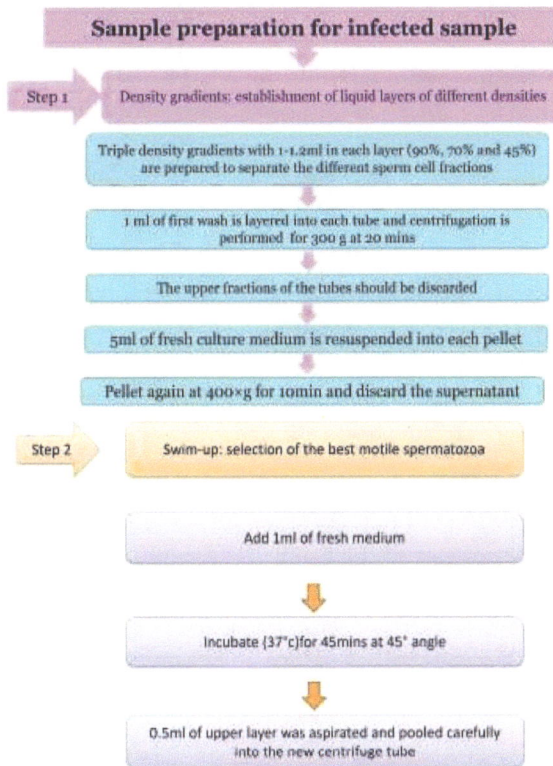

Sample preparation for infected sample

Step 1 — Density gradients: establishment of liquid layers of different densities

Triple density gradients with 1-1.2ml in each layer (90%, 70% and 45%) are prepared to separate the different sperm cell fractions

1 ml of first wash is layered into each tube and centrifugation is performed for 300 g at 20 mins

The upper fractions of the tubes should be discarded

5ml of fresh culture medium is resuspended into each pellet

Pellet again at 400×g for 10min and discard the supernatant

Step 2 — Swim-up: selection of the best motile spermatozoa

Add 1ml of fresh medium

Incubate {37°c}for 45mins at 45° angle

0.5ml of upper layer was aspirated and pooled carefully into the new centrifuge tube

Processing of Surgically Retrieved Sperms

Different surgical sperm retrieval methods have been developed to retrieve spermatozoa from either the epididymis or the testis from the azoospermic men. Surgically-retrieved sperms can be used for intracytoplasmic sperm injection (ICSI) and/or cryopreservation [40]. The commonly performed surgical methods are Percutaneous epididymal sperm aspiration (PESA), Testicular epididymis sperm aspiration (TESA), Testicular epididymis sperm extraction (TESE), and Microdissection testicular epididymis sperm extraction (microTESE). Based on the procedure, the sperm processing technique will be varying. A detailed semen processing procedure for PESA is described in flowchart 10; for all other surgical procedures, the processing method is described in flowchart 11 [41].

Flowchart 10: Processing of PESA Sample.

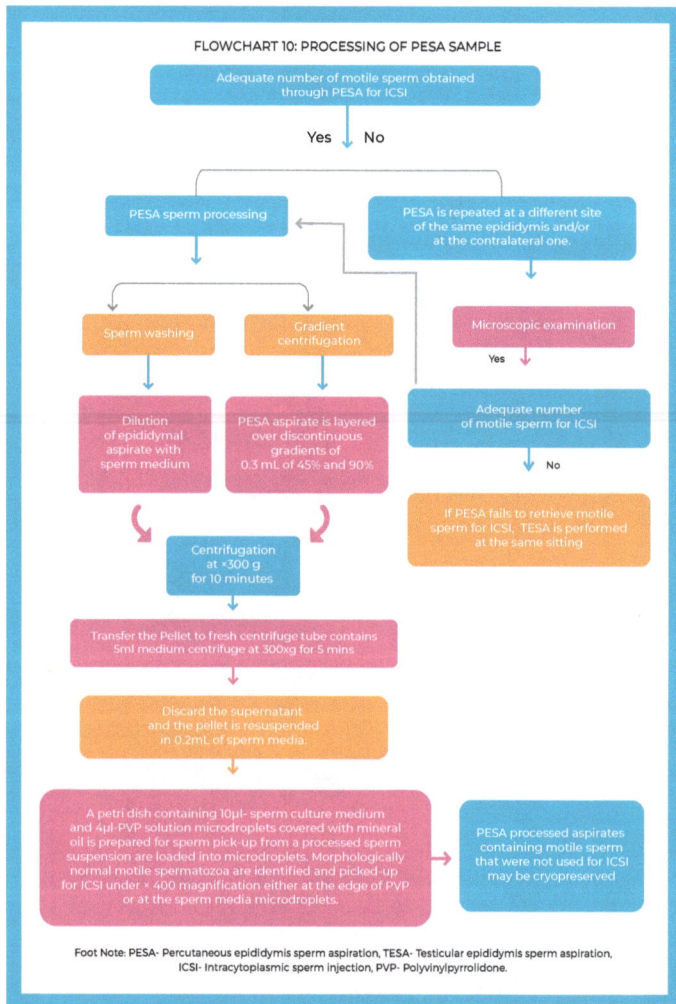

FLOWCHART 10: PROCESSING OF PESA SAMPLE

Foot Note: PESA- Percutaneous epididymis sperm aspiration, TESA- Testicular epididymis sperm aspiration, ICSI- Intracytoplasmic sperm injection, PVP- Polyvinylpyrrolidone.

Flowchart 11: Processing of Surgically Retrieved Sperm Sample.

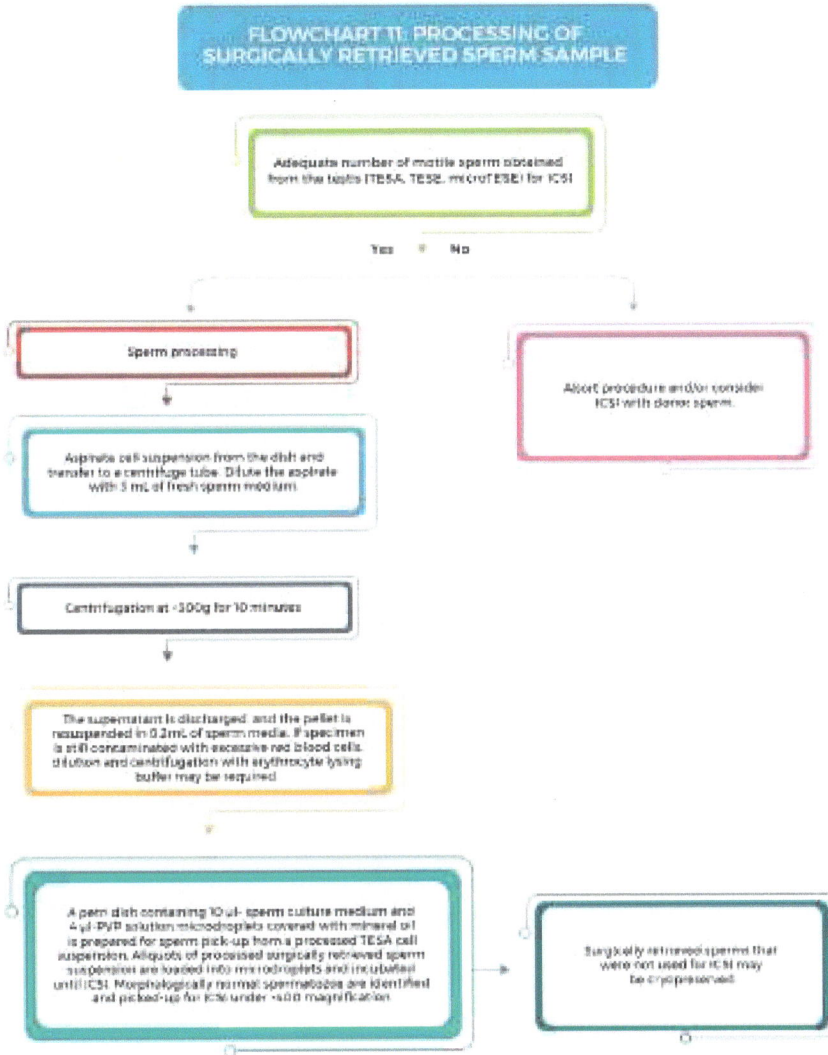

FLOWCHART 11: PROCESSING OF SURGICALLY RETRIEVED SPERM SAMPLE

Adequate number of motile sperm obtained from the testis (TESA, TESE, microTESE) for ICSI

Yes   No

Sperm processing

Abort procedure and/or consider ICSI with donor sperm.

Aspirate cell suspension from the dish and transfer to a centrifuge tube. Dilute the aspirate with 5 mL of fresh sperm medium.

Centrifugation at ~300g for 10 minutes

The supernatant is discharged and the pellet is resuspended in 0.2mL of sperm media. If specimen is still contaminated with excessive red blood cells, dilution and centrifugation with erythrocyte lysing buffer may be required

A petri dish containing 10 µl- sperm culture medium and 4 µl-PVP solution microdroplets covered with mineral oil is prepared for sperm pick-up from a processed TESA cell suspension. Aliquots of processed surgically retrieved sperm suspension are loaded into microdroplets and incubated until ICSI. Morphologically normal spermatozoa are identified and picked-up for ICSI under ~400 magnification.

Surgically retrieved sperm that were not used for ICSI may be cryopreserved

Foot Note: TESA- Testicular epididymis sperm aspiration, TESE- Testicular epididymis sperm extraction, ICSI- intracytoplasmic sperm injection, microTESE- microdissection Testicular epididymis sperm extraction PVP- Polyvinylpyrrolidone

CONCLUSION

Semen assessment and processing are crucial steps in the diagnosis and management of male subfertility. Though SA is not a definite tool in the diagnosis of the male factor, it still provides much-needed vital information on male factor and aids in a better understanding of the male factor. Thus, it has become an inevitable procedure in the field of medically assisted reproduction.

As far as preparation techniques are concerned, an array of methods is performed worldwide. The choice of the method should focus on various factors, including the quality of the sample and the insemination method. It should yield a maximum quantity of quality sperm and also enable the use of even a severely impaired sample for the management of male fertility. Based on the literature evidence the density gradient centrifugation method was found to be effective. Therefore, knowledge of the various assessment and processing techniques is essential for all personnel involved in MAR.

CONSENT FOR PUBLICATION

Not applicable

CONFLICT OF INTEREST

The author declares no conflict of interest, financial or otherwise.

ACKNOWLEDGEMENT

Declared none.

REFERENCES

[1]   World Health Organization. WHO laboratory manual for the examination and processing of human semen. 2021.

[2]   Molnar C, Gair J. Concepts of Biology. 1st Canadian Edition., 2015.

[3]   Esteves SC, Miyaoska R. Sperm physiology and assessment of spermatogenesis kinetics *in vivo*. Handbook of Fertility. Academic Press 2015; pp. 383-96.
[http://dx.doi.org/10.1016/B978-0-12-800872-0.00034-2]

[4]   Hedger MP. The Immunophysiology of Male Reproduction. Knobil and Neill's Physiology of Reproduction 2015; pp. 805-92.
[http://dx.doi.org/10.1016/B978-0-12-397175-3.00019-3]

[5]   Vasan SS. Semen analysis and sperm function tests: How much to test? Indian J Urol 2011; 27(1): 41-8.
[http://dx.doi.org/10.4103/0970-1591.78424] [PMID: 21716889]

[6]   Talwar P. Talwar P. Semen Analysis: Nuts and Bolts. 2016. Available from: http://indianfertilitysociety.org/pdf/nexus_2.%20mailer.pdf

[7]   Agarwal A, Gupta S, Sharma R. Andrological evaluation of male infertility. Switzerland: Springer International Publishing 2016.
[http://dx.doi.org/10.1007/978-3-319-26797-5]

[8]   Jung JH, Kim MH, Kim J, *et al.* Treatment of leukocytospermia in male infertility: a systematic review. World J Mens Health 2016; 34(3): 165-72.
[http://dx.doi.org/10.5534/wjmh.2016.34.3.165] [PMID: 28053945]

[9]   Monclus MA, Fornes MW. Sperm conjugation in mammal reproductive function: Different names for

the same phenomenon? Mol Reprod Dev 2016; 83(10): 884-96.
[http://dx.doi.org/10.1002/mrd.22636] [PMID: 26970336]

[10] Moskovtsev SI, Librach CL. Methods of sperm vitality assessment InSpermatogenesis. Totowa, NJ: Humana Press 2013; pp. 13-9.

[11] Björndahl L, Söderlund I, Johansson S, Mohammadieh M, Pourian MR, Kvist U. Why the WHO recommendations for eosin-nigrosin staining techniques for human sperm vitality assessment must change. J Androl 2004; 25(5): 671-8.
[http://dx.doi.org/10.1002/j.1939-4640.2004.tb02839.x] [PMID: 15292094]

[12] Agarwal A, Gupta S, Sharma R. Eosin-Nigrosin staining procedure. Andrological Evaluation of Male Infertility. Cham: Springer 2016; pp. 73-7.
[http://dx.doi.org/10.1007/978-3-319-26797-5_8]

[13] Talwar P, Hayatnagarkar S. Sperm function test. J Hum Reprod Sci 2015; 8(2): 61-9.
[http://dx.doi.org/10.4103/0974-1208.158588] [PMID: 26157295]

[14] Park YS, Lee SH, Choi HW, Lee HS, Lee JS, Seo JT. Abnormal human sperm parameters contribute to sperm DNA fragmentation in men with varicocele. World J Mens Health 2018; 36(3): 239-47.
[http://dx.doi.org/10.5534/wjmh.180014] [PMID: 30079641]

[15] http://hotrosinhsanbenhvienbuudien.com.vn/ky-thuat/vo-sinh-nam-ky-thuat-hos-test/

[16] Vasan SS. Semen analysis and sperm function tests: How much to test?. Indian journal of urology: IJU: journal of the Urological Society of India 2011; 27(1): 41.

[17] Grimes DA, Lopez LM. "Oligozoospermia," "azoospermia," and other semen-analysis terminology: the need for better science. Fertil Steril 2007; 88(6): 1491-4.
[http://dx.doi.org/10.1016/j.fertnstert.2007.04.013] [PMID: 17582404]

[18] Talluri TR, Mal G, Ravi SK. Biochemical components of seminal plasma and their correlation to the fresh seminal characteristics in Marwari stallions and Poitou jacks. Vet World 2017; 10(2): 214-20.
[http://dx.doi.org/10.14202/vetworld.2017.214-220] [PMID: 28344405]

[19] Kothari RP, Chaudhari AR. Zinc levels in seminal fluid in infertile males and its relation with serum free testosterone. J Clin Diagn Res 2016; 10(5): CC05-8.
[http://dx.doi.org/10.7860/JCDR/2016/14393.7723] [PMID: 27437207]

[20] Krause W, Bohring C. Why do we determine α□glucosidase activity in human semen during infertility work□up? Andrologia 1999; 31(5): 289-94.

[21] Gupta S, Singh M, Dhaka Y, Kumar P. Functional and biochemical attributes of bull semen with reference to freezability at different stages of semen processing.

[22] Chen Z, Toth T, Godfrey-Bailey L, Mercedat N, Schiff I, Hauser R. Seasonal variation and age-related changes in human semen parameters. J Androl 2003; 24(2): 226-31.
[http://dx.doi.org/10.1002/j.1939-4640.2003.tb02666.x] [PMID: 12634309]

[23] Lu JC, Yue RQ, Feng RX, Kong LZ, Xu YC. Accuracy evaluation of the depth of six kinds of sperm counting chambers for both manual and computer-aided semen analyses. Int J Fertil Steril 2016; 9(4): 527-33.
[PMID: 26985341]

[24] Talarczyk-Desole J, Berger A, Taszarek-Hauke G, Hauke J, Pawelczyk L, Jedrzejczak P. Manual vs. computer-assisted sperm analysis: can CASA replace manual assessment of human semen in clinical practice? Ginekol Pol 2017; 88(2): 56-60.
[http://dx.doi.org/10.5603/GP.a2017.0012] [PMID: 28326513]

[25] Agarwal A, Sharma RK. Automation is the key to standardized semen analysis using the automated SQA-V sperm quality analyzer. Fertil Steril 2007; 87(1): 156-62.
[http://dx.doi.org/10.1016/j.fertnstert.2006.05.083] [PMID: 17081526]

[26] Hawaldar R, Sodani S. Study of semen quality by SQA (QwikCheck Gold). Indian Journal of

Pathology and Oncology 2018; 5(2): 290-3.

[27]   Henkel RR, Schill WB. Sperm preparation for ART. Reprod Biol Endocrinol 2003; 1: 108.
        [http://dx.doi.org/10.1186/1477-7827-1-108] [PMID: 14617368]

[28]   Gangrade BK, Agarwal A. Sperm Processing and Selection Techniques in an IVF/ICSI.

[29]   Beydola T, Sharma RK, Lee W, Agarwal A, Rizk B, Aziz N. Sperm preparation and selection
        techniques Male Infertility Practice. New Delhi: Jaypee Brothers Medical Publishers 2013; pp. 244-51.

[30]   Kaneko S, Sato H, Kobanawa K, Oshio S, Kobayashi T, Iizuka R. Continuous-step density gradient
        centrifugation for the selective concentration of progressively motile sperm for insemination with
        husband's semen. Arch Androl 1987; 19(1): 75-84.
        [http://dx.doi.org/10.3109/01485018708986803] [PMID: 3426341]

[31]   Malvezzi H, Sharma R, Agarwal A, Abuzenadah AM, Abu-Elmagd M. Sperm quality after density
        gradient centrifugation with three commercially available media: a controlled trial. Reprod Biol
        Endocrinol 2014; 12(1): 121.
        [http://dx.doi.org/10.1186/1477-7827-12-121] [PMID: 25466430]

[32]   Makker K, Agarwal A, Sharma RK. Magnetic activated cell sorting (MACS): utility in assisted
        reproduction.

[33]   Shirota K, Yotsumoto F, Itoh H, *et al.* Separation efficiency of a microfluidic sperm sorter to minimize
        sperm DNA damage. Fertil Steril 2016; 105(2): 315-21.e1.
        [http://dx.doi.org/10.1016/j.fertnstert.2015.10.023] [PMID: 26551440]

[34]   Nagy ZP, Varghese AC, Agarwal A, Eds. In vitro Fertilization: A Textbook of Current and Emerging
        Methods and Devices. Springer 2019; p. 305.
        [http://dx.doi.org/10.1007/978-3-319-43011-9]

[35]   Beydola T, Sharma RK, Lee W, *et al.* Sperm preparation and selection techniques Male Infertility
        Practice. New Delhi: Jaypee Brothers Medical Publishers 2013; pp. 244-51.

[36]   Windt ML, Bouic PJ, Lombard CJ, Menkveld R, Kruger TF. Antisperm antibody tests: traditional
        methods compared to ELISA. Arch Androl 1989; 23(2): 139-45.
        [http://dx.doi.org/10.3109/01485018908986836] [PMID: 2686572]

[37]   AS V. Dhama K, Chakraborty S, Abdul Samad H, K Latheef S, Sharun K, Khurana SK, Tiwari R,
        Bhatt P, Chaicumpa W. Role of antisperm antibodies in infertility, pregnancy, and potential for
        contraceptive and antifertility vaccine designs: research progress and pioneering vision. Vaccines
        (Basel) 2019; 7(3): 116.
        [http://dx.doi.org/10.3390/vaccines7030116]

[38]   Barazani Y, Stahl PJ, Nagler HM, Stember DS. Management of ejaculatory disorders in infertile men.
        Asian J Androl 2012; 14(4): 525-9.
        [http://dx.doi.org/10.1038/aja.2012.29] [PMID: 22580636]

[39]   Rivera-Egea R, Samochin TV, Escrivá MM. Processing of Sperm Samples in HIV-Positive Patients In
        in vitro Fertilization. Cham: Springer 2019; pp. 335-45.

[40]   Esteves SC, Miyaoka R, Agarwal A. Sperm retrieval techniques for assisted reproduction. Int Braz J
        Urol 2011; 37(5): 570-83.
        [http://dx.doi.org/10.1590/S1677-55382011000500002] [PMID: 22099268]

[41] Kumari S, Kacker G. Live birth from TESA-ICSI in a modified natural cycle– A case report. Fertility Science and Research 2018; 5(1): 33.
[http://dx.doi.org/10.4103/fsr.fsr_26_18]

CHAPTER 3

Sperm Freezing

Mohammad A. Ibrahim[1,*]

[1] *Department of Theriogenology, Faculty of Veterinary Medicine, Alexandria University, Alexandria Governorate 5424041, Egypt*

Abstract: Sperm freezing is one of the reproductive technologies responsible for increasing domestic animals' fertility and preserving wild animal fertility over the last 100 years. This chapter will shed light on the current understanding of the effect of cryopreservation on sperm cells, advances in cryopreservation in terms of cryoprotectants and freezing protocols for bovines, caprines, ovines, equines, and camelids. Moreover, this chapter will shed light on other technologies needed to be adopted for different animal species.

Keywords: Cryopreservation, Cryoprotectants, Fertility, Sperm.

INTRODUCTION

Sperm freezing is essential for the distribution of superior genetics among domestic animal species and the preservation of wild animal species. Sperm freezing enables increasing sperm longevity and availability to inseminate animals in another part of the world or even in the future when the original donor is already dead. However, this at the expense that sperm cells thawed after freezing undergo alternation in their functionality, and intracellular and plasma membrane structures, which leads to compromised fertilizing ability.

Advances in sperm cryopreservation go around improving post-thawing sperm parameters to be as close to pre-freezing sperm parameters. In this context, a substantial amount of work has been dedicated to testing different cryoprotectants, additives, cooling rates, and freezing dose. It is worth noting that a freezing protocol that is suitable for one animal species does not essentially work for another. This is could be attributed to many reasons but primarily due to the different composition of sperm cell plasma membrane of different animal species. The cryotolerance of the sperm plasma membrane is determined according to membrane cholesterol to phospholipid ratios. In this context, species can be

---

[*] **Corresponding author Mohammad A. Ibrahim:** Theriogenology Department, Faculty of Veterinary Medicine, Alexandria University, Alexandria Governorate 5424041, Egypt; Tel: 00201553330486; E-mail: mohammad.ibrahim@ alexu.edu.eg

ordered, according to their plasma membrane cholesterol to phospholipid ratio, hence sperm cryotolerance is as following: human (0.99-0.83), rabbit (0.88), bull (0.40-0.45), ram (0.38), stallion (0.36), rooster (0.30), and boar (0.20-0.26) [1 - 5].

EFFECT OF CRYOPRESERVATION ON SPERM CELLS

As with any plasma membrane, the spermatozoa plasma membrane is built up of lipids and proteins. Lipids are in the form of phospholipids and cholesterol. The phospholipids are arranged as a dynamic bilayer. Each phospholipid is composed of a phospho-head group and two fatty acyl chains. The fatty acyl chains are varying in length and structure ranging from 14 to 22 carbons and contain 0–6 unsaturated bonds. Cholesterol is composed of four lipophilic carbon rings (associated with the fatty acyl chains) and a carbon side chain (associated with the head groups of the phospholipid). Therefore, cholesterol can fill in any gaps created in the membrane's core because fatty acyl chains have different carbon lengths or different degrees of unsaturation. Cholesterol, therefore, helps to stabilize the membrane at body temperature. When the spermatozoa undergo cooling, the lipid parts undergo a transition state called a phase transition, where they transform from being in a fluid state into a solid (gel) state. Cellular plasma membranes do not undergo the phase transition at a single temperature but over a range of temperatures. This is because the phase transition of each fatty acyl chain occurs at a different temperature. As lipids undergo the phase transition coalesces into 'icebergs', they leave their protein partners, which in turn coalesce to form protein clumps within the membrane, which significantly alters their functional capacities. As the temperature is reduced further, more and more lipid species undergo the phase transition and coalesce in these 'icebergs' of solid lipid, leaving a constantly smaller portion of the fluid membrane. Finally, upon reaching the lowest temperature of the membrane range of transition, the entire membrane is in the 'gel' state. At −5 °C, the extracellular solute, the cells, and intracellular water are still not frozen but in a supercooled state. Between −5 to −15°C, the intracellular water is still in a supercooled state, but the extracellular medium starts to form ice crystals [6]. Due to concentration differences, the intracellular water will diffuse to the extracellular medium and freeze. If the cooling speed beyond -15°C is slow, almost all intracellular water will diffuse out to an external medium leading to dehydration and shrinkage of the cell, and ultimately a hyperosmotic shock. However, if the cooling speed is fast, ice crystal formation will start due to the presence of intracellular water that did not get out [7]. Therefore, sperm cooling and freezing rate should be fast sufficiently to circumvent spermatozoa dehydration and shrinkage but slow adequately to dodge intracellular ice crystals formation [7]. Moreover, cooling, and freezing temperatures lead to the compromise of sperm ATP-dependent ion channels of potassium, sodium, magnesium, and calcium leading to depolarization and an increase in permeability of plasma membrane and mitochondria, leading to premature capacitation, cell death, peroxidation, loss of lipids from the plasma membrane, and release of reactive oxygen species (ROS) [8].

The objective of cryopreservation protocol is the survival of sperm cells during freezing and after thawing. During thawing, sperm cells undergo rehydration, osmotic stress, and ultimately plasma membrane disruption. Therefore, optimizing sperm cryopreservation consists of optimizing three main components: sperm cryoprotectant, cooling/freezing rate, and thawing rate.

Cryoprotectants are used to decrease cryoinjury during freezing and thawing. Cryoprotectants are classified according to their ability to penetrate the sperm cell into permeating or non-permeating cryoprotectants. *Permeating cryoprotectant* is a type of cryoprotectant able to penetrate the sperm cell membrane, consequently, they change the viscosity of the cytoplasm and decrease the concertation of intracellular electrolytes. Therefore, upon subjecting sperm cells to freezing temperatures, they will cause dehydration of the sperm cell and reduce the formation of intracellular ice and minimize the extent of osmotic shrinkage of sperm cells [9, 10]. Glycerol is generally the most used permeating cryoprotectant in mammalian spermatozoa cryopreservation. Also, ethylene glycol and dimethylacetamide have been used as penetrating cryoprotectants [11]. Glycerol and other permeating cryoprotectants can penetrate the sperm cells, therefore it is relatively toxic to the sperm cells. Since cytoplasm viscosity of sperm cell differs between animal species [12], therefore, glycerol concentrations tolerability depend on animal species. For example, bull and boar semen is relatively tolerate higher concentration of glycerol compared with stallions [13]. *Non-permeating cryoprotectants are a* type of cryoprotectants that do not penetrate the sperm cell membrane. It acts extracellularly by stabilizing the plasma membrane, decreasing the freezing point of the extracellular compartment, therefore, lessening formations of ice crystals around the sperm cells [14]. Non-permeating cryoprotectants include sugars such as lactose or trehalose or dextran, and proteins [15]. The protein of animal sources has been used as a cryoprotectant, as chicken egg yolk and fat-free skimmed milk are widely utilized for different animals' sperm cryopreservation. However, there is an increasing interest to use animal-free cryoprotectants to avoid risks for viral or other disease transmissions.

As previously mentioned, since sperm cell wall composition is different between species, hence does each animal spices sperm prefer a particular freezing medium? what is the ideal concentration of the ingredients? The next sections will discuss the current understanding of cryopreservation protocols for different animal species.

Bull Sperm Cryopreservation

Different types of extenders have been used to decrease the cold shock effect on bovine sperm to enhance sperm parameters and fertility after thawing. Chicken egg yolk has been widely integrated into bovine semen extenders at a rate of 20%. The egg yolk cryoprotectant effect is mainly due to the presence of low-density lipoproteins (LDL) that coat the sperm plasma membrane and maintain components of the sperm plasma membrane during freezing and thawing [16, 17].

It is believed that seminal plasma proteins bound to sperm cause destabilization of the cell wall by causing cholesterol and phospholipid efflux, hence sperm become more susceptible to cold shock during freezing (Thérien *et al.*, 1998) (Manjunath *et al.*, 1988), (Desnoyers *et al.*, 1992; Manjunath *et al.*, 1994). (Thérien *et al.*, 1999). The LDL bind to proteins of seminal plasma, in addition, it cements the plasma membrane of the sperm; therefore, it increases the sperm cryotolerance during freezing.

Alternately to egg yolk, milk-based cryoprotectants have been used for cryopreservation of bovine sperm cells. The protective effect of these cryoprotectants is attributed to casein [18]. Casein binds to seminal plasma proteins, therefore preventing sluffing of lipids from the sperm plasma membrane (Bergeron *et al.*, 2007).

However, the use of animal-based products in semen extenders raised recently many concerns due to the risk of bacterial or xenobiotic contamination, presence of endotoxins, interference with sperm evaluation, variability of egg yolk composition from batch to batch which make quality control more difficult, and integrity of sperm cell more adversely affected [19]. Based on these issues and others, a growing interest raised in searching for alternatives for animal-based semen cryoprotectants for the artificial insemination industry. Therefore, animal protein-free media has been introduced [20].

Plant-based cryoprotectants were introduced as an alternative for milk and egg yolk-based extenders. Soybean contains lecithin used as a plant-based extender, providing adequate protection for sperm during cryopreservation while lessening the hazard of disease transmission. Many comparisons have been made between animal protein-based extenders (*e.g.* Triladyl, BullXcel, Laciphos) and animal protein-free extenders (*e.g.* OptiXcell, AndroMed, Biociphos plus). Even in vitro sperm parameters showed some difference, the pregnancy rate, 56-day, or 60-day nonreturn rate did not significantly differ [20 - 24].

Bull Sperm Freezing using Protein-free Extenders

According to observation, the plasma membrane of sperm cells loses its cholesterol during freezing [25]. Therefore, adding cholesterol to the freezing medium has improved post-thawing sperm parameters [26, 27]. Bull sperm was recently successfully cryopreserved exclusive of integrating any animal or plant protein [28]. This was done *via* the treatment of bull semen with cholesterol-cyclodextrin complex (CC) followed by dilution in an extender containing glycerol and tris salts (TG). The post-thawing sperm parameters were comparable to that frozen in exogenous protein-based extenders. Moreover, the protein profile of fresh sperm was comparable to that frozen in CC+TG. Therefore, the CC+TG extender could be a promising extender [28].

Bull Sperm Cryopreservation using Nanoparticles and Nanovesicles

Nanoparticles and nanovesicles (liposomes and exosomes) have been incorporated and their effect on decreasing cryoinjury has been investigated. Nanoparticles, derived from natural herbs or metals, act by decreasing oxidative stress, decreasing cell apoptosis, and enhancing plasma membrane integrity. Liposomes (synthetic nanovesicles) are a spherical single lipid bilayer, produced by sonication of lipid suspension (sonicated liposome) [29] or extrusion by passing lipid suspension through extruders have nano-sized pores (extruded liposome) [30]. Liposomes act by integrating into the sperm plasma membrane and repairing the impaired parts due to freezing and thawing. Exosomes are membrane-bounded nano-extracellular vesicles that are released from cells by exocytosis. Exosomes (*e.g.*, oviduct-derived) decease sperm cryoinjury by decreasing ROS and lipid peroxidation that takes place during freezing and thawing by delivering bioactive components such as antioxidant enzymes, miRNA, and mRNA [31 - 35]

Bull Sperm Lyophilization

Sperm lyophilization or freeze-drying was introduced as early as mid of 20th century [36]. However, the lack of reproducibility to fertilize an oocyte and production of full-term calves led to the loss of interest in sperm lyophilization [37, 38]. However, interest has been revived after the breakthrough work done on mouse sperm, in which live mice pups were produced using lyophilized sperm *via* ICSI [39]. Lyophilization of bull sperm may be a good substitute for conventional cryopreservation, due to fewer maintenance costs, easier transport, avoid hassle due to improper temperature control [40]. Some studies have stated the successful freeze-drying of the bovine spermatozoa, oocyte fertilization, and production of blastocysts [41 - 43]. However, efforts to enhance the technique are still needed to produce the first live calves from freeze-dried sperm.

Bull Sperm Encapsulation

Sperm encapsulations involve mixing a suspension of sperm with sodium alginate solution then mixing with ions (*e.g.*, barium ions) to form alginate gel beads. The bead suspension is then filled into semen straws and undergoes freezing using programmable freezers or conventional freezing in liquid nitrogen vapor using styrofoam boxes. Sperm encapsulation could help better protection of sperm during cryopreservation. In addition, it helps steady the release of sperm from the alginate gel beads, which may achieve a better conception rate. Pregnancy rates using encapsulated cryopreserved sperm did not differ significantly from conventionally cryopreserved sperm [44]. Bovine spermatozoa alginate encapsulation may be a promising technique for increasing sperm release and survival inside the uterus prolonging. However, the technology needs to be further optimized to be conventionally applicable in the field.

Bull Sperm multi-thermal Gradient Freezing

In a conventical slow freezing system, ice grows with uncontrolled velocity and morphology. In Multi-thermal gradient freezing, ice crystal propagation in semen is controlled during freezing by moving sperm samples through a gradient of linear cooling temperature so that ice crystal formation is accurately managed in terms of ice crystals morphology and formation velocity [45]. This method showed a better result when freezing sex-sorted sperm [46].

Prediction of Bovine Sperm Freezability

Prediction of bull frozen semen fertility is important for advancing the cattle artificial insemination industry. One important parameter that determines the fertility of bull frozen sperm is the ability to survive the cryodamage. Conventional semen analysis is not sufficient in predicting sperm freezability and it involves many steps such as semen collection, processing, freezing, and fertility trials. That is not convenient in terms of costs and time. Recently, proteomic studies have been done to identify biomarkers that are associated with cryotolerance of the bovine sperm. Bovine seminal ribonuclease protein and seminal plasma protein (BSP-5) were associated with high freezability semen, while tubulins, glucose-6-phosphate isomerase, peroxiredoxin-5, spermadhesin-1, gelsolin, sperm equatorial segment protein 1, ATP synthase, calmodulin, glyceraldehyde-3-phosphate dehydrogenase, and secretoglobin family 1D were associated with low freezability semen [47, 48]. Indeed, proteomic biomarkers of sperm freezability represent a promising area of research, and translation of these findings into the field use could provide a tool for the identification of semen samples worth processing for cryopreservation.

Bovine Sperm Selection before Cryopreservation

The selection of bulls according to their freezability could lead to the concentration of these bull genetics into the gene pool, which could be associated with selection for other undesirable traits. Bovine competent sperm selection before cryopreservation provides a better alternative rather than bull selection. Sperm collection could be done through sperm migration (swimming-up), sperm filtration (*e.g.*, using Sephadex beads), colloid centrifugation (*e.g.*, single layer centrifugation) [49, 50]. Recently, a sperm selection method based on rheotaxis and thermotaxis has been reported [51]. In this method, the sperm are incubated in a slowly rotating fluid (rheotaxis), which has a temperature gradient between the bottom at 25°C to the upper layers at 30°C (thermotaxis). Motile viable sperm will migrate to the upper layers, where they can be harvested. This method improved the post-thawing sperm parameters significantly in comparison with control [51].

Buck (Goat) Semen Cryopreservation

Goat's semen contains phospholipase enzymes of a bulbourethral origin that interfere with cryopreservation diluents that contain egg yolk or milk. One enzyme is the egg yolk-coagulating enzyme that acts on egg yolk lecithin to hydrolyze it into lysolecithin and fatty acids [52]. Also, bulbourethral gland lipase (BUSgp60) hydrolyzes skim milk triglycerides into fatty acid production like oleic acid [53]. Lysolecithin and oleic acid are toxic to the buck's sperm by causing chromatin decondensation and premature acrosomal reaction [54, 55]. Therefore, seminal plasma should be separated from buck's semen before freezing using milk or egg yolk-based dilutants [56]. Another alternative to seminal plasma removal and sperm washing is to use a lower level of egg yolk 2.5% rather than 20% [57, 58]. Plant-based diluents like soybean lecithin extenders showed to be better than egg yolk-based diluents [59, 60].

Many additives have been incorporated into caprine semen extenders to decrease cryoinjury and improve post-thawing semen fertility. Most of the additives act by decreasing sperm plasma membrane lipid peroxidation and release of the sperm-toxic ROS. Substances as antioxidants like vitamin E and butylated hydroxytoluene, IGF-1, vitamin C, and EDTA have been shown to improve post-thawing acrosomal and plasma membrane integrity, and motility [61 - 63].

Recently, nanoparticles additives have been incorporated into goat semen extenders. Nanoparticles of lecithin integrated at a rate of 2% into a tris-based extender, have been shown to improve motility, viability, and decrease apoptosis [64]. Also, nano-formulations of herbal extracts of curcumin, mint, and thyme have been shown to improve plasma membrane integrity and motility, while decreasing apoptosis and chromatin decondensation [65].

One of the interesting areas of research in semen cryopreservation is the prediction of semen freezability. Markers of goat's semen freezability beyond conventional sperm quality parameters [66] were investigated using molecular markers. A single nucleotide polymorphism (SNP) in the gene encoding for caprine heat shock protein 70 (HSP70) was correlated with better sperm freezability in goats [67]. Therefore, SNPs could be contemplated as early indicators for semen superiority and quality of goat-bucks.

Ram Semen Cryopreservation

Generally, ram semen cryopreservation protocols follow the same protocols used for buck goat semen, except the fact that ram's seminal plasma does not adversely interact with egg yolk or milk-based extenders [68]. To enhance the post-thaw fertility of ram frozen semen, many additives to ram semen extenders have been investigated. Casein, palm oil, and coconut oil have been compared to egg yolk in semen extenders, however, post-thawing sperm parameters were superior in presence of egg yolk was [69]. Natural or lyophilized low-density lipoproteins have enhanced post-thaw sperm motility [70]. Soybean lecithin in ram semen extenders gave good results comparable to egg yolk with the advantage of being a plant-based extender to avoid possible hazards of animal-based extenders [71, 72]. Treatment of ram semen with cyclodextrins pre-loaded with cholesterol improved cryosurvival and longevity [73]. Adding trehalose, a disaccharide that consists of two molecules of glucose, to semen extenders have been investigated in different animal species. In rams, adding trehalose to semen extenders has been shown to enhance post-thawing sperm parameters [74 - 77]. Trehalose cryoprotective effect could be credited to many theories. First, glass transition in trehalose occurs at higher temperatures (-30°C) than glycerol (-60°C) that leads to extracellular vitrification formation about sperm cells at an appropriate rate that reduces ice crystal formation. Second, through hydrogen bonds, trehalose replaces the water, shell, and therefore reduce cryoinjury. Third, it modifies the sperm plasma membrane fluidity and therefore decreases the effect of cryoinjury. Fourth, trehalose decreases sperm cell dehydration, therefore decreasing rapid alteration of cell volume and subsequent damage. Fifth, trehalose showed antioxidant properties. Sixth, trehalose might block acrosomal reactions induced by freezing and thawing [78 - 81].

Antifreeze proteins (AFPs) were discovered in Antarctic fish, which it uses to survive through freezing weather [82]. Adding AFPs to freezing medium has been shown to improve ram's post-thawing acrosome integrity of sheep spermatozoa

[83]. AFPs act by modifying ice crystal formation and prohibiting recrystallization [84]. Due to the high cost of natural AFPs, synthetic AFPs have been designed [85]. Synthetic AFP as cyclohexanetriol was found to enhance ram spermatozoa cryosurvival by decreasing apoptosis induced by the freezing and thawing [86].

Nanowater (NW) has been explored in semen extender during freezing of ram semen [87]. NW is generated by subjecting the water to cold plasma treatment (declusterization). During this procedure, water molecules clusters are divided down into single molecules each measuring one nanometer [88]. Semen diluent prepared with NW was correlated with a considerable enhancement in the frozen-thawed ram semen fertilizing ability and live lamb production from inseminated ewes [87]. The beneficial effect of NW on cyrosurvival of ram sperm cells may be due to improved stability of the mitochondrial and plasma membranes evident by lower aspartate aminotransferase and alkaline phosphatase [87].

Stallion Semen Cryopreservation

Advances in cryopreservation of stallion sperm go around improving and discovering new cryoprotectants, and sperm selection before cryopreservation. This is to increase the quality of frozen semen and to allow freezing retiring aged stallions or elite stallions with bad sperm freezability [89].

Due to the high volume of stallion ejaculates, seminal plasma should be removed before cryopreservation. The conventional method is to centrifuge semen and resuspend sperm pellet into the extender. However, centrifugation could be detrimental to sperm. Therefore, filtering the semen through a synthetic hydrophilic membrane has been used recently [90].

Sperm selection before cryopreservation is crucial to get rid of the non-viable sperm cell. Dead sperm cells are known to release a toxic substance that affects the other viable sperm population [91, 92]. Density gradient centrifugation is one of the methods used clinically for sperm selection. This method is based on viable, motile, and normal sperm cells that are denser, therefore, are separated from the lighter dead, immotile, and abnormal sperm cells when centrifuged through a dense colloid medium [93 - 95]. Deep horn insemination with frozen-thawed subjected to density gradient centrifugation did not improve pregnancy rates [96]. However, density gradient centrifugation may be useful for artificial insemination using ejaculates from a stallion with poor fertility [97].

Like other animal species, egg yolk, skimmed milk, and glycerol were universally incorporated into stallion semen freezing media. Also, plant-based extenders have been shown to provide cryoprotection for stallion sperm comparable to animal-based extenders [98]. However, a multicentric fertility trial is still needed to validate the use of plant-based extenders for stallion frozen semen. Other additives and constituents have been investigated to have an optimized formula for stallion cryopreservation. Amides as dimethylformamide and methylformamide were shown to improve post-thawing semen quality in bad freezer stallion [99]. The addition of amino acids such as betaine, proline, glutamine has been shown to improve post-thawing stallion semen parameters [100, 101]. Adding cholestanol-loaded-cyclodextrin to stallion semen freezing medium has improved post-thawing sperm parameters [102 - 104].

Freeze-drying or sperm lyophilization has been reported in stallion [105]. Freeze-drying of sperm is better than conventional freezing using liquid nitrogen due to fewer maintenance costs, easier way of transportation, and less born for contamination [106]. Freeze-drying of sperm despite being a promising tool, still needs a lot of optimizations in non-laboratory animals. The first live non-laboratory animal to be produced using freeze-dried sperm after ICSI was reported in equine [107]. This will encourage further research and optimization of freeze-drying in different animals.

Identification of freezability biomarkers and prediction of stallion sperm freezability represent great importance to identify good and poor freezability ejaculates before starting the cryopreservation. The conventional sperm parameters as motility and sperm membrane integrity are not enough to predict if a particular ejaculate will satisfactorily survive the cryoinjury [108]. It is found that good and poor freezability ejaculates differ in their ROS production and mitochondrial membrane potential [109]. Aquaporins (AQPs), a family of ubiquitous water channels, were found to be associated with stallion sperm freezability, especially AQP3, AQP7, AQP9, and AQP11 [110, 111]. Recently, classification and regression trees (CART) analysis were used to predict good and poor stallion ejaculates retrospectively. CART analysis was used to determine if there is a relationship between the percentage of ejaculates acceptable for AI and age and breed of the stallion as well as season [112]. It showed that more acceptable frozen-thawed stallion ejaculates depend on explanatory variables of stallion age and breed.

Cryopreservation of Camel-bull Semen

Cryopreservation of camel semen represents many challenges due to low ejaculate volume and sperm concentration, and high semen viscosity which means that sperm need to be liberated *via* liquefaction to hinder cryoprotect accessibility to sperm cells [113].

Efficient liquefaction is achieved by diluting the semen sample into a Tris-Citrat--Fructose extender at a rate of 1:5–1:10 according to semen viscosity and applying gentle aspiration with a sterile transfer pipette for about 30–60 min [113, 114].

Green buffer, an egg yolk-based extender, and a SHOTOR extender are often used for camel semen cryopreservation. SHOTOR is a Tris-based extender that contains 20% egg yolk, Tris, citric acid, glucose, fructose, penicillin G sodium, and streptomycin sulfate. Like many other species sperm cryopreservation, glycerol, and ethylene glycol are generally incorporated into camel semen freezing medium. Using green buffer, 3% glycerol or 6% ethylene glycol offered the best cryoprotection for camel sperm [115]. Using SHOTOR, dromedary camel semen cryopreservation could be enhanced using dimethyl formamide 6% rather than glycerol or ethylene glycol as a cryoprotectant agent [116].

The fertility of cryopreserved camel semen has not reached an acceptable rate to be offered commercially [113, 117, 118]. Therefore, much work needs to be done to optimize the fertility of camel cryopreserved sperm. In this endeavor, cholesterol-loaded cyclodextrins were used to pre-treat camel sperm before cryopreservation [119]. Despite enhancing post-thawing sperm parameters, it was not an advantage for *in vitro* or *in vivo* fertilization [120]. Another methodology was tried to enhance post-thawing camel semen fertility by sperm selection using colloid centrifugation. Using single layer centrifugation before sperm cryopreservation, camel sperm post-thawing parameters [114], and in vitro fertilization ability were significantly enhanced [121]. Therefore, colloid centrifugation represents a promising tool to incorporate into camel semen cryopreservation protocols.

New Technologies to be Adopted for Animal Sperm Cryopreservation

Freezing Small Numbers of Sperm

Semen samples are usually frozen into 250-500 μl volume containing millions of sperm. However, conventional cryopreservation methods are not convenient for freezing a small number of sperm cells in micro volumes (less than 100 μl). Therefore, biological and non-biological carriers have been used for this purpose [122]. Empty zona pellucida has been used as a bio-carrier for freezing a small number of sperm cells that are used to achieve successful pregnancy in humans [123, 124]. Individual human sperm cells were also cryopreserved in micro-vessels using an ultra-rapid cooling approach without cryoprotectant using polydimethylsiloxane chips in microfluidics [125]. Indeed, these methods were applied to human sperm cells, however, optimizing these methods to animal sperm cells may provide a useful tool to achieve pregnancy and live birth from rare, poor, or little sperm samples.

Solid Surface Vitrification (SSV)

Human sperm cell suspension mixed with cryoprotectant was loaded into thin capillaries, which allowed them to encounter a cold surface (−180°C). The vitrified droplets were transferred into a cryovial for storage in liquid nitrogen [126]. This method provided a rapid and practical way to cryopreserve human sperm cells with a comparable, if not better, post-thawing quality compared with the conventional cryopreservation method. This could open the door for cryopreservation of animal sperm cells using vitrification.

Sperm Separation via Electrophoresis

Before cryopreservation or after thawing, viable, motile, and normal sperm separation from dead, immotile, and abnormal sperm cell population is done through density gradient centrifugation. This method is widely adopted in the field. However, it was reported that this method was associated with apoptotic-like changes and oxidative DNA damage [127, 128]. Alternatively, sperm separation was done *via* electrophoresis, which was associated with free DNA damage [129]. This method is based on that high-quality sperm cells carry a negative charge; therefore, it migrates toward the anode in an electric field. This method accomplishes sperm separation within 20-30 minutes. This technique has been adopted for human sperm cells [129], and the first live baby produced using electrophoretically isolated spermatozoa was reported [130]. The development and adoption of this method for the separation of animal sperm cells will be invaluable [91].

CONCLUSION

Sperm freezing has been progressively improved over the last 20 years leading to optimized fertility in different animal species. However, the ultimate goal to achieve post-thawing sperm fertility similar to that of freshly ejaculated sperm is yet to be achieved. There are many promising areas needed to be explored to enhance the freezing of sperm. These areas include understanding the effect of freezing on the sperm cells, formulating the perfect freezing medium, and enhancing the method of frozen sperm cell delivery. With the progressive advances in artificial intelligence technologies and algorithms, the researcher should try these technologies to adjust sperm freezing.

CONSENT FOR PUBLICATION

Not applicable

CONFLICT OF INTEREST

The author declares no conflict of interest, financial or otherwise.

ACKNOWLEDGEMENT

Declared none.

REFERENCES

[1]   Parks JE, Lynch DV. Lipid composition and thermotropic phase behavior of boar, bull, stallion, and rooster sperm membranes. Cryobiology 1992; 29(2): 255-66.
[http://dx.doi.org/10.1016/0011-2240(92)90024-V] [PMID: 1582232]

[2]   Darin-Bennett A, White IG. Influence of the cholesterol content of mammalian spermatozoa on susceptibility to cold-shock. Cryobiology 1977; 14(4): 466-70.
[http://dx.doi.org/10.1016/0011-2240(77)90008-6] [PMID: 560945]

[3]   Parks JE, Hammerstedt RH. Development changes occurring in the lipids of ram epididymal spermatozoa plasma membrane. Biol Reprod 1985; 32(3): 653-68.
[http://dx.doi.org/10.1095/biolreprod32.3.653] [PMID: 3158356]

[4]   Mack SR, Everingham J, Zaneveld LJD. Isolation and partial characterization of the plasma membrane from human spermatozoa. J Exp Zool 1986; 240(1): 127-36.
[http://dx.doi.org/10.1002/jez.1402400116] [PMID: 3021893]

[5]   Parks JE, Arion JW, Foote RH. Lipids of plasma membrane and outer acrosomal membrane from bovine spermatozoa. Biol Reprod 1987; 37(5): 1249-58.
[http://dx.doi.org/10.1095/biolreprod37.5.1249] [PMID: 2832008]

[6]   Gao D, Critser JK. Mechanisms of cryoinjury in living cells. ILAR J 2000; 41(4): 187-96.
[http://dx.doi.org/10.1093/ilar.41.4.187] [PMID: 11123179]

[7]   Yeste M. Sperm cryopreservation update: Cryodamage, markers, and factors affecting the sperm freezability in pigs. Theriogenology 2016; 85(1): 47-64.
[http://dx.doi.org/10.1016/j.theriogenology.2015.09.047] [PMID: 26506124]

[8]   Amann RP, Pickett BW. Principles of cryopreservation and a review of cryopreservation of stallion spermatozoa. J Equine Vet Sci 1987; 7: 145-73.
[http://dx.doi.org/10.1016/S0737-0806(87)80025-4]

[9]   Holt WV. Basic aspects of frozen storage of semen. Anim Reprod Sci 2000; 62(1-3): 3-22.
[http://dx.doi.org/10.1016/S0378-4320(00)00152-4] [PMID: 10924818]

[10]   Lovelock JE, Polge C. The immobilization of spermatozoa by freezing and thawing and the protective action of glycerol. Biochem J 1954; 58(4): 618-22.
[http://dx.doi.org/10.1042/bj0580618] [PMID: 13230012]

[11]   Yeste M, Rodríguez-Gil JE, Bonet S. Artificial insemination with frozen-thawed boar sperm. Mol Reprod Dev 2017; 84(9): 802-13.
[http://dx.doi.org/10.1002/mrd.22840] [PMID: 28608609]

[12]   Hammerstedt RH, Keith AD, Snipes W, Amann RP, Arruda D, Griel LC Jr. Use of spin labels to evaluate effects of cold shock and osmolality on sperm. Biol Reprod 1978; 18(4): 686-96.
[http://dx.doi.org/10.1095/biolreprod18.4.686] [PMID: 656535]

[13]   Moore AI, Squires EL, Bruemmer JE, Graham JK. Effect of cooling rate and cryoprotectant on the cryosurvival of equine spermatozoa. J Equine Vet Sci 2006; 26: 215-8.
       [http://dx.doi.org/10.1016/j.jevs.2006.03.003]

[14]   Hammerstedt RH, Graham JK, Nolan JP. Cryopreservation of mammalian sperm: what we ask them to survive. J Androl 1990; 11(1): 73-88.
       [PMID: 2179184]

[15]   Hezavehei M, Sharafi M, Kouchesfahani HM, *et al.* Sperm cryopreservation: A review on current molecular cryobiology and advanced approaches. Reprod Biomed Online 2018; 37(3): 327-39.
       [http://dx.doi.org/10.1016/j.rbmo.2018.05.012] [PMID: 30143329]

[16]   Medeiros CMO, Forell F, Oliveira ATD, Rodrigues JL. Current status of sperm cryopreservation: why isn't it better? Theriogenology 2002; 57(1): 327-44.
       [http://dx.doi.org/10.1016/S0093-691X(01)00674-4] [PMID: 11775978]

[17]   Bergeron A, Crête M-H, Brindle Y, Manjunath P. Low-density lipoprotein fraction from hen's egg yolk decreases the binding of the major proteins of bovine seminal plasma to sperm and prevents lipid efflux from the sperm membrane. Biol Reprod 2004; 70(3): 708-17.
       [http://dx.doi.org/10.1095/biolreprod.103.022996] [PMID: 14613896]

[18]   Bergeron A, Manjunath P. New insights towards understanding the mechanisms of sperm protection by egg yolk and milk. Mol Reprod Dev 2006; 73(10): 1338-44.
       [http://dx.doi.org/10.1002/mrd.20565] [PMID: 16868925]

[19]   Layek SS, Mohanty TK, Kumaresan A, Parks JE. Cryopreservation of bull semen: Evolution from egg yolk based to soybean based extenders. Anim Reprod Sci 2016; 172: 1-9.
       [http://dx.doi.org/10.1016/j.anireprosci.2016.04.013] [PMID: 27509873]

[20]   Aires VA, Hinsch K-D, Mueller-Schloesser F, Bogner K, Mueller-Schloesser S, Hinsch E. In vitro and in vivo comparison of egg yolk-based and soybean lecithin-based extenders for cryopreservation of bovine semen. Theriogenology 2003; 60(2): 269-79.
       [http://dx.doi.org/10.1016/S0093-691X(02)01369-9] [PMID: 12749940]

[21]   Murphy EM, O'Meara C, Eivers B, Lonergan P, Fair S. Comparison of plant- and egg yolk-based semen diluents on in vitro sperm kinematics and in vivo fertility of frozen-thawed bull semen. Anim Reprod Sci 2018; 191: 70-5.
       [http://dx.doi.org/10.1016/j.anireprosci.2018.02.010] [PMID: 29496341]

[22]   van Wagtendonk-de Leeuw AM, Haring RM, Kaal-Lansbergen LMTE, den Daas JHG. Fertility results using bovine semen cryopreserved with extenders based on egg yolk and soy bean extract. Theriogenology 2000; 54(1): 57-67.
       [http://dx.doi.org/10.1016/S0093-691X(00)00324-1] [PMID: 10990347]

[23]   Arifiantini RI, Yusuf TL. Developing of Tris Soy Milk Diluent for Frisian Holstein Bull Frozen Semen. Hayati J Biosci 2010; 17: 91-4.
       [http://dx.doi.org/10.4308/hjb.17.2.91]

[24]   Muiño R, Fernández M, Peña AI. Post-thaw survival and longevity of bull spermatozoa frozen with an egg yolk-based or two egg yolk-free extenders after an equilibration period of 18 h. Reprod Domest Anim 2007; 42(3): 305-11.
       [http://dx.doi.org/10.1111/j.1439-0531.2006.00784.x] [PMID: 17506810]

[25]   Bailey JL, Bilodeau JF, Cormier N. Semen cryopreservation in domestic animals: a damaging and capacitating phenomenon. J Androl 2000; 21(1): 1-7.
       [PMID: 10670514]

[26]   Yu Y, Chipot C, Cai W, Shao X. Molecular dynamics study of the inclusion of cholesterol into cyclodextrins. J Phys Chem B 2006; 110(12): 6372-8.
       [http://dx.doi.org/10.1021/jp056751a] [PMID: 16553456]

[27]   Purdy PH, Graham JK. Effect of cholesterol-loaded cyclodextrin on the cryosurvival of bull sperm.

Cryobiology 2004; 48(1): 36-45.
[http://dx.doi.org/10.1016/j.cryobiol.2003.12.001] [PMID: 14969680]

[28]   Anzar M, Rajapaksha K, Boswall L. Egg yolk-free cryopreservation of bull semen. PLoS One 2019;
       14(10): e0223977.
       [http://dx.doi.org/10.1371/journal.pone.0223977] [PMID: 31613900]

[29]   Graham JK, Foote RH. Effect of several lipids, fatty acyl chain length, and degree of unsaturation on
       the motility of bull spermatozoa after cold shock and freezing. Cryobiology 1987; 24(1): 42-52.
       [http://dx.doi.org/10.1016/0011-2240(87)90005-8] [PMID: 3816287]

[30]   Röpke T, Oldenhof H, Leiding C, Sieme H, Bollwein H, Wolkers WF. Liposomes for cryopreservation
       of bovine sperm. Theriogenology 2011; 76(8): 1465-72.
       [http://dx.doi.org/10.1016/j.theriogenology.2011.06.015] [PMID: 21820724]

[31]   Saadeldin IM, Khalil WA, Alharbi MG, Lee SH. The Current Trends in Using Nanoparticles,
       Liposomes, and Exosomes for Semen Cryopreservation. Animals (Basel) 2020; 10(12): 10.
       [http://dx.doi.org/10.3390/ani10122281] [PMID: 33287256]

[32]   Jahanbin R, Yazdanshenas P, Amin Afshar M, Mohammadi Sangcheshmeh A, Varnaseri H, Chamani
       M, *et al.* Effect of zinc nano-complex on bull semen quality after freeze-thawing process. Anim Prod
       2015; 17: 371-80.

[33]   Mousavi SM, Towhidi A, Zhandi M, *et al.* Comparison of two different antioxidants in a nano
       lecithin-based extender for bull sperm cryopreservation. Anim Reprod Sci 2019; 209: 106171.
       [http://dx.doi.org/10.1016/j.anireprosci.2019.106171] [PMID: 31514920]

[34]   Khalil WA, El-Harairy MA, Zeidan AEB, Hassan MAE. Impact of selenium nano-particles in semen
       extender on bull sperm quality after cryopreservation. Theriogenology 2019; 126: 121-7.
       [http://dx.doi.org/10.1016/j.theriogenology.2018.12.017] [PMID: 30551018]

[35]   Franchi A, Moreno-Irusta A, Domínguez EM, Adre AJ, Giojalas LC. Extracellular vesicles from
       oviductal isthmus and ampulla stimulate the induced acrosome reaction and signaling events
       associated with capacitation in bovine spermatozoa. J Cell Biochem 2020; 121(4): 2877-88.
       [http://dx.doi.org/10.1002/jcb.29522] [PMID: 31692037]

[36]   Polge C, Smith AU, Parkes AS. Revival of spermatozoa after vitrification and dehydration at low
       temperatures. Nature 1949; 164(4172): 666.
       [http://dx.doi.org/10.1038/164666a0] [PMID: 18143360]

[37]   Saacke RG, Almquist JO. Freeze-drying of bovine spermatozoa. Nature 1961; 192: 995-6.
       [http://dx.doi.org/10.1038/192995a0] [PMID: 14495721]

[38]   Meryman HT, Kafig E. Special Article Freeze-drying of Bovine Spermatozoa. Reproduction 1963; 5:
       87.
       [http://dx.doi.org/10.1530/jrf.0.0050087]

[39]   Wakayama T, Yanagimachi R. Development of normal mice from oocytes injected with freeze-dried
       spermatozoa. Nat Biotechnol 1998; 16(7): 639-41.
       [http://dx.doi.org/10.1038/nbt0798-639] [PMID: 9661196]

[40]   Sathe S, Shipley CF. Cryopreservation of Semen. Bovine Reproduction 2014; pp. 662-70.

[41]   Keskintepe L, Pacholczyk G, Machnicka A, *et al.* Bovine blastocyst development from oocytes
       injected with freeze-dried spermatozoa. Biol Reprod 2002; 67(2): 409-15.
       [http://dx.doi.org/10.1095/biolreprod67.2.409] [PMID: 12135874]

[42]   Martins CF, Báo SN, Dode MN, Correa GA, Rumpf R. Effects of freeze-drying on cytology,
       ultrastructure, DNA fragmentation, and fertilizing ability of bovine sperm. Theriogenology 2007;
       67(8): 1307-15.
       [http://dx.doi.org/10.1016/j.theriogenology.2007.01.015] [PMID: 17383718]

[43]   Hara H, Abdalla H, Morita H, Kuwayama M, Hirabayashi M, Hochi S. Procedure for bovine ICSI, not

sperm freeze-drying, impairs the function of the microtubule-organizing center. J Reprod Dev 2011; 57(3): 428-32.
[http://dx.doi.org/10.1262/jrd.10-167N] [PMID: 21325738]

[44] Perteghella S, Gaviraghi A, Cenadelli S, *et al.* Alginate encapsulation preserves the quality and fertilizing ability of Mediterranean Italian water buffalo (*Bubalus bubalis*) and Holstein Friesian (*Bos taurus*) spermatozoa after cryopreservation. J Vet Sci 2017; 18(1): 81-8.
[http://dx.doi.org/10.4142/jvs.2017.18.1.81] [PMID: 27456772]

[45] Arav A, Yavin S, Zeron Y, Natan D, Dekel I, Gacitua H. New trends in gamete's cryopreservation. Mol Cell Endocrinol 2002; 187(1-2): 77-81.
[http://dx.doi.org/10.1016/S0303-7207(01)00700-6] [PMID: 11988314]

[46] Arav A, Saragusty J. Directional freezing of sperm and associated derived technologies. Anim Reprod Sci 2016; 169: 6-13.
[http://dx.doi.org/10.1016/j.anireprosci.2016.02.007] [PMID: 26879097]

[47] Gomes FP, Park R, Viana AG, *et al.* Protein signatures of seminal plasma from bulls with contrasting frozen-thawed sperm viability. Sci Rep 2020; 10(1): 14661.
[http://dx.doi.org/10.1038/s41598-020-71015-9] [PMID: 32887897]

[48] Ryu D-Y, Song W-H, Pang W-K, Yoon S-J, Rahman MS, Pang M-G. Freezability biomarkers in bull epididymal spermatozoa. Sci Rep 2019; 9(1): 12797.
[http://dx.doi.org/10.1038/s41598-019-49378-5] [PMID: 31488871]

[49] Morrell JM, Rodriguez-Martinez H. Practical applications of sperm selection techniques as a tool for improving reproductive efficiency. Vet Med Int 2010; 2011: 894767.
[PMID: 20814435]

[50] Januskauskas A, Lukoseviciute K, Nagy S, Johannisson A, Rodriguez-Martinez H. Assessment of the efficacy of Sephadex G-15 filtration of bovine spermatozoa for cryopreservation. Theriogenology 2005; 63(1): 160-78.
[http://dx.doi.org/10.1016/j.theriogenology.2004.04.002] [PMID: 15589282]

[51] Nagata MB, Egashira J, Katafuchi N, *et al.* Bovine sperm selection procedure prior to cryopreservation for improvement of post-thawed semen quality and fertility. J Anim Sci Biotechnol 2019; 10: 91.
[http://dx.doi.org/10.1186/s40104-019-0395-9] [PMID: 31807306]

[52] Iritani A, Nishikawa Y. Studies on the egg yolk coagluting enzyme in goat semen IV. 1963.

[53] Pellicer-Rubio MT, Magallon T, Combarnous Y. Deterioration of goat sperm viability in milk extenders is due to a bulbourethral 60-kilodalton glycoprotein with triglyceride lipase activity. Biol Reprod 1997; 57(5): 1023-31.
[http://dx.doi.org/10.1095/biolreprod57.5.1023] [PMID: 9369166]

[54] Purdy PH. A review on goat sperm cryopreservation. Small Rumin Res 2006; 63: 215-25.
[http://dx.doi.org/10.1016/j.smallrumres.2005.02.015]

[55] Leboeuf B, Restall B, Salamon S. Production and storage of goat semen for artificial insemination. Anim Reprod Sci 2000; 62(1-3): 113-41.
[http://dx.doi.org/10.1016/S0378-4320(00)00156-1] [PMID: 10924822]

[56] Cabrera F, González F, Batista M, Calero P, Medrano A, Gracia A. The effect of removal of seminal plasma, egg yolk level and season on sperm freezability of canary buck (Capra hircus). Reprod Domest Anim 2005; 40(3): 191-5.
[http://dx.doi.org/10.1111/j.1439-0531.2005.00544.x] [PMID: 15943690]

[57] Cseh S, Faigl V, Amiridis GS. Semen processing and artificial insemination in health management of small ruminants. Anim Reprod Sci 2012; 130(3-4): 187-92.
[http://dx.doi.org/10.1016/j.anireprosci.2012.01.014] [PMID: 22333390]

[58] Ibrahim M, El-Bawab I, El-Amrawi G. Evaluation of the protective effect of duck egg yolk during cooling of buck semen. Alex J Vet Sci 2012; 37: 87-100.

[http://dx.doi.org/10.5455/ajvs.41000]

[59]   Vidal AH, Batista AM, da Silva ECB, *et al.* Soybean lecithin-based extender as an alternative for goat sperm cryopreservation. Small Rumin Res 2013; 109: 47-51.
[http://dx.doi.org/10.1016/j.smallrumres.2012.07.022]

[60]   Chelucci S, Pasciu V, Succu S, *et al.* Soybean lecithin-based extender preserves spermatozoa membrane integrity and fertilizing potential during goat semen cryopreservation. Theriogenology 2015; 83(6): 1064-74.
[http://dx.doi.org/10.1016/j.theriogenology.2014.12.012] [PMID: 25595356]

[61]   Saraswat S, Jindal S, Kharche S, Rout P, Ranjan R, Priyadharsini R. Role of antioxidant additives in the protection of DNA integrity of buck spermatozoa with RAPD assay. Indian J Anim Sci 2013; 84: 3.

[62]   Padilha RT, Magalhães-Padilha DM, Cavalcante MM, *et al.* Effect of insulin-like growth factor-I on some quality traits and fertility of cryopreserved ovine semen. Theriogenology 2012; 78(4): 907-13.
[http://dx.doi.org/10.1016/j.theriogenology.2012.04.005] [PMID: 22541171]

[63]   Kalthur G, Raj S, Thiyagarajan A, Kumar S, Kumar P, Adiga SK. Vitamin E supplementation in semen-freezing medium improves the motility and protects sperm from freeze-thaw-induced DNA damage. Fertil Steril 2011; 95(3): 1149-51.
[http://dx.doi.org/10.1016/j.fertnstert.2010.10.005] [PMID: 21067726]

[64]   Nadri T, Towhidi A, Zeinoaldini S, *et al.* Lecithin nanoparticles enhance the cryosurvival of caprine sperm. Theriogenology 2019; 133: 38-44.
[http://dx.doi.org/10.1016/j.theriogenology.2019.04.024] [PMID: 31055160]

[65]   Ismail AA, Abdel-Khalek AE, Khalil WA, *et al.* Effects of mint, thyme, and curcumin extract nanoformulations on the sperm quality, apoptosis, chromatin decondensation, enzyme activity, and oxidative status of cryopreserved goat semen. Cryobiology 2020; 97: 144-52.
[http://dx.doi.org/10.1016/j.cryobiol.2020.09.002] [PMID: 32916165]

[66]   Dorado J, Muñoz-Serrano A, Hidalgo M. The effect of cryopreservation on goat semen characteristics related to sperm freezability. Anim Reprod Sci 2010; 121(1-2): 115-23.
[http://dx.doi.org/10.1016/j.anireprosci.2010.04.182] [PMID: 20462708]

[67]   Nikbin S, Panandam JM, Yaakub H, Murugaiyah M, Sazili AQ. Novel SNPs in heat shock protein 70 gene and their association with sperm quality traits of Boer goats and Boer crosses. Anim Reprod Sci 2014; 146(3-4): 176-81.
[http://dx.doi.org/10.1016/j.anireprosci.2014.03.001] [PMID: 24674824]

[68]   Ntemka A, Tsakmakidis I, Kiossis E, Milovanović A, Boscos C. Current status and advances in ram semen cryopreservation. J Hell Vet Med Soc 2018; 69: 911-24.
[http://dx.doi.org/10.12681/jhvms.18014]

[69]   Del Valle I, Souter A, Maxwell WM, Muiño-Blanco T, Cebrián-Pérez JA. Function of ram spermatozoa frozen in diluents supplemented with casein and vegetable oils. Anim Reprod Sci 2013; 138(3-4): 213-9.
[http://dx.doi.org/10.1016/j.anireprosci.2013.02.022] [PMID: 23561943]

[70]   Moustacas VS, Zaffalon FG, Lagares MA, *et al.* Natural, but not lyophilized, low density lypoproteins were an acceptable alternative to egg yolk for cryopreservation of ram semen. Theriogenology 2011; 75(2): 300-7.
[http://dx.doi.org/10.1016/j.theriogenology.2010.08.016] [PMID: 20934209]

[71]   Masoudi R, Sharafi M, Zare Shahneh A, *et al.* Effect of dietary fish oil supplementation on ram semen freeze ability and fertility using soybean lecithin- and egg yolk-based extenders. Theriogenology 2016; 86(6): 1583-8.
[http://dx.doi.org/10.1016/j.theriogenology.2016.05.018] [PMID: 27374423]

[72]   Emamverdi M, Zhandi M, Zare Shahneh A, Sharafi M, Akbari-Sharif A. Optimization of ram semen

cryopreservation using a chemically defined soybean lecithin-based extender. Reprod Domest Anim 2013; 48(6): 899-904.
[http://dx.doi.org/10.1111/rda.12183] [PMID: 23701189]

[73]   Mocé E, Purdy PH, Graham JK. Treating ram sperm with cholesterol-loaded cyclodextrins improves cryosurvival. Anim Reprod Sci 2010; 118(2-4): 236-47.
[http://dx.doi.org/10.1016/j.anireprosci.2009.06.013] [PMID: 19615833]

[74]   Quan GB, Hong QH, Hong QY, Yang HY, Wu SS. The effects of trehalose and sucrose on frozen spermatozoa of Yunnan semi-fine wool sheep during a non-mating season. Cryo Lett 2012; 33(4): 307-17.
[PMID: 22987242]

[75]   Aisen EG, Medina VH, Venturino A. Cryopreservation and post-thawed fertility of ram semen frozen in different trehalose concentrations. Theriogenology 2002; 57(7): 1801-8.
[http://dx.doi.org/10.1016/S0093-691X(02)00653-2] [PMID: 12041684]

[76]   Bucak MN, Ateşşahin A, Varişli O, Yüce A, Tekin N, Akçay A. The influence of trehalose, taurine, cysteamine and hyaluronan on ram semen Microscopic and oxidative stress parameters after freeze-thawing process. Theriogenology 2007; 67(5): 1060-7.
[http://dx.doi.org/10.1016/j.theriogenology.2006.12.004] [PMID: 17280711]

[77]   Jafaroghli M, Khalili B, Farshad A, Zamiri MJ. The effect of supplementation of cryopreservation diluents with sugars on the post-thawing fertility of ram semen. Small Rumin Res 2011; 96: 58-63.
[http://dx.doi.org/10.1016/j.smallrumres.2010.11.010]

[78]   Lv C, Wu G, Hong Q, Quan G. Spermatozoa Cryopreservation: State of Art and Future in Small Ruminants. Biopreserv Biobank 2019; 17(2): 171-82.
[http://dx.doi.org/10.1089/bio.2018.0113] [PMID: 30499684]

[79]   Ahmad E, Naseer Z, Aksoy M, *et al.* Trehalose enhances osmotic tolerance and suppresses lysophosphatidylcholine-induced acrosome reaction in ram spermatozoon. Andrologia 2015; 47(7): 786-92.
[http://dx.doi.org/10.1111/and.12329] [PMID: 25269572]

[80]   Crowe JH, Crowe LM, Oliver AE, Tsvetkova N, Wolkers W, Tablin F. The trehalose myth revisited: introduction to a symposium on stabilization of cells in the dry state. Cryobiology 2001; 43(2): 89-105.
[http://dx.doi.org/10.1006/cryo.2001.2353] [PMID: 11846464]

[81]   Aboagla EM, Terada T. Trehalose-enhanced fluidity of the goat sperm membrane and its protection during freezing. Biol Reprod 2003; 69(4): 1245-50.
[http://dx.doi.org/10.1095/biolreprod.103.017889] [PMID: 12801983]

[82]   DeVries AL, Wohlschlag DE. Freezing resistance in some Antarctic fishes. Science 1969; 163(3871): 1073-5.
[http://dx.doi.org/10.1126/science.163.3871.1073] [PMID: 5764871]

[83]   Payne SR, Oliver JE, Upreti GC. Effect of antifreeze proteins on the motility of ram spermatozoa. Cryobiology 1994; 31(2): 180-4.
[http://dx.doi.org/10.1006/cryo.1994.1021] [PMID: 8004998]

[84]   Rubinsky B, Arav A, Devries AL. The cryoprotective effect of antifreeze glycopeptides from antarctic fishes. Cryobiology 1992; 29(1): 69-79.
[http://dx.doi.org/10.1016/0011-2240(92)90006-N] [PMID: 1606831]

[85]   Chou KC. Energy-optimized structure of antifreeze protein and its binding mechanism. J Mol Biol 1992; 223(2): 509-17.
[http://dx.doi.org/10.1016/0022-2836(92)90666-8] [PMID: 1738160]

[86]   Quan GB, Wu SS, Lan ZG, Yang HY, Shao QY, Hong QH. The effects of 1,4-cyclohexanediol on frozen ram spermatozoa. Cryo Lett 2013; 34(3): 217-27.

[PMID: 23812311]

[87]   Murawski M, Schwarz T, Grygier J, *et al.* The utility of nanowater for ram semen cryopreservation. Exp Biol Med (Maywood) 2015; 240(5): 611-7.
[http://dx.doi.org/10.1177/1535370214557219] [PMID: 25491414]

[88]   Mystkowska J, Dąbrowski J, Kowal K, Niemirowicz K, Car H. Physical and chemical properties of deionized water and saline treated with low-pressure and low-temperature plasma. Chemik 2013; 67: 719-24.

[89]   Alvarenga MA, Papa FO, Ramires Neto C. Advances in Stallion Semen Cryopreservation. Vet Clin North Am Equine Pract 2016; 32(3): 521-30.
[http://dx.doi.org/10.1016/j.cveq.2016.08.003] [PMID: 27726989]

[90]   Alvarenga MA, Papa FO, Carmo MT, Kievitsbosch T, Castro Chaves MMB, Ramires Neto C. Methods of Concentrating Stallion Semen. J Equine Vet Sci 2012; 32: 424-9.
[http://dx.doi.org/10.1016/j.jevs.2012.06.003]

[91]   Gibb Z, Aitken RJ. Recent Developments in Stallion Semen Preservation. J Equine Vet Sci 2016; 43: S29-36.
[http://dx.doi.org/10.1016/j.jevs.2016.06.006]

[92]   Ball BA, Vo AT, Baumber J. Generation of reactive oxygen species by equine spermatozoa. Am J Vet Res 2001; 62(4): 508-15.
[http://dx.doi.org/10.2460/ajvr.2001.62.508] [PMID: 11327456]

[93]   Morrell JM, Dalin AM, Rodriguez-Martinez H. Comparison of density gradient and single layer centrifugation of stallion spermatozoa: yield, motility and survival. Equine Vet J 2009; 41(1): 53-8.
[http://dx.doi.org/10.2746/042516408X322139] [PMID: 19301582]

[94]   Johannisson A, Morrell JM, Thorén J, Jönsson M, Dalin AM, Rodriguez-Martinez H. Colloidal centrifugation with Androcoll-E prolongs stallion sperm motility, viability and chromatin integrity. Anim Reprod Sci 2009; 116(1-2): 119-28.
[http://dx.doi.org/10.1016/j.anireprosci.2009.01.008] [PMID: 19217226]

[95]   Macías García B, Morrell JM, Ortega-Ferrusola C, *et al.* Centrifugation on a single layer of colloid selects improved quality spermatozoa from frozen-thawed stallion semen. Anim Reprod Sci 2009; 114(1-3): 193-202.
[http://dx.doi.org/10.1016/j.anireprosci.2008.08.025] [PMID: 18842364]

[96]   Cerny KL, Hughes S, Campos JR, Coleman RJ, Troedsson MHT, Squires EL. Fertility of Mares Inseminated With Frozen-Thawed Semen Processed by Single Layer Centrifugation Through a Colloid. J Equine Vet Sci 2012; 32: 289-91.
[http://dx.doi.org/10.1016/j.jevs.2011.09.075]

[97]   Morrell JM, Mari G, Kútvölgyi G, *et al.* Pregnancies following artificial insemination with spermatozoa from problem stallion ejaculates processed by single layer centrifugation with Androcoll-E. Reprod Domest Anim 2011; 46(4): 642-5.
[http://dx.doi.org/10.1111/j.1439-0531.2010.01721.x] [PMID: 21114793]

[98]   Aspland AM, Satake N, Cawdell-Smith AJ, Collins CW. Assessing novel, plant-based cryopreservation extenders for freezing stallion semen. J Equine Vet Sci 2014; 34: 54-5.
[http://dx.doi.org/10.1016/j.jevs.2013.10.032]

[99]   Alvarenga MA, Papa FO, Landim-Alvarenga FC, Medeiros ASL. Amides as cryoprotectants for freezing stallion semen: a review. Anim Reprod Sci 2005; 89(1-4): 105-13.
[http://dx.doi.org/10.1016/j.anireprosci.2005.07.001] [PMID: 16099609]

[100]   Koskinen E, Junnila M, Katila T, Soini H. A preliminary study on the use of betaine as a cryoprotective agent in deep freezing of stallion semen. Zentralbl Veterinärmed A 1989; 36(2): 110-4.
[http://dx.doi.org/10.1111/j.1439-0442.1989.tb00710.x] [PMID: 2501949]

[101]   Trimeche A, Yvon JM, Vidament M, Palmer E, Magistrini M. Effects of glutamine, proline, histidine

and betaine on post-thaw motility of stallion spermatozoa. Theriogenology 1999; 52(1): 181-91.
[http://dx.doi.org/10.1016/S0093-691X(99)00120-X] [PMID: 10734416]

[102] Moraes EA, Matos WCG, Graham JK, Ferrari WD Jr. Cholestanol-loaded-cyclodextrin improves the quality of stallion spermatozoa after cryopreservation. Anim Reprod Sci 2015; 158: 19-24.
[http://dx.doi.org/10.1016/j.anireprosci.2015.04.004] [PMID: 26005214]

[103] Spizziri BE, Fox MH, Bruemmer JE, Squires EL, Graham JK. Cholesterol-loaded-cyclodextrins and fertility potential of stallions spermatozoa. Anim Reprod Sci 2010; 118(2-4): 255-64.
[http://dx.doi.org/10.1016/j.anireprosci.2009.08.001] [PMID: 19762177]

[104] Oliveira CH, Vasconcelos AB, Souza FA, *et al.* Cholesterol addition protects membrane intactness during cryopreservation of stallion sperm. Anim Reprod Sci 2010; 118(2-4): 194-200.
[http://dx.doi.org/10.1016/j.anireprosci.2009.08.011] [PMID: 19758774]

[105] Choi YH, Love CC, Varner DD, Hinrichs K. Equine blastocyst development after intracytoplasmic injection of sperm subjected to two freeze-thaw cycles. Theriogenology 2006; 65(4): 808-19.
[http://dx.doi.org/10.1016/j.theriogenology.2005.04.035] [PMID: 16095679]

[106] Tedder RS, Zuckerman MA, Goldstone AH, *et al.* Hepatitis B transmission from contaminated cryopreservation tank. Lancet 1995; 346(8968): 137-40.
[http://dx.doi.org/10.1016/S0140-6736(95)91207-X] [PMID: 7603227]

[107] Choi YH, Varner DD, Love CC, Hartman DL, Hinrichs K. Production of live foals *via* intracytoplasmic injection of lyophilized sperm and sperm extract in the horse. Reproduction 2011; 142(4): 529-38.
[http://dx.doi.org/10.1530/REP-11-0145] [PMID: 21846810]

[108] Kuisma P, Andersson M, Koskinen E, Katila T. Fertility of frozen-thawed stallion semen cannot be predicted by the currently used laboratory methods. Acta Vet Scand 2006; 48: 14.
[http://dx.doi.org/10.1186/1751-0147-48-14] [PMID: 16987393]

[109] Yeste M, Estrada E, Rocha LG, Marín H, Rodríguez-Gil JE, Miró J. Cryotolerance of stallion spermatozoa is related to ROS production and mitochondrial membrane potential rather than to the integrity of sperm nucleus. Andrology 2015; 3(2): 395-407.
[http://dx.doi.org/10.1111/andr.291] [PMID: 25294093]

[110] Delgado-Bermúdez A, Noto F, Bonilla-Correal S, *et al.* Cryotolerance of Stallion Spermatozoa Relies on Aquaglyceroporins rather than Orthodox Aquaporins. Biology (Basel) 2019; 8(4): 85.
[http://dx.doi.org/10.3390/biology8040085] [PMID: 31726707]

[111] Bonilla-Correal S, Noto F, Garcia-Bonavila E, Rodríguez-Gil JE, Yeste M, Miro J. First evidence for the presence of aquaporins in stallion sperm. Reprod Domest Anim 2017; 52 (Suppl. 4): 61-4.
[http://dx.doi.org/10.1111/rda.13059] [PMID: 29052325]

[112] Aurich J, Kuhl J, Tichy A, Aurich C. Efficiency of Semen Cryopreservation in Stallions. Animals (Basel) 2020; 10(6): 1033.
[http://dx.doi.org/10.3390/ani10061033] [PMID: 32545785]

[113] Skidmore JA, Malo CM, Crichton EG, Morrell JM, Pukazhenthi BS. An update on semen collection, preservation and artificial insemination in the dromedary camel (Camelus dromedarius). Anim Reprod Sci 2018; 194: 11-8.
[http://dx.doi.org/10.1016/j.anireprosci.2018.03.013] [PMID: 29572045]

[114] Malo C, Crichton EG, Morrell JM, Pukazhenthi BS, Skidmore JA. Single layer centrifugation of fresh dromedary camel semen improves sperm quality and in vitro fertilization capacity compared with simple sperm washing. Reprod Domest Anim 2017; 52(6): 1097-103.
[http://dx.doi.org/10.1111/rda.13036] [PMID: 28801955]

[115] Malo C, Crichton EG, Skidmore JA. Optimization of the cryopreservation of dromedary camel semen: Cryoprotectants and their concentration and equilibration times. Cryobiology 2017; 74: 141-7.
[http://dx.doi.org/10.1016/j.cryobiol.2016.11.001] [PMID: 27818266]

[116] El-Badry DA, Mohamed RH, El-Metwally HA, Abo Al-Naga TR. The effect of some cryoprotectants on dromedary camel frozen-thawed semen. Reprod Domest Anim 2017; 52(3): 522-5.
[http://dx.doi.org/10.1111/rda.12934] [PMID: 28176384]

[117] Akbar S, Hassan S, Ahmad M. Studies on semen processing, cryopreservation and artificial insemination in dromedary camel. International Journal of Animal Science 2018; 2: 1018.

[118] Al-Bulushi S, Manjunatha BM, Bathgate R, Rickard JP, de Graaf SP. Artificial insemination with fresh, liquid stored and frozen thawed semen in dromedary camels. PLoS One 2019; 14(11): e0224992.
[http://dx.doi.org/10.1371/journal.pone.0224992] [PMID: 31697770]

[119] Crichton EG, Pukazhenthi BS, Billah M, Skidmore JA. Cholesterol addition aids the cryopreservation of dromedary camel (Camelus dromedarius) spermatozoa. Theriogenology 2015; 83(2): 168-74.
[http://dx.doi.org/10.1016/j.theriogenology.2014.09.005] [PMID: 25312816]

[120] Crichton EG, Malo C, Pukazhenthi BS, Nagy P, Skidmore JA. Evaluation of cholesterol- treated dromedary camel sperm function by heterologous IVF and AI. Anim Reprod Sci 2016; 174: 20-8.
[http://dx.doi.org/10.1016/j.anireprosci.2016.08.013] [PMID: 27622555]

[121] Malo C, Crichton EG, Morrell JM, *et al.* Colloid centrifugation of fresh semen improves post-thaw quality of cryopreserved dromedary camel spermatozoa. Anim Reprod Sci 2018; 192: 28-34.
[http://dx.doi.org/10.1016/j.anireprosci.2018.02.005] [PMID: 29525206]

[122] AbdelHafez F, Bedaiwy M, El-Nashar SA, Sabanegh E, Desai N. Techniques for cryopreservation of individual or small numbers of human spermatozoa: a systematic review. Hum Reprod Update 2009; 15(2): 153-64.
[http://dx.doi.org/10.1093/humupd/dmn061] [PMID: 19109313]

[123] Cohen J, Garrisi GJ. Micromanipulation of gametes and embryos: Cryopreservation of a single human spermatozoon within an isolated zona pellucida. Hum Reprod Update 1997; 3(5): 453.
[http://dx.doi.org/10.1093/humupd/3.5.453] [PMID: 9528910]

[124] Walmsley R, Cohen J, Ferrara-Congedo T, Reing A, Garrisi J. The first births and ongoing pregnancies associated with sperm cryopreservation within evacuated egg zonae. Hum Reprod 1998; 13 (Suppl. 4): 61-70.
[http://dx.doi.org/10.1093/humrep/13.suppl_4.61] [PMID: 10091058]

[125] Zou Y, Yin T, Chen S, Yang J, Huang W. On-chip cryopreservation: a novel method for ultra-rapid cryoprotectant-free cryopreservation of small amounts of human spermatozoa. PLoS One 2013; 8(4): e61593.
[http://dx.doi.org/10.1371/journal.pone.0061593] [PMID: 23646110]

[126] Satirapod C, Treetampinich C, Weerakiet S, Wongkularb A, Rattanasiri S, Choktanasiri W. Comparison of cryopreserved human sperm from solid surface vitrification and standard vapor freezing method: on motility, morphology, vitality and DNA integrity. Andrologia 2012; 44 (Suppl. 1): 786-90.
[http://dx.doi.org/10.1111/j.1439-0272.2011.01267.x] [PMID: 22211943]

[127] Brum AM, Sabeur K, Ball BA. Apoptotic-like changes in equine spermatozoa separated by density-gradient centrifugation or after cryopreservation. Theriogenology 2008; 69(9): 1041-55.
[http://dx.doi.org/10.1016/j.theriogenology.2008.01.014] [PMID: 18378291]

[128] Aitken RJ, Finnie JM, Muscio L, *et al.* Potential importance of transition metals in the induction of DNA damage by sperm preparation media. Hum Reprod 2014; 29(10): 2136-47.
[http://dx.doi.org/10.1093/humrep/deu204] [PMID: 25141857]

[129] Fleming SD, Ilad RS, Griffin A-MG, *et al.* Prospective controlled trial of an electrophoretic method of sperm preparation for assisted reproduction: comparison with density gradient centrifugation. Hum Reprod 2008; 23(12): 2646-51.
[http://dx.doi.org/10.1093/humrep/den330] [PMID: 18765400]

[130] Ainsworth C, Nixon B, Jansen RPS, Aitken RJ. First recorded pregnancy and normal birth after ICSI using electrophoretically isolated spermatozoa. Hum Reprod 2007; 22(1): 197-200.
[http://dx.doi.org/10.1093/humrep/del351] [PMID: 16971383]

CHAPTER 4

Controlled Ovarian Stimulation

Firas Al-Rshoud[1,*] and **Tamara Darwish**[1]

[1] *Department of Obstetrics and Gynecology, Faculty of Medicine, Hashemite University, P.O Box 330127, Zarqa 13133, Jordan*

Abstract: For conception to occur, whether naturally or via assisted reproductive technology (ART), three things are required; a healthy sperm for fertilization, an ovum released from both ovaries (in the fallopian tubes for natural pregnancy or in the laboratory in IVF/ICSI), and a healthy place for the embryo to implant (a healthy receptive endometrial cavity), so any defect in any of the aforementioned pregnancy requirements will affect the achievement of pregnancy in both types; the natural one or that resulting from ART [1]. Ovarian stimulation is used to overcome any ovulatory disorder or in ART to collect as many eggs as possible (10-15 eggs at least) [2] to maximize the chances of pregnancy. Assessment of ovarian reserve is the first and most important step in ovarian stimulation, as this will indicate the proper method and dose needed for ovulation induction. We test for ovarian reserve using Biomarkers (FSH and AMH) and physical markers (antral follicles count(AFC)) by transvaginal ultrasound(TVS). Both AMH and AFC have similar accuracy in assessing ovarian reserve and are both superior to FSH as they have no cyclic variations [3]. This chapter will discuss all aspects of the physiology of the ovulation process and ovarian stimulation.

Keywords: Anovulation, ART, Ovarian stimulation.

THE OVARY

The human ovary is composed of the following; a) the hilum; b) the outer cortex; c) the central medulla. The hilum comprises the blood supply, nerves, and hilus cells. At the embryonic stage, the ovary is part of the mesovarium. The ovarian cortex consists of two parts, the outer layer (the tunica albuginea) and the inner one (ovarian follicles are implanted in stromal tissue). Further, mesonephric is the origin of the central medulla.

---

* **Corresponding author Firas Al-Rshoud:** Department of Obstetrics and Gynecology, Faculty of Medicine, Hashemite University, P.O Box 330127, Zarqa 13133, Jordan; Tel: 00962777682626; E-mail: reshoud75@yahoo.com

Embryology and Ovarian Reserve

It is well known that in average cases the primitive ectoderm is the origin site for human germ cells after they develop and migrate to the gonadal ridges (the migration takes place within 5 weeks of gestation). At this time window, the gonadal ridges and mesonephric ducts formulate. The process of migration entails the proliferation of germ cells followed by mitosis. Importantly, the variation of different gonads into either ovary or testis is controlled by genes-related sex chromosomes. However, at this early stage, rapid mitotic germ cell multiplication takes place in which 6-7 million oogonia develop(in the ovarian cortex), which is the highest number or the peak number, keeping in mind that this takes place by 20 weeks gestation. After delivery, the number of eggs decreases with age reaching puberty with less than one million and menopause with around 1000 eggs [4].

Ovarian Physiology

The cyclic activity in women is their reproductive capacity; a characteristic that mirrors the development of dominant follicles. In normal conditions, in each menstrual cycle, a single dominant follicle is produced by the ovaries, in which, this follicle plays a major role during the follicular phase; the production of estradiol. Afterward, ovulation takes place. The dominant follicle transforms into the corpus luteum, through which progesterone production takes place (luteal phase). Together, estradiol and progesterone are needed to facilitate the implantation of the human embryo. Hence, a thorough understanding of the life cycle of dominant follicles will facilitate an understanding of female fertility. This section will briefly discuss the following aspects: Follicular structure and function, the process of folliculogenesis, ovulation, and luteogenesis.

FOLLICULAR STRUCTURE AND FUNCTION

Folliculogenesis resembles the maturation of the ovarian follicle. This follicle is comprised of a shell composed of somatic cells consisting of the immature oocyte and represents the transformation of small primordial follicles into large preovulatory follicles and atresia of the remaining ones. This takes place monthly during the menstrual cycle. Folliculogenesis is divided into two stages; the first stage is gonadotropin-independent while the second stage is gonadotropin-dependent. The first stage represents the development and differentiation of the oocyte and it is organized by the production of growth factors through paracrine mechanisms. The second stage is follicular size increment until it reaches around 25-30 mm, and is controlled by FSH, LH, along with growth factors.

The Process of Folliculogenesis, Ovulation, and Luteogenesis

This process entails the maturation of the follicle from primordial through the secondary follicle.

Primordial

The primordial phase takes place at 18–22 weeks intrauterine, at which the ovarian cortex has its peak number of follicles (4 to 5 million in average case), however, the highest number of follicles could reach 6 to 7 million. These primordial follicles consist of immature oocytes separated from ovarian stroma by the basal lamina and supported by granulosa cells. These cells are dormant, and they can stay at the dormant stage for up to 50 years. Thus, it explains the fact that ovarian cycle length precludes this period. Hence, before birth, the follicles are reduced to around 180,000 by puberty. However, by quality, only 400 follicles reach the preovulatory phase, and by menopause, around 1,000 follicles remain. Thus, it is plausible that early and late menopause are bound by the number of populations at birth, however, no definite evidence is reported. Of note, when primordial cells 'wake up', the initial recruitment phase would take place mediated by hormones and growth factors.

The Primary Phase

In this phase, the activation of the ovarian follicle takes place. The granulosa cells that are around the primordial follicles change their morphological flat features to a cuboidal structure, signposting the launching of the primary follicle, and genes are then transcribed. Besides, paracrine mechanisms begin to initiate the signaling between the follicle and the oocyte. The oocyte and the follicle develop and grow to nearly 0.1 mm in diameter.

The primary follicle during this stage creates and develops follicle-related stimulating hormone (FSH) receptors, in addition, the oocyte will be surrounded by the zona pellucida (A glycoprotein polymer), to form a buffer zone between the oocytes and granulosa cells. The zona pellucida will be kept with the oocyte when the ovulation process is completed as it releases enzymes that enhance the ability of the sperm to penetrate the ovum.

The Secondary Follicle

At this stage, the oocyte-secreted signals will gather around the theca cells; the outermost layer of the follicle, the basal lamina, and undergo cytodifferentiation and will be surrounded by these cells to emerge as the theca externa and theca interna. Consequently, vascular capillary supply will be established between these two thecal layers marking the beginning of blood supply to the follicle. Fully developed oocytes enclosed by a zona pellucida signal the fact that the late-term secondary follicle is developed in which also the antrum begins to develop.

OVARIAN STIMULATION IN GENERAL

Ovarian stimulation otherwise called ovulation induction is a stimulation of the ovulation process using medical interventions as indicated, and it is used to induce ovarian follicles developments. In other words, it is a process in which ovarian follicular development, as well as the maturation of oocytes, is induced through medical interventions. The method of ovarian stimulation depends on the cause of anovulation and the number of eggs needed as well. Ovarian stimulation is not an intervention that is free of disadvantages, such as Multiple pregnancies, Ovarian hyperstimulation(OHSS), and other side effects, so we need to choose the best method for each individualized patient to have a safe and successful ovarian stimulation.

WHO Classification

The WHO classifies women with ovulatory disorders into the following groups:

Group I ovulation disorders (hypogonadotropic hypogonadal): This entity is caused by hypothalamic-pituitary axis failure and comprises 10% of women with ovulatory disorders. These women usually present with amenorrhea whether secondary or primary associated with low serum levels of estrogen and gonadotrophins.

Group II ovulation disorders (normo-gonadotropic, normo-estrogenic): This entity is caused by a dysfunction of the hypothalamic-pituitary axis and is evident among 85% of women with ovulatory disorders. This category includes women with PCOS and those with amenorrhea due to hyperprolactinemia.

Group III ovulation disorders (hyper-gonadotropic, hypoestrogenic). This entity is caused by ovarian insufficiency and comprises around 5% of women with ovulation disorders.

Indications for Ovarian Stimulation

Ovulation stimulation reverses anovulation and stimulates ovulation among women who suffer from oligoovulation for example but not limited to women with polycystic ovary syndrome (PCOS), hypogonadotropic hypogonadism (HH), primary ovarian insufficiency (POI), and hyperprolactinemia. Importantly, diagnosing such conditions carries not only fertility implications, but also has detrimental health implications including bone health, sexual health, long-term cardio-metabolic risks, and wellbeing.

The use of gonadotrophin for ovarian stimulation plays an integral role in the treatment of infertility. Studies characterizing pituitary regulation of gonadal function in the early 1900s laid the foundation for the development of gonadotrophin preparations for ovarian stimulation [5]. In natural cycles, GnRH stimulates the secretion of FSH and LH from the anterior pituitary gland [6]. Follicle development in the ovary and the selection of a dominant follicle are regulated by FSH and LH. Ovulation is then induced by a mid-cycle surge of LH. In the 1940s, HMG, which contains a 1:1mixture of FSH and LH activity, was first extracted from postmenopausal urine and became commercially available a decade later [7].

Anovulatory Infertility

Amenorrhea otherwise known as the absence of the menstrual cycle can exist as a transient, or permanent condition which stems from dysfunctional hypothalamus, pituitary, ovaries, uterus, or vagina. In addition, it is categorized into primary (absence of menarche by age 15 years or older) or secondary (absence of menstrual cycle for more than three months in females who had a regular cycle, or six months among females who had irregular menses. However, any amenorrhea that lasts for three or more months in duration and oligomenorrhea (fewer than nine menstrual cycles per year or cycle length greater than 35 days) requires investigation. The etiologic and diagnostic considerations for oligomenorrhea are the same as for secondary amenorrhea. Importantly, pregnancy has to be ruled out before starting the investigations, then a sound method to investigate the primary or secondary amenorrhea in women should take place to start a fundamental base for controlling the menstrual cycle and other related implications related to the hypothalamus, pituitary, ovary, and uterus. Hence, tracing the site with the dysfunctional problem is highly important because it will form the underlying ground for possible treatment options. It is highly important to rule out common causes of secondary amenorrhea such as hypothalamic-related amenorrhea problems or polycystic ovary syndrome (PCOS), and anatomic or pathologic causes. In addition to that, a thorough history should be taken considering the past and current medical history, familial and personal risk factors.

Hypogonadotrophic - Hypogonadism

This pathology is a consequence of a total or partial absence of GnRH-induced LH and FSH secretion. The causes for this pathology are either congenital or idiopathic featured by a problem in secreting the GnRH hormone from the hypothalamus or a defect in the GnRH action on the pituitary. The most common cause of this type of anovulation is stress on the pituitary [8].

The idiopathic cases of HH are related to heterogeneous situations, however, there is a potential contribution of multiple genetic defects or epigenetic changes. For example, the defective embryonic migration of GnRH from the nasal placode to the hypothalamus results in Kallmann syndrome which is featured by anosmia or hyposmia. Around 50% of the cases of Kallmann syndrome are associated with other comorbidities such as neural hearing defects, cleft palate, renal agenesis, and dental defects. Loss-of-function mutations of KISS1 and KISS1-R genes have also been reported in patients with idiopathic HH. However, acquired reasons for HH were nervous system malignancies, and infiltrative diseases including; hemochromatosis sarcoidosis, Wegener's granulomatosis, and histiocytosis), infections (meningitis, encephalitis, and tuberculosis), radiotherapy to the brain, head trauma, and the usage of the following medications: glucocorticoids, narcotics, and chemotherapy. Impairment correlates strongly to the age of the patient, the extent of trauma, and the dose/type of radiotoxic/chemotoxic medications.

Polycystic Ovary Syndrome (PCOS) and Premature Ovarian Insufficiency PCOS

This entity is the most prevalent hormonal disorder among women of childbearing age. It comprises menstrual cycle abnormalities, obesity, androgen excess, subfertility problems along its psychological impact on the wellbeing of women. This entity also carries increased future risks for cardiovascular diseases, endometrial cancer, and metabolic syndrome. It stems from a genetic origin that becomes evident by insulin excess. Insulin excess results from insulin resistance driven by obesity [9].

PCOS is diagnosed according to the present criteria, the Rotterdam criteria(ESHRE/ASRM 2003) if two of the following are present:

1. Clinical (acne and hirsutism) or biochemical hyperandrogenism.

2. Anovulation/oligoovulation.

3. Polycystic ovaries on ultrasound.

Treatment options are tailored according to the presentation which varies according to age. PCOS per se accounts for up to 90% of cases with anovulatory infertility. Obesity is still considered the primary cause of ovulatory dysfunction. Not only that, obesity hinders monitoring ovulation via ultrasound. LH hypersecretion, present in around 40% of cases on the other hand is to be blamed for ovulatory disorders among PCOS women with normal BMI.

In this section, we will discuss the available treatment options for women diagnosed with PCOS who present with subfertility.

Women diagnosed with PCOS may conceive spontaneously or with ovulation induction. Many factors play a role in their subfertility other than obesity including the metabolic, endocrine, and inflammatory aspects of this entity and their effect on follicles' quality and the survival of the developing embryo. Management options for women with PCOS presenting with subfertility include weight loss, oral anti-estrogens, gonadotrophin therapy, and laparoscopic ovarian drilling. The fact that hyperinsulinemia is the core of this entity raises the question of the need to use insulin-sensitizing agents in its management. None of these modalities has been proven to be superior to others in any RCT. However, agreed upon is the need for careful monitoring of ovulation as PCOS patients are at risk of developing both; OHSS and multiple pregnancies.

Insulin Sensitizing Agents

The biguanide metformin affects both; hepatic insulin secretion and enhances its sensitivity at the cellular level. Although many RCTs failed to prove their effectiveness in the treatment of anovulatory infertility, their reproductive effects are still promising.

Antiestrogens (Clomiphine Citrate and Letrozole)

Clomiphene citrate has long been the first-line therapy in women with anovulatory subfertility. CC induces ovulation in up to 80% of candidates and up to 70% are expected to get pregnant. Treatment usually commences on the second day of the cycle and continues for five days. However, starting treatment any day between the second and fifth day of the cycle is deemed acceptable. Starting dose must be 50mg/ day, this could be titrated up to 100 mg/day with careful monitoring according to the response to prevent OHSS. If ovulation fails to occur with the 100mg/ day dose, CC should be discontinued. On the other hand, if an exuberant response was noted on the 50mg/day dose, it should be decreased down to 25 mg/day. If pregnancy fails to happen after 6-9 ovulatory cycles, IVF should be the next line of treatment.

Gonadotrophin Therapy

This is prescribed to women who are resistant to antiestrogen therapy. Cumulative pregnancy rate after 6 cycles reaches up to 70%. Starting with a dose as low as

25-50 IU FSH or HMG should be done to prevent hyperstimulation. If this dose is deemed inadequate an increment of 25-50 IU should be applied to start day 14 and continue for 7 days. Although prolonged, this ensures no superovulation or multiple pregnancies. Ovulation is usually triggered using the 5000 unit of hCG once a 17 mm follicle is the appearance on ultrasound. On the contrary, if two or more follicles with a size of 14 were seen on ultrasound, hCG should be withheld and the couple should be advised to abstain from sexual intercourse to prevent OHSS. If conception fails t occur after 6 cycles in women younger than the age of 25 and 12 cycles in those older than 25, this method should be abandoned in favor of IVF.

Laparoscopic Ovarian Drilling

In women with PCOS who have normal BMI a surgical alternative for ovulation induction exists; laparoscopic ovarian drilling. This procedure involves making 4 holes with diathermy of 40 watts in each ovary. Studies have proved this technique to be as effective as gonadotrophin therapy with no risk of OHSS and multiple pregnancies. This surgery, however, must be performed by a trained laparoscopist.

POI

Premature ovarian insufficiency (POI), previously known as premature ovarian failure, this entity affects 1% of women under the age of 40. The name has been changed as follicular function fluctuates in around 50% of those women and 5-10% of them conceive spontaneously at some point of their reproductive life. Its diagnosis is based on the presence of primary/secondary amenorrhea, elevated gonadotropin levels, and hypoestrogenism. Its effects go beyond subfertility to include climacteric symptoms, cardiovascular risks, and osteoporosis.

Many causes have been postulated to cause POI, including genetic, infectious, autoimmune, and metabolic ones. However, idiopathic POI remains to be the most prevalent among all accounting for up to 90% of cases.

Women are born with a set number of follicles that reach up to 300000 around the time of puberty. Of which, only 500 oocytes are released during childbearing age. Consequently, POI could be a result of a decreased pool since the beginning or an accelerated rate of loss of follicles.

Many genetic diseases have been associated with POI including X-linked disorders, fragile X syndrome, deletions, translocations, and autosomal dominant disorders. The most common X-linked disorder related to POI is Turner syndrome

(XO). It's a rare entity in which women have streak ovaries due to follicular atresia.

Fragile X syndrome is caused by the FMR1 premutation and is found in 5% of women with POI. Moreover, viral and bacterial infections may lead to POI. In specific, CMV has been linked to POI.

Around 15% of women with POI have been found to have autoimmune disorders reflecting a possible linkage between them both. These women should be investigated for the presence of type 1 diabetes, thyroid disorders, and many others.

Ovarian insufficiency could also be caused by chemotherapy, surgery, or radiotherapy.

Diagnosis and Investigations

The diagnosis of POI could be challenging as ovarian function is unstable and fluctuates. In general, two or more readings of gonadotrophins and estrogen with 6 weeks apart are needed to confirm the diagnosis.

Recently, many tests have been investigated for their usefulness in the diagnosis, including AFC and AMH. A low level of AMH/decreased AFC could be used to predict POI. However Estradiol level and FSH level remain to be the diagnostic ones of choice.

Once diagnosed, karyotyping of the patient should be done to rule out the genetic basis of this disease and autoimmune disorders should also be excluded based on clinical presentation.

Albeit not a must, pelvic ultrasound showing small ovaries with no follicles could be used to support the diagnosis.

A DEXA scan must also be done to keep a baseline bone density to counsel the patient regarding her risk of fracture and the need for HRT.

Subfertility and POI

For the time being, no treatment option has been postulated for women with POI suffering from subfertility. For those who are karyotypically normal, resumption of follicular activity could happen to lead to spontaneous pregnancy. IVF with donor oocyte and adoption seem to be the only available options currently available to help these women experience motherhood.

Ovarian Stimulation (COS) *in Vitro*-Fertilization (IVF) Treatment

In Vitro-Fertilization (IVF) Treatment

The first step for *in Vitro* fertilization (IVF) treatment is controlled ovarian stimulation(COS), the choice of COS protocol and the gonadotropins type and dose (FSH/LH) depends on many factors: age of the female, ovarian reserve, Body mass index(BMI) and any risk factors for OHSS. Three main principles in COS, the first one is to suppress the Hypothalamic-Pituitary-Ovarian axis to prevent premature luteinizing hormone (LH) surge, then to stimulate multiple follicular growths and the last one is to enhance the final maturation of the eggs [10]. For patients less than 40 years with a normal ovarian reserve and no risk factors for OHHS, GnRH agonist long protocol is the most widely used in COS. For older patients with poor ovarian reserve or in any patient at risk for OHHS, GnRH antagonists are the ideal choice [11]. The advantage of GnRH antagonists is a shorter duration of ovarian stimulation so, less total gonadotropins dose used in COS and more importantly you can trigger the final maturation of the eggs using GnRH agonist to avoid the risk of OHSS. Short GnRH agonist protocol has got less popular as it is inferior for both mentioned protocols (GnRH agonist long protocol and GnRH antagonists) in terms of IVF pregnancy rate and higher incidence of OHSS, for these reasons most of the ART clinics in the world do not use this protocol for many years. Mild stimulation protocol (using oral ovulation medications) has a role in special cases, such as patients with a very poor ovarian reserve and in patients with very high AMH and AFC(PCOS patients) to avoid the OHSS risk in the later, as OHSS a life-threatening condition if severe or critical stage. The ovarian stimulation on IVF treatment should be tailored according to the case and the confounding factors to maximize the pregnancy chances.

Complications of Ovarian Stimulation: OHSS, M.P

Ovarian stimulation is a rather safe interference. However, it carries the risk of both; multiple pregnancies (M.P) and Ovarian hyperstimulation syndrome OHSS. The risk of multiple pregnancies is most prevalent among women with PCOS, hence careful monitoring of follicular maturation cannot be overemphasized. It is estimated that around 5-10% of women are managed with CC [12], and around 25% of those induced with gonadotrophins will have multiple pregnancies [13].

Ovarian hyperstimulation syndrome occurs in almost one-third of IVF cycles, 3-8% of which are graded to be severe/critical. The risk noticeably increases if the pregnancy was achieved. OHSS is classified according to the chart below. It has been suggested that elevated levels of LH and hCG associated with follicular stimulation lead to increased levels of VEGF which in turn results in an inflammatory reaction, increased permeability, thrombosis, and third spacing, all of which constitute the basis for the development of OHSS.

Risk factors for the development of OHSS include the previous history of OHSS, PCOS, increased AFC, increased AMH, using an agonist cycle, multiple pregnancies, and achieving pregnancy.

OHSS is classified according to its timing into early and late OHSS. Early OHSS usually occurs within 1 week of hCG administration, is due to the exaggerated ovarian response, and generally carries a good prognosis. Late OHSS on the contrary occurs more than 10 days after the administration of hCG, is usually endogenously induced, and carries a prolonged/worse prognosis.

Monitoring of the course of OHSS should be done using the following biochemical parameters; CBC, electrolytes, osmolality, LFT, and ultrasound. hCG should also be checked if appropriate to confirm/exclude pregnancy.

Treatment is usually maintained on an outpatient basis with careful management of fluid intake (oral upon thirst only), avoidance of NSAIDS to prevent kidney injury, and a twice-weekly assessment of waist circumference and clinical symptoms. Criteria for admission include those with severe pain, nausea, fluid disturbances in the form of >1000 cc positive fluid balance or those with < 1000cc urine/day, those worsening as outpatients, and cases classified as critical OHSS.

Women who conceive should be given LMWH all through the first trimester to prevent the occurrence of VTEs.

CONCLUSION

Pregnancy to occur, whether naturally or via assisted reproductive technology (ART), needs three main things; a healthy sperm for fertilization, an ovum released from both ovaries (in the fallopian tubes for natural pregnancy or in the laboratory in IVF/ICSI), and a healthy place for the embryo to implant (a healthy receptive endometrial cavity). Ovulation stimulation is a medical intervention via oral or more commonly used injectable medications to reverse anovulation (hypogonadotropic hypogonadism (HH), primary ovarian insufficiency (POI), and hyperprolactinemia) and stimulates ovulation among women who suffer from oligoovulation (women with polycystic ovary syndrome (PCOS)). Ovarian stimulation is a rather safe interference. However it carries the risk of both; multiple pregnancies (M.P) and Ovarian hyperstimulation syndrome which is in a severe form of a life-threatening condition, for this reason, this intervention should be provided by a specialized person to individualize this medical intervention.

CONSENT FOR PUBLICATION

Not applicable

CONFLICT OF INTEREST

The author declares no conflict of interest, financial or otherwise.

ACKNOWLEDGEMENT

Declared none.

REFERENCES

[1]   Al Rshoud F, Kilani R, Al Asali F, Assi O, Ikhlaif N, Maaita W. The effect of endometrial polyp resection by office hysteroscopy during *in vitro* fertilization/intracytoplasmic sperm injection treatment cycle on pregnancy rate a series of 25 cases. J Hum Reprod Sci 2011; 14(4): 439-2.

[2]   Sunkara SK, Rittenberg V, Raine-Fenning N, Bhattacharya S, Zamora J, Coomarasamy A. Association between the number of eggs and live birth in IVF treatment: an analysis of 400 135 treatment cycles. Hum Reprod 2011; 26(7): 1768-74.
[http://dx.doi.org/10.1093/humrep/der106] [PMID: 21558332]

[3]   Barbakadze L, Kristesashvili J, Khonelidze N, Tsagareishvili G. The correlations of anti-mullerian hormone, follicle-stimulating hormone and antral follicle count in different age groups of infertile women. Int J Fertil Steril 2015; 8(4): 393-8.

[4]   Mamsen LS, Brøchner CB, Byskov AG, Møllgard K. The migration and loss of human primordial germ stem cells from the hind gut epithelium towards the gonadal ridge. Int J Dev Biol 2012; 56(10-12): 771-8.

[5]   Lunenfeld, 2004; Practice Committee of American Society for Reproductive Medicine 2008.

[6]   Baerwald AR, Adams GP, Pierson RA. Ovarian antral folliculogenesis during the human menstrual cycle: a review. Hum Reprod Update 2012; 18(1): 73-91.

[7]   Leao, Esteves. Practice Committee of American Society for Reproductive Medicine 2014; 2008

[8]   Valsamakis G, Chrousos G, Mastorakos G. Stress, female reproduction and pregnancy. Psychoneuroendocrinology 2019; 100: 48-57.

[9]   Osibogun O, Ogunmoroti O, Michos ED. Polycystic ovary syndrome and cardiometabolic risk: Opportunities for cardiovascular disease prevention. Trends Cardiovasc Med 2020; 30(7): 399-404.

[10]   Allahbadia GN, Morimoto Y. Ovarian Stimulation Protocols. India: Springer 2016.

[11]   Hohmann FP, Macklon NS, Fauser BC. A randomized comparison of two ovarian stimulation protocols with gonadotropin-releasing hormone (GnRH) antagonist cotreatment for *in Vitro* fertilization commencing recombinant follicle-stimulating hormone on cycle day 2 or 5 with the standard long GnRH agonist protocol. J Clin Endocrinol Metab 2003; 88(1): 166-73.
[http://dx.doi.org/10.1210/jc.2002-020788] [PMID: 12519847]

[12]   Garthwaite H, Stewart J, Wilkes S. Multiple pregnancy rate in patients undergoing treatment with clomifene citrate for WHO group II ovulatory disorders: a systematic review. Hum Fertil (Camb) 2021; 1-10.

[13]   Tadokoro N, Vollenhoven B, Clark S, *et al.* Cumulative pregnancy rates in couples with anovulatory infertility compared with unexplained infertility in an ovulation induction programme. Hum Reprod 1997; 12(9): 1939-44.
[http://dx.doi.org/10.1093/humrep/12.9.1939] [PMID: 9363711]

CHAPTER 5

Recent Approaches in Intrauterine Insemination in Livestock

Ömer Uçar[1,*]

[1] *Muğla Sıtkı Koçman University, Milas Veterinary Faculty, Division of Reproduction & A.I., Milas-Muğla, Turkey*

Abstract: In the present chapter, the recent studies on artificial insemination (AI) of livestock (large and small) animals via the intrauterine route are summarized. For this, intrauterine inseminations especially in cattle, horse, camel, buffalo, sheep, goat, dog, and cat species will be considered in detail. Brief data on inseminations in other species (pig, turkey, hen, honeybees, silkworm) is also given.

Once semen sample is collected from sexually mature and clinically healthy breeding male animals, it has to be kept alive (mainly by dilution and cooling) first and then used either immediately (fresh) or stored (chilled/frozen) until being used in oestrous females. During insertion of the semen (particularly stored ones), it is necessary to place the inseminate in the vicinity of the ovum (Graafian follicle) as close as possible. Hence, fertility results of insemination are always higher when semen is deposited intrauterine. However, this is not as easy as we would expect, due mainly to the anatomical structure of females (cat, dog, sheep, and hen) or viability of semen (fish and camel).

To overcome this female- or male-originated limiting factors of fertility, various novel approaches have been reported towards acceptable rates of fertility outcome. Hence, these recent insemination techniques are outlined herein.

Keywords: Animal, Artificial insemination, Breeding, Intrauterine, Livestock, Review.

INTRODUCTION

In a broad sense, reproduction aims to sustain the life of living creatures via bearing viable and productive offspring [1]. In animals, the duration of natural mating takes a very short time, only a couple of seconds (cattle, sheep, horse, hen) or so (camel, cat, turkey, fish). The exceptions are mating-lock ('tie-to-tie') in

* **Corresponding author Ömer Uçar**: Muğla Sıtkı Koçman University, Milas Veterinary Faculty, Department of Clinical Vet. Sci., Division of Reproduction & A.I., 48200-Milas, Muğla, Turkey; Tel.: 00 90 252 2115829; E-mail: omeurucar@mu.edu.tr

Islam M. Saadeldin (Ed.)

dogs (5-25 minutes long), a long mating period in camels (generally 7-15 minutes or occasionally day-long), or several matings (by 7-10 males) during 'mating fly' in honeybees (a couple of minutes) [2, 3]. Alternatively, however, the AI technologies are also practiced largely in livestock animals (mostly in cow, ewe, doe, sow, mare as well as in turkey, fish, honeybees, and silkworm). The most common practice of AI is seen in cattle (Figs. **1a**, **1B**) due mainly to achieving genetic improvement in calves (Figs. **1c**, **1d**) by spreading especially males' desirable genetic characteristics (high milk yield and heavier calves born) and preventing mating-related diseases (Brucellosis, IBR/IPV, Tuberculosis, *etc.*) [4]. AI is one of the fastest ways to control diseases and increase individual (offspring) genetic characteristics on a national scale [5].

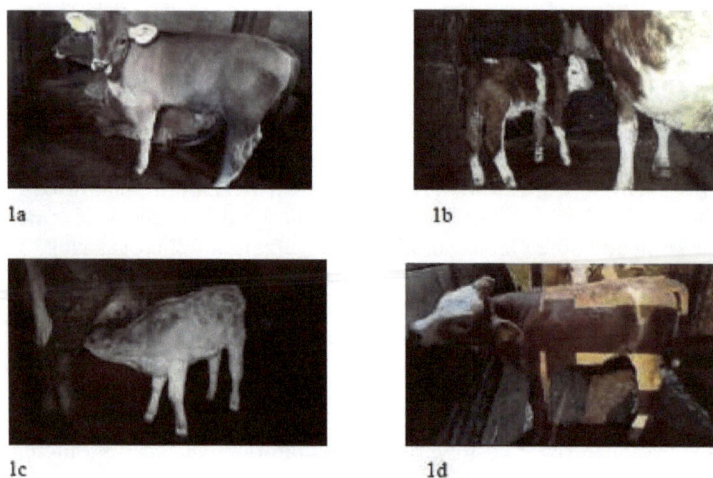

1a

1b

1c

1d

Fig. (1). AI caves in Brown Swiss (1a), Simmental (1b), or their crossed breeds (1c and 1d, resp.).

Conventional AI technologies include mostly the rectovaginal transcervical route in large animals (cattle, buffalo, camel), while the transcervical route is used both in large (horse, cattle) and small animals (dog). However, surgical methods of AI are also practiced mainly in small animals (dog, cat, sheep, goat).

Herein, a brief explanation of AI, relevant species, and people involved are given below.

DESCRIPTION OF AI

In short, AI may be described as the placement of an adequate dose (inseminate) of viable semen collected from the sire male animal to the relevant part of the genital tract of the oestrous female by using appropriate insemination techniques and methods [6, 7]. Globally, it is most commonly applied in cattle in numerous countries (USA, UK, Denmark, Germany, Russia, Turkey, Israel, Japan, Australia, *etc.*) [2 - 4]. For guiding the routine conduct of AI practices in livestock animals [3], reproductive physiological values (oestrus, ovulation time, viabilities of sperm and ovum, optimal insemination time) of different species (cow, mare, ewe, doe, sow, bitch, queen, hen/turkey) are given in Table **1**.

Table 1. Reproductive physiological values of various livestock species*.

Species	Physiological values in livestock			
	Oestrus	ovulation time	Viabilities of gametes	optimal insemination time
			Sperm	Ovum, h	

Species	Oestrus	ovulation time	Sperm	Ovum, h	optimal insemination time
Cattle	12-18 h	10-12 h after oestrus ends	24-36 h	20-24	The second half of oestrus or final 1/3 period
Horse	4-7 day	1-2 days after oestrus begins	72-96 h	6-8	1 day before ovulation or near ovulation
Sheep	30-36 h	24-30th h after oestrus begins	30-48 h	16-24	16-24th h after oestrus begins
Goat	30-36 h	24-30th h after oestrus begins	30-48 h	16-24	16-24th h after oestrus begins
Pig	60-72 h	36-48th h after oestrus begins	36-48 h	8-10	16-24th h after oestrus begins
Dog	9 day	36-48th h after oestrus begins	4-6 day	48-72	48th and 96th h after oestrus begins
Cat	6 day	48-72nd h after mating	24-36 h	8-16	nearly 40 h after hCG administration
Poultry (a hen)	none	Every day at certain times	30 d in the female genital tract	1-2	Once a week (afternoon)

*Slightly modified from [3].

Species Concerned

AI can be performed mainly in ruminants (cattle, buffalo, sheep, and goats), pigs, and horses as well as other species such as bees and silkworms. Further, it is also performed in certain scale fish (rainbow trout) [1 - 3, 8, 9]. However, AI is prohibited especially in thoroughbred (British) horses and AI foals are not registered officially [1, 4]. An exception to this is the disease control for dourine (venereal trypanosomiasis) and infectious anemia by equine AI in some eastern European countries [4].

Personnel Involved

In the field, AI is administered by a Veterinarian or Veterinary Health Technician (sometimes farmers/breeders or their employees, UK) as authorized people [4]. Anatomy and physiology knowledge of these people is expected to be high enough to support their insemination skills [9].

STRUCTURE OF SPERM AND A SUMMARY OF SEMEN HANDLING (FREEZE-THAWING)

Sperm Structure

For long-term storage (in liquid nitrogen, LN_2) semen samples from bulls are considered somewhat cryo-resistant against the deteriorating effects of freeze-thawing. However, semen from dogs, stallions, pigs, and rams is known to be vulnerable due mainly to low high unsaturated/saturated fatty acid ratio and low cholesterol content. Especially in the latter case, dilution rate is also a matter for cryosurvival, where high dilution rates lead to the leakage of cellular structural elements of the sperm cells, such as albumin content [8].

Semen Freeze-thawing

Considering the actual effect of freeze-thawing on the structure of individual sperm cells, the long-term storage process may be summarized in two words: freezing=dehydration, while thawing=rehydration [10]. Specifically, the two most critical "injuring" factors are the ice formation/nucleation, especially occurring during the release of the latent heat below zero degrees C (between -5 to -45 °C, or more precisely between -15 and -25 °C), and critical osmotic gradient, as occurring particularly across the outer (plasma) membrane (PM) or the inner membranes (of inner and outer acrosomal, nuclear and mitochondrial ones) [8].

Further, semen packed are usually frozen at a freezing rate of 100 (for ampoules and pellets) to -170 °C/min⁻¹ (for straws), while thawing occurs at much faster rates: 1000-2000 °C/min during thawing by a conventional method, *i.e.* at body temperature (around 37 °C). For semen, the optimal freezing rate varies between -10 and -80 ^{C}/min⁻¹ for the critical temperature zone, between -5 and -45 °C. On the other hand, for thawing, a rate of 1000 °C /min occurs between -70 to zero °C. This means that keeping the freezing rate (being rather slow, bearing resultant smaller-sized ice) in mind, a 10-20 times higher rate of change in the mass formation of cells occurs (during the thawing). Much faster availability of free extracellular water osmotically enforces the membranes to get in at greater (bulky) sizes. Surely, this kind of great gradient change in osmotic pressure disrupts the membranes since they cannot easily/readily conduct the movement (conductivity) of water in and out on all occasions.

To add, for glycerolisation (glycerol, as being the most common cryoprotectant), surely, lower concentrations of less than 5% would be less toxic (*e.g.*, for acrosome reaction-AR, a concentration of 2% (v/v, final) was non-toxic for dog sperm) [8, 10]. But lesser concentrations need a lower (slower) freezing rate, especially during the initial stages of the freezing curve (earlier than -5 or -15 °C).

Moreover, the PM of sperm cells of many species (*e.g.*, dog, stallion, boar, fox, ram, *etc.*) has high ratios of unsaturated/saturated fatty acids (making the cells vulnerable to oxidative stress, which usually occurs during the membrane damage and/or especially freeze-thawing) and low cholesterol content (related with the elasticity/lipid phase transition of membranes). Considering the "instinctive" high sensitivity of membranes against the oxidative stress (especially following high rates of dilution and improper freeze-thawing steps) and low elasticity of membranes against the osmotic gradient that inevitably occurs during freezing, it would be quite hard to survive for the cells afterward (the following cryopreservation) [10].

Above all, minimizing or avoiding the ice nucleation (with smaller-sized crystals) or avoiding the release of the latent heat (the sigmoid part of the freezing curve that usually occurs during the initial freezing) would be one of the best approaches to reaching a high cryo-survival [8].

So, what could be presumed for optimal cryo-survival (and ultimately the highest fertility outcome possible) is that if we; i) choose "good freezer" individuals if possible, ii) practice efficient and fast initial evaluation, iii) extend the semen in optimal diluent at a favourable rate of dilution, iv) use glycerol (or other cryoprotectants, from an organic origin) at low concentrations and to be added in increments at +4 °C, v) slow down the initial freezing (or more precisely along with cooling to and equilibrating at 4 °C, that would avoid the cold shock, starting at +17 °C, more efficiently) and thawing rate (much slower than minus 1000-2000 °C per min) somehow, we would gain pretty much higher survival (and ultimate fertility) rates, as suggested in the literature [8, 10].

Surely, effective oestrus detection, promptly timed intrauterine insemination by an expert inseminator, and proper management (feeding-housing) of mother animals are other prerequisites for optimum fertility results.

INTRAUTERINE AI TECHNOLOGIES USED

Brief History of Reproductive Technologies

The first biotechnological studies in history started in the 1780s by the Italian Scientist L. Spallanzani, as he used fresh semen taken from a male dog for AI, resulting in two live puppies. In the 20th century, agricultural production and livestock breeding were both international imperatives for human well-being and animal welfare across the world. In farms, animals (horses, cattle, sheep), biotechnological techniques (mainly AI), the first attempts began in the early 1900s (firstly in Russia and then Turkey) and were adapted by the national cooperatives (in Denmark and USA). Subsequently, AI has become widespread in Europe (*e.g.*, UK, France, and the Netherlands) and America and now it is widely used all over the world. The AI is practiced very widely in Denmark, Japan, and Israel, while at least half of the dairy farms prefer AI [4]. Beef producers seem to prefer natural mating, due mainly to difficulties in handling, oestrus detection, and the purpose of production (beef instead of milk). However, oestrus synchronisation (mainly by synthetic progestogens and prostaglandins), skillful cooperation between Vets and farmers along with close monitoring of oestrous females would all allow efficient management and higher fertility to be achieved within the herd.

On the other hand, laboratory studies in reproductive biotechnology started to gain momentum with the discovery of long-term storage medium liquid nitrogen after 1935 and glycerol, a cryoprotectant, in 1949. In the following years, especially after the 1960s, scientific studies (including capacitation, acrosome reaction-AR, packaging semen, diluent optimization, development of freeze-thawing protocols) for many species dramatically increased [8, 10]. After the 1980s, embryo transfer (ET), cloning, and transgenic animal production have all been succeeded around the beginning of the second millennium. Except for the ET with constant use (mainly in the USA and Europe), other advanced techniques remain rather limited in use.

The AI is the best known biotechnological method used in cattle breeding since the early 1900s [3]. This method is mostly used in cows compared to other farm animals to reach healthy individuals with higher yield properties (milk and meat). The insemination of the cows using bull semen allows gaining or sustaining superior genetic characteristics of calves. It also provides information about the general animal health and actual farm conditions for management and feeding.

The main purpose here is to spread the genetic characteristics of the healthy and high-breeding bull. However, to maintain the purity of the breed, it is necessary to continue the AI in the herd continuously and persistently. Breeding bulls should be changed every 2 years, but if they remain longer (risk of inbreeding), they may cause loss of calf crop and yield. This method, which is very common in America and Europe, is widely used in herd-based large dairy farms.

Main Intrauterine Insemination Techniques

In livestock animals, various (mainly four) AI techniques are used in practice, as follows:

a. Rectovaginal intrauterine AI (cattle, buffalo, camel),
b. Transvaginal intrauterine AI (horse, cattle),
c. Laparoscopic (non-surgical) intrauterine AI (sheep, goat), and
d. Laparotomic (surgical) intrauterine AI (dog, cat).

In various livestock species, some details (insemination technique, semen deposition site, inseminate volume, sperm concentration) of the AI techniques are given in Table **2**. Considering the insemination techniques, different routes of transferring semen (vaginal, speculum, rectovaginal, laparoscopic) into the female genital tract are illustrated herein.

Table 2. Technical details of AI in various livestock species*.

Species	Technical details of AI					
	insemination techniques	semen deposition site	inseminate volume, ml		sperm concentration (inseminate)	
			Diluted semen	Frozen semen	Diluted semen x10^6	Frozen semen x10^6
Cattle	Recto-Vaginal	Corpus uteri	1	0.25- 0.50	10	25
Horse	Vaginal	Corpus uteri	10-30	2- 4	10-250	250-500
Sheep	Speculum	Cervix	0.1-0.2	0.25	50-250	250-500
	Laparoscopic	Corpus	0.1-0.2	0.25	10-20	25-50
Goat	Speculum	Cervix	0.1-0.2	0.25	50-250	250-500
	Laparoscopic	Corpus uteri	0.1-0.2	0.25	10-20	25-50
Pig	Vaginal	Corpus uteri	20- 80	0.5-4	1000	1000
Dog	Vaginal	Cranial vagina (Cornu uteri)	2-4	0.25- 0.50	200	800
	Laparoscopic	Corpus uteri	0.25- 0.50	0.25- 0.50	20	100

Technical details of AI					
Cat	Vaginal	Cranial vagina (Cornu uteri)	0.1	0.1	50	100
	Laparoscopic	Corpus uteri	0.1- 0.25	0.1- 0.25	10	20
Hen	Vaginal	Vagina	0.1	0.1	200	1000

*Modified from Sönmez [2012].

AI PRACTICES IN DIFFERENT SPECIES

In the field, AI is practiced mainly in livestock animals (cattle, buffalo, sheep, horse, pig, dog). However, other farm animals (such as rabbits, cats, turkeys) are also considered but with less intensity for AI use. Herein, some details of AI practices are given based on the order of most frequent usage according to species.

AI in Cows

Cows are the main animals to practice AI in the world because of their genital anatomical structure allowing quite simple and easy insemination by expert personnel and their importance for the global food supply chain (mainly milk, meat and calve) [2 - 4]. Today, milk is important as a source of animal protein commonly used for meeting human requirements. The milk in consumption is mostly obtained from breeding (high-yielding and healthy) cows [8]. The AI industry is well developed in this species [6, 7] both for manufacturing freezing equipment (France, USA, Canada, Germany, China), commercial semen production from bull sires at strictly ruled, well-organized breeding centers (run by cooperatives or private companies).

The main technique for AI in the field is the rectovaginal-transcervical route as the deposition site for the inseminate. Frozen/thawed bull semen is used routinely for this purpose.

However, new techniques are either equipped with a mobile USG (with a rectal probe) and/or using the transvaginal route only (by-passing the rectal handling) with the help of speculum and vaginal lighting. The success rates are quite comparable. AI fertility could well be increased by proper management and feeding [11 - 13], normal/good BCS (body condition score) [14, 15], antibiotic therapy [16 - 18], vitamin-mineral supplements [19, 20], hormonal support (oestrus synchronisation or ovulatory hormones) [14, 16 - 18, 21].

Herein, numerous factors directly or indirectly that are inter-related with the outcome of AI (fertility, calf health, milking/fattening performance, *etc.*) should be considered with some detail [3, 4]. In the field practice, low fertility (infertility) is a major problem in the dairy industry, and it is a 'herd problem' rather than a female's disorder (affecting presumably half of the herd population). Hence, a comprehensive approach to solving numerous herd problems (*e.g.*, animal care-management disorders, feeding disorders, genetic defects, immunological factors, pathological cases, genital infections, and insemination errors) is required [9]. To overcome these critical herd problems, some issues will be dealt with herein as follows: rectal palpation, pregnancy detection, oestrus detection, semen, insemination, feeding, management, health services, reproductive health, and education and organization.

The Importance of Rectal Palpation (Experience And Timing)

This basic palpation method has been increasingly used in the field for various purposes (for inflammation, anomaly, pregnancy, heat detection, and control) due to the increasing culture breed cow population and high yield expectation, especially after the 1950s, as well as being cheap and practical. Before the rectal examination, technical personnel should have a good knowledge of anatomy and physiology, as well as gain the necessary experience on the abattoir material with written, digital, and visual media opportunities. During the examination, the cow should not be scared or stressed, and maximum attention should be paid to the hygiene and duration of the examination. Although the method is a very common infield practice, it is recommended to be supported by ultrasound to be more reliable today.

For efficient rectal palpation, theoretical background and practical experience are required. Other than undergraduate training, field practice under the supervision of an expert Vet would be advisable for gaining a sufficient start-point.

The Importance of Pregnancy Detection (Experience, Method, and Timing)

Although pregnancy determinations in different periods (early: 2.5 months, middle: 2.5-5 months, advanced: 5-7.5 months, and last: 7.5-9 months and above) is mostly performed rectally in the field, nowadays, patients' preference and reliability It is recommended to be performed with ultrasound support in the early (30-45 days before) and advanced (between 5.5.-6.5 months), which are not accessible. In the examination, it is not aimed to detect pregnancy for only 1-2 minutes. Examination of other internal findings (such as anomaly, uterine/ovary diseases, abomasum displacement) and pregnancy findings should be evaluated together.

For defining the stage of pregnancy, professional experience is required (3-5 thousand animals to be checked). This is particularly so for rectal palpation. Now, record-keeping especially in large enterprises with high milk yield and USG examination by experienced technical personnel may also be used effectively for pregnancy detection.

The Importance of Oestrus Detection (Experience, Method, and Timing)

Almost half of the oestrus symptoms are missed in dairy cattle enterprises [1]. This necessitates strict and continuous oestrus monitoring especially for adult and non-pregnant individuals. Oestrus is defined as the female accepting the male to mate due to various psychological, physiological and hormonal factors. In cows that show year-round oestrus seasonally, oestrus is observed at the average intervals of 3 weeks (1-2 days less in heifers). Once the permission for mounting starts, it lasts for about 18-19 hours and ovulation occurs approximately 6-8 hours after the symptoms of oestrus ends. In cows, compared to other farm animals, the oestrus symptoms should be interpreted correctly to catch the correct time for insemination (Table **1**). To achieve ideal pregnancy, the technique of oestrus detection is surely critical due to shorter oestrus duration and ovulation occurring after the heat in cattle.

Undoubtedly, the internal symptom of oestrus is the presence of a developed follicle (with 1-2.5 cm Graaf follicle in diameter, Fig. **2**) and the tonicity (increased contractility of the myometrium), are critical for high fertility. For this, external symptoms of permission to mount and vaginal discharge should also be present. The oestrus detection may be conducted by recto-vaginal palpation and regular observation are quite easy (sufficient experience needed). Also, pedometer, ultrasound examination could be used (especially in large enterprises, but these tools require labor and expense. In this respect, combined (holistic) approach as observation (pedometer), palpation, and conductivimeter along with using teaser bull are effective for pinpointing the right time to AI.

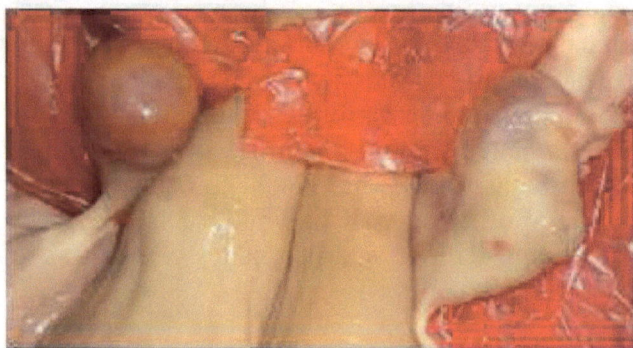

Fig. (2). Graaff follicle (left) along with regressive corpus luteum (right) on the ovaries of a cow in oestrus.

The Importance of Semen (Handling, Storage, Thawing, and Breed Chosen)

In cattle, frozen/thawed semen (mostly packed in French 0.25 ml straws) is used for AI since the early 1970s with considerable success [8]. Bull semen has at least 50% progressive motility post-thaw and 30% abnormality at maximum. Straws with less than 30% post-thawing motility are not used, but in case of a doubt, two doses of inseminating should be used to achieve acceptable fertility following AI.

Frozen semen is transported and/or stored within the LN_2 tanks of varying sizes (*e.g.*, 4-10 L for semen tank, 25-50 L for storage tank). Initial assessment of native semen (with more than 70% live-motile and less than 30% abnormal sperm morphology), or following pre-freeze handling (dilution, glycerolisation, equilibration) packaging semen in 0.25 or 0.50 ml French type straws are used mostly. Straws containing diluted semen of the same breed species (with high-yielding, non-relative, healthy sire males) are stored within the LN_2 tanks, which should at least be filled twice a week and kept in a clean, shadow area without high airflow. Thawing at/around body temperature; 35 °C for 60 sec. (*e.g.*, dog) or 37 °C for 30 sec. (*e.g.*, bull) is mostly used in practice [8].

The Importance of Insemination (Experience, Site, Number, and Timing)

Experience in insemination requires basic physiological knowledge (Table **1**) and an officially proven AI certificate at first. It is critically important that this experience for the AI is expected to be improved by regular self-learning on the abattoir material as well as practicing under the supervision of an experienced colleague in the field [6, 7]. Now, millions of cattle are inseminated with ST in the USA, Europe, Turkey, Japan, and many countries in the world.

The principle for the insemination site is the most possible vicinity of the ovum. In farm animals (cow, mare) this is mostly the corpus uteri of females [3]. If not, intracervical (ewe) or deep intravaginal (hen, ewe, bitch, queen) routes would also be possible, but with possible lower fertility. Against this, second insemination (12 h apart in cows) would be advised to compensate the site obstacle if present anatomically or when oestrus synchronisation is preferred (24 h apart in cows).

The number of inseminations is generally one (cow, hen), two (bitch, cow, mare), daily (bitch, mare), every two days (mare, bitch), or weekly (a hen) [1 - 3]. Handling the semen (inseminate) as fresh/native, diluted, or frozen-thawed may change both the occasion and site of inseminations (bitch, ewe).

Fort timing inseminations, it should be well before (bitch) or before (cow, mare, ewe, doe) ovulation (Tables **1** and **2**) [3]. Cows are normally inseminated in the middle (or at the beginning of the second half) of oestrus (9-15th h of 12-18 h of oestrus period) and the presence of Graaf follicle (1-2 cm in diameter) is expected mostly palpated on the left or right ovaries (Fig. **1**) for ideal insemination resulting pregnancy.

The Importance of Feeding (Storage, Silage, Supplements, Watering, Undernutrition, BCS)

Ideal management-feeding provides animal welfare and productivity. Indeed, feeding affects the calf yield (30% of the whole) along with management (40%), genetics (15%), hygiene (10%), and climate (5%) [13].

Storage of forage and concentrated feed should be away from direct contact with pollutants (feces, urine, *etc.*), moisture, water, wind, and sun. Providing a dry, clean shadow/protected area from animal (including rodents) and human contact is recommended against contamination as such and keep the feedstuff safe [22].

Using silage material supplemented with molasses, urea, salt, *etc.* as needed is highly recommended especially for high-yielding dairy cows and/or during harsh climate periods (long/heavy winter and dry seasons) [15]. Among nutrients, sufficient energy protein along with vitamin-mineral contents of daily rations are critically important for normal functions (especially digestive and reproductive systems) of breeding females and sire males. Daily, continuous, and clean water supply (10% or more of the live weight of a cow, as a daily requirement) is needed especially during the post-partum (lactation) period Undernutrition or low BCS would lead either to lower fertility or even anoestrus in ruminants [10].

The Importance of Management (Manure Disposal, Air Conditioning, Welfare, Culture, Exercise, Scratching)

As the most critical factor in cattle breeding, management highly affects the calf yield (40% of the whole) [6, 7, 9, 13].

For dairy cattle farms especially with indoor breeding, the long-term presence of feces on the walking grounds is one of the main obstacles for inner air quality. Its continuous presence also worsens the odor of the environment, increases the moisture, and prevents animals from easy walking especially when it is frozen and interfered with routine work of automatic manure cleaning machines. Poor indoor air quality and long-term presence of feces inevitably affect udder, lung, uterus, and skin health. Indeed, *E. coli* is the major bacterium of endometritis and mastitis. It would also increase predisposition to infections including repeat breeding cases when it is present in excess [11, 12, 22].

Also, for housing, overcrowded herds and/or cold winter season would both decrease the sufficient oxygen supply and air conditioning of carbon dioxide, ammonium and methane gases. Undoubtedly, animals need oxygen to survive including reproduction. To minimize/prevent potential diseases as well as heat stress, proper management should be practiced, dry and clean bedding, non-slippery walking ground, sufficient area of resting in and outside the barn are all pre-requisites to support animal welfare. Sufficient straching tools equally distributed at reach within the barn are also vital against the possible infestation of parasites severely threatening skin health [6, 7, 9, 22].

Above all, animal keeping (breeding) culture of private owners or executives running the enterprises is also needed to both keep cooperation with stakeholders of the sector (caretakers, Veterinarians, Government officials, marketing people, *etc.*) sustain profitable running the livestock (especially high-yielding dairy) enterprise. These all include working with high-yielding (profitable) animals under proper management conditions with sufficient welfare practices needed [11, 12].

The Importance of Health Services (Routine Controls, Vaccination, Culling)

Using AI not only improves yield characteristics but also has an important function in preventing some infectious (Brucella, IBR/IPV, Tuberculosis, *etc.*) and genetically transmitted diseases such as Leucosis, BLAD, CVM, *etc.* [1, 2, 6, 7]. These undesirable microbial and/or congenital diseases may be prevented by using antimicrobial substances added to sperm and/or local (intrauterine administrations) [17, 18] prior determination of genetic disease in semen both by strict laboratory and managerial protocols [2, 6, 7].

Routine controls of Veterinarians allow early detection or even prevention of potential health problems (general health status of the herd or other specific disorders of individuals such as hypocalcemia, endometritis, repeat breeding, *etc.*) [17, 18].

Regular Vaccination programs (against Brucellosis, IBR/IPV, Anthrax, FMD, Leptospirosis, mixed vaccines including Coronavirus and Clostridia infections, *etc.*) are needed for herd health [11, 22].

Culling annually (10-15% of the total herd) is needed for the exclusion of low-profit animals (especially elder individuals, repeat breeders or those with serious disorders such as fractures, paralysis, advanced age, and bad-tempered behaviours, *etc.*).

The Importance of Reproductive Health (General Health, Libido, Semen, Breeding Soundness, Conception, Pregnancy, Birth, Abortus, Repeat Breeding, Inbreeding)

Undoubtedly, semen collection and successful insemination, pregnancy, delivery (calving, foaling, lambing, *etc.*), and post-partum recovery, suckling, *etc.* all require healthy individuals, as male and female partners. For this, general health of both genders must be normal both clinically (no apparent abnormal symptoms/discharge, fever, anorexia, *etc.*), genetically (CVM, BLAD, *etc.*), and microbiologically (*e.g.*, vaccinated against Brucellosis, IBR/IPV, FMD, *etc.*) [3, 6, 7].

For males, sexually mature individuals (*e.g.*, 9-16 months for cattle) and regular breeding soundness examinations are critical. Also, sufficient libido upon stimulus, and regular weekly semen collections (normal spermiogram) without any serious clinical and/or microscopic problems are required. For females, annual regular oestrus followed by conception within acceptable time limits, sustainable embryogenesis (less than 3% abortus in the herd), and successful delivery following insemination/mating are expected. Proper precautions (such as proper vaccinations and sufficient veterinary/advisory health services) must be taken to prevent the unwanted early termination of pregnancy that would critically interfere with the main profits of keeping mother animals [3, 6, 7].

Repeat breeding (especially in high-milk-yielding cows) has to be monitored closely, especially after the early lactation period (postpartum day 55 onwards). Heavy load of milk yield also suppresses the success rate (by 1-3% fertility loss following the AI with every 5 liters of extra milk) [11]. Normally, within 85 days postpartum, subsequent conception should take place. Otherwise, even with normal health and with the lack of abnormal vaginal discharge in middle-aged females, failure of inseminations 2-3 times or more (especially beyond 90-120 days postpartum) is considered *repeat breeding*. This critical reproductive disorder must be solved by a proper clinical approach (mainly by hormonal and/or antimicrobial support, energy balancing of daily ration, or dealing with insemination failures (semen viability, the experience of relevant/technical personnel, *etc.*) or other management-feeding problems. On the other hand, *inbreeding* (due to a mating relative/genetically close partners) should also be prevented strictly by regular changing of sire bulls (every 6 months or 2 years for frozen semen and mating bulls, respectively). Otherwise, peculiar visibility of (semi)lethal genes in newborns (*e.g.*, calves, lambs) would lead inevitably to congenital abnormalities (such as dwarfism) within the herd [1].

The Importance of Education and Organization (Continuous Training, Certification, Farmer/Breeder Associations, Feedbacks, Financial Supports)

Unfortunately, the educational level of the food chain (animal) sector is rather low especially in underdeveloped or even developing countries [11, 23, 24]. So, sufficient training of farmers and then their certification seem urgent and inevitable to more skillful animal husbandry and profitable sector. Financial support of breeders/breeders and their relevant associations would provide a fluent stream of data transfer, feedback, control, and support.

AI in Horses

In horses, AI is not as common as in cattle. This is mainly due to the jelly nature of native semen and poor freezability of stallion spermatozoa (compared to bull's semen), seasonal nature (photoperiod dependency) of the sexual cycle (April-June in Northern hemisphere), difficulty in finding follicles, variation in ovulation time and breed (Arabian mares may be inseminated by AI [1], but it is forbidden in English thoroughbred mares [4].

The AI technique in oestrous mares is the (vaginal) transcervical route as the deposition site for the inseminate. Frozen/thawed (or diluted/chilled) semen is used routinely for this purpose [2]. Fertility rates may change due mainly to lower post-thawing semen traits (vulnerable against cold shock over cooling for pre-freeze handling) [8] and to the insemination time such that foaling oestrus (6-11th day postpartum) as compared to the subsequent (second) oestrus would lead lower fertility (40 to 70%, respectively [1]. Also, it is expected that showing a couple of prior oestrus during the seasonal breeding season would eventually increase the conception rate. Later matings/inseminations in season (towards the end of breeding season) lead to higher fertility [2].

AI in Buffaloes

Similar to cattle, buffalo cows are inseminated through the rectovaginal intrauterine route. But, pregnancy rates in buffaloes are somewhat lower than in cattle possibly due mainly to species differences (*e.g.* breeding type, season, long oestrus, *etc.*) [25].

AI in Camels

In female camels, rectovaginal intrauterine AI is preferred under the sitting posture of the female given sedatives earlier. But the results of inseminations have not been encouraging yet, due mainly to low initial values of sperm motility (50-55%) as well as the likely excessive exudates of the cervix [Ptaszynska, 2001]. For this, more efficient techniques to increase sperm viability and accessible cervix by enlightening the vaginal tract (using a speculum and headlight) to reach a more visible orifice and easier passage of the cervical channel. These would allow efficient intrauterine insemination in a shorter time along with higher inseminate volume and concentration that surely support higher fertility.

AI in Ewes

In ewes, the cervical obstacle is the major barrier preventing effective intrauterine insemination. However, intracervical inseminations are routinely applicable with fresh/diluted semen [1].

AI in Dogs

Intracervical and intrauterine inseminations are possible but need expertise [26]. Laparoscopic or laparotomic inseminations may also prefer in certain circumstances (inappropriate sizes of mating partners or upon the client's request), although surgical interventions are not preferred (presumably due to ethical issues) in most cases even their higher fertility outcome [8].

CONCLUSION AND FUTURE RECOMMENDATIONS

In history, AI has been considered an effective practice for animal breeding [27]. Further, to date, it became an inevitable biotechnological tool for the prevention of diseases (*e.g.* Brucellosis, IBR/IPV, FMD, *etc.*) on a national scale [10] in animal husbandry and sustainable healthy breeding [28].

AI will become increasingly popular especially in cattle as a result of awareness for higher rates of calving weight, milk yield, feeding (fattening) performance, and preventing venereal diseases (*e.g.* Brucella). However, the widespread use of adequate management-feeding and Veterinary medical services are also pre-requisites for modern livestock breeding.

Considering fertility parameters in cattle, as the major species for AI, it is briefly recommended that [13];

i. Conception rates should not be below 50% following the first inseminations,

ii. Pregnancy Index (insemination number per conception) should not exceed 2,

iii. Open/Service period (Calving-Insemination/Conception) should not exceed 120 days,

iv. Calving interval should not exceed 400 days.

CONSENT FOR PUBLICATION

Not applicable

CONFLICT OF INTEREST

The author declares no conflict of interest, financial or otherwise.

ACKNOWLEDGEMENT

Declared none.

REFERENCES

[1] İleri İK. Suni tohumlama teknikleri. Evcil Hayvanlarda Reprodüksiyon ve Sun'i Tohumlama Ed. Ak K, Pabuccuoğlu S, Birler S. İstanbul Üniversitesi Veteriner Fakültesi Masaüstü Yayımcılık Ünitesi. Ders Notu No: 138, İstanbul: Türkiye 2002, s.133-45. 2002.

[2] Ptaszynska M. Compendium of Animal Reproduction, 6th Revised Edn, INTERVET Int bv, Netherland. 2001.

[3] Sönmez M. Reprodüksiyon, Suni Tohumlama ve Androloji Ders Notları Fırat Üniv Vet Fak. Elazığ: Türkiye 2012.

[4] Arthur GH, Noakes DE, Pearson H. Artificial insemination (Noakes DE) Veterinary Reproduction and Obstetrics. 6th ed. London, UK: Bailliere Tindall 1993; pp. 567-81.

[5] Uçar Ö. 2004 Akrozom reaksiyonunun sperma kalitesi, muhafazası ve suni tohumlama başarısındaki rolü. Kafkas Univ Vet Fak Derg 2004; 10: 117-24. [derleme].

[6] Daşkın A. İneklerde Uygulamalı Suni Tohumlama Aydan Web Ofset. Ankara: Türkiye 2005.

[7] Daşkın A. Sığırcılık İşletmelerinde Reprodüksiyon Yönetimi ve Suni Tohumlama Aydan Web Ofset. Ankara: Türkiye 2005.

[8] Uçar Ö. Acrosome Reaction and Cryopreservation of Dog Spermatozoa PhD Thesis, University of Bristol, Bristol, England 2000.

[9] Gökçen H. İneklerde İnfertilite, Nobel Tıp Kitabevleri. İstanbul: Türkiye 2008.

[10] Uçar Ö, Ünal Y, Yıldız S. Ruminantlarda yetersiz beslenmenin sindirimsel ve metabolik adaptasyonlar ve üreme üzerine etkileri. Kafkas Univ Vet Fak Derg 2004; 10: 227-41. [çeviri].

[11] Uçar Ö, Bülbül T, Saatcı M. The effects of insufficient management and nutrition on reproductive performance in dairy cows: Struggle between modern and traditional breeding. ICABGEH 20 (IV. International Congress on Domestic Animal Breeding, Genetics and Husbandry. 2020; August 12-14, 2020.

[12] Uçar Ö. Kafkas Üniversitesi Veteriner Fakültesi Dölerme ve Suni Tohumlama Anabilim Dalına getirilen hayvanların bireysel özellikleri, geliş zamanları ve uygulama yaklaşımları yönünden değerlendirilmesi. Atatürk Üniv Vet Bilim Derg 2006; 1(3-4): 39-50.

[13]   İleri İK. Suni tohumlama ve gebelik şansını artırıcı uygulamalar. IV. Ulusal Reprodüksiyon ve Suni Tohumlama Bilim Kongresi, 25-28 Ekim, 2007. Manavgat-Antalya, Türkiye. Sayfa 22-26.

[14]   Uçar Ö, Özkanlar S, Kaya M, Özkanlar Y, Şenocak MG, Polat H. Ovsynch synchronisation programme combined with vitamins and minerals in underfed cows: biochemical, hormonal and reproductive traits. Kafkas Univ Vet Fak Derg 2011; 2011(17): 963-70.

[15]   Çolak A, Uçar Ö. Sütçü ineklerde Vücut Kondisyon Skoru (VKS) ile fertilite ilişkisi. Bültendif Veteriner Bülten, 2007:28, Ağustos, sayfa 6-9 (derleme).

[16]   Öztürkler Y, Uçar Ö, Yıldız S, Güngör Ö. The effect of hCG and Gentamicin administration related to artificial insemination following oestrus synchronisation upon the calving rates in repeat breeder cows. Kafkas Univ Vet Fak Derg 2001; 7: 207-11. a

[17]   Öztürkler Y, Uçar Ö, Lehimcioğlu NL. İneklerde sun'i tohumlamayı takiben intrauterin ilaç uygulamasının gebelik oranları üzerine etkisi. Kafkas Univ Vet Fak Derg 2001; 7: 197-200. b

[18]   Öztürkler Y, Uçar Ö. Treatment of infertile cows: efficacy of $PGF_{2\alpha}$ and Gentamicin their combination in a PRID regime. Turk J Vet Anim Sci 2006; 30: 521-5.

[19]   Kaçar C, Kamiloğlu NN, Uçar Ö, Arı UÇ, Pancarcı ŞM, Güngör Ö. İneklerde β-karoten + E vitamini uygulamasıyla kombine edilen Ovsynch ve Cosynch senkronizasyon programlarının gebelik oranı üzerine etkisi. Kafkas Univ Vet Fak Derg 2008; 14: 45-50.

[20]   Uslu BA, Mis L, Gülyüz F, *et al.* Is there a relationship between serum minerals (Ca, Mg) and trace elements (Cu, Fe, Mn, Zn) at mating and pregnancy rates in fat-tailed Morkaraman sheep? Indian J Anim Res 2017; 51(2): 256-62.
[http://dx.doi.org/10.18805/ijar.v0iOF.7258]

[21]   Polat B, Colak A, Kaya M, Ucar O. Stimulation of delayed puberty in heifers by using a PRID regime. Rev Med Vet (Toulouse) 2009; 160: 149-53.

[22]   Yavuz F, Çolak A, Akbulut Ö, *et al.* Erzurum İli Büyükbaş Hayvan Islahı Projesi "Çiftçi El Kitabı". Editör: Yavuz, F. (Atatürk Üniv. Erzurum Valiliği ve Tarım ve Köyişleri Bakanlığı işbirliğiyle), Mega Ofset Matbaacılık, Erzurum, 4. Baskı, Mart. 2011.

[23]   Aydın MK, Keskin M. Muğla ilinde süt sığırı yetiştiriciliğinin mevcut durumu, bazı verim ve yapısal özellikleri. Lalahan Hay. Araşt. Enst Derg 2019; 59(2): 57-63.

[24]   Aydın MK, Keskin M. Muğla ilinde küçükbaş hayvan yetiştiriciliğinin yapısal özellikleri. Mediterr Agric Sci 2018; 31(3): 317-23.
[http://dx.doi.org/10.29136/mediterranean.428577]

[25]   Küçükkebapçı M, Aslan S. Evcil dişi mandaların üreme özellikleri (Derleme). Lalahan Hayvanc Arast Enst Derg 2002; 42(2): 55-63.

[26]   Kalkan O, Uçar Ö. Köpeklerde spermanın alınması, saklanması ve suni tohumlamada kullanılmasına kısa bir bakış. Cumhuriyet Üniv Sağl Bil Enst Derg 2016; 1(2): 24-35.

[27]   Tekin N. Hayvan yetiştiriciliğinde reprodüktif biyotekniklerin yer ve önemi IV Ulusal Reprodüksiyon ve Suni Tohumlama Bilim Kongresi, 25-28 Ekim. Manavgat-Antalya, Türkiye. Sayfa 2007; pp. 19-21.

[28]   Pabuccuoğlu S. Reprodüktif biyoteknolojinin dünü bugünü yarını, IX Ulusal Reprodüksiyon ve Suni Tohumlama Bilim Kongresi, 5-9 Eylül. Hatay Türkiye 2018; pp. 7-14.

CHAPTER 6

Artificial Oocyte Activation

Thomas Ebner[1,*]

[1] *Department of Gynecology, Obstetrics and Gynecological Endocrinology, Johannes Kepler University, Altenberger Strasse 69, 4040 Linz, Austria*

Abstract: Against all expectations, the presence of a carefully selected normal spermatozoon does not guarantee oocyte activation/fertilization. In contrast, some ICSI cycles will have to face no or low fertilization in several consecutive cycles. Both sperm- and oocyte-derived problems may account for such a dilemma. In case of physiological activation problems, any artificial increase in ooplasmic calcium could rescue the fertilization process. Such approaches are summarized under the term artificial oocyte activation (AOA). AOA can be achieved by modified ICSI techniques, piezoelectrical manipulation, or chemical stimuli. Amongst these approaches, the latter is the currently most accepted one in IVF laboratories around the world and particularly the Ca^{2+}-ionophores ionomycin and calcimycin are the most extensively studied agents. Recently, a ready-to-use ionophore (A23187) has been introduced which is CE-marked and as such will assist in the standardization of AOA techniques. There is growing evidence that for proper indications usage of AOA can be considered quite safe. This conclusion is based on studies on morphokinetics, chromosome segregation, and gene expression. More importantly, available neonatal and neurodevelopmental data are reassuring. However, since artificial oocyte activation rarely results in physiological Ca^{2+} oscillations and is not beneficial for all patients with a suspected activation deficiency these techniques should not be used without profound indication.

Keywords: Activation failure, Calcimycin, Fertilization failure, ICSI, Ionomycin, Ionophore.

INTRODUCTION

In a recent report, the ESHRE Special Interest Group of Embryology and Alpha Scientists in Reproductive Medicine [1] stated that in a clinical laboratory setting it is highly recommended that Performance Indicators (PI) are evaluated. On the one hand, this would allow for the systematic analysis of the lab´s contribution to patient care and, on the other, these PIs could represent an important instrument within an existing Quality Management System. More importantly, particular Key

---
[*] **Corresponding author Thomas Ebner**: Department of Gynecology, Obstetrics and Gynecological Endocrinology, Johannes Kepler University, Altenberger Strasse 69, 4040 Linz, Austria; Tel.: ++43 05768084 24600, Fax.: ++43 05768084 24604; E-mail: Thomas.Ebner@Kepleruniklinikum.at

Performance Indicators (KPI) such as ICSI degeneration rate, fertilization rate, or blastocyst development rate would prove useful to perform internal and external quality control. By ascertaining such KPIs embryologists could not only set minimum competency limits but also aspirational goals which would assist benchmarking with others.

According to these figures [1] normal (two polar bodies and two pronuclei) fertilization rate after conventional IVF should be between 60% (competency value) and 75% (benchmark value). Obviously, higher fertilization rates are expected for ICSI (65%-80%). These rational expectations are based on the mere fact that with the ICSI technique a single spermatozoon is directly injected into the ooplasm [2] which avoids some of the physiological barriers of the cumulus-oocyte-complex. ICSI, however, bears the risk that gametes are accidentally chosen for injection which shows deficits in its genetic constellation, phospholipase C zeta (PLCζ) content, protamine ratio, the integrity of centrosome, and/or DNA methylation. Since such a scenario would have an impact on further outcomes, embryologists tend to select sperms considered for usage in the most physiological way [3]. Such selection processes aim for sperms with optimal morphology at higher magnification (IMSI), high sperm head birefringence, completed maturation (so-called pICSI), and/or DNA intactness.

However, even in the presence of a presumably functional spermatozoon that has been processed and identified by a "physiological" selection method oocyte activation and fertilization cannot be taken for granted. In fact, approximately 10% of all ICSI cycles show fertilization problems and 3% result in complete fertilization failure. As a consequence, some patients suffer from repeated fertilization failure which leads to cycle cancellation in the presence of normal sperm parameters and good ovarian response.

For affected couples, artificial oocyte activation (AOA) is the method of choice to restore fertilization and in fact, sometimes is the only way to achieve embryo transfer and eventually pregnancy. It should be kept in mind that AOA techniques should not be applied routinely since they are no physiological approaches at all.

PHYSIOLOGICAL AND ARTIFICIAL OOCYTE ACTIVATION

Before one could further focus on AOA and understand its mechanisms it is of importance to bring physiological oocyte activation to mind. In contrast to the natural conception or conventional IVF ICSI bypasses several physiological steps such as cumulus cell penetration, binding to the zona protein 3 receptor, active passage through the perivitelline space, and Izumo 1 to Juno binding once the membranes of the gametes fuse. A sperm entering the oocyte causes an elaborated reaction in the same. In detail, time-lapse sequences of early preimplantation development revealed that the first sign of oocyte activation is the extrusion of the second polar body (on average 3.3h post-injection) followed by the formation and juxtapositioning of the pronuclei (6.3h and 8.5h, respectively) [4].

These events observed at light-microscopical resolution parallel physiological processes at the molecular level. Of note, the sperm-derived oocyte activation factor, an enzyme known as PLCζ, enters the ooplasm upon fusion or injection. It then binds to membrane-bound phosphatidylinositol-biphosphate (PIP2) which in turn cleaves into two molecules, diacylglycerol (DAG) and inositol-3-phosphate (IP3). While DAG is involved in zona reaction to prevent polyspermy the latter compound is driving oocyte activation.

The produced IP3 interacts with its receptors located at the membrane of the smooth endoplasmic reticulum (sER) whereupon the release of Ca^{2+} ions from inside the sER is initiated. This Ca^{2+} flux is modulated in a time-dependent manner so that it is literally presented in an oscillating mode. sER Ca^{2+}-depletion is unlasting and in order to avoid complete loss of calcium the oocyte uses Ca^{2+} sensors that monitor available Ca^{2+} within the sER and in case a limited or even drastic depletion of the same is detected calcium entry from outside the egg is triggered to refill sER storages. In order to maintain the oscillatory activity of calcium over a longer period of time mechanisms have to exist which ensure restoration of the Ca^{2+}- level to baseline after every single peak. This is mostly done *via* ion pumps, exchanger proteins, or calcium uptake into mitochondria (for review see [5]). Finally, oscillations cease once the two pronuclei are formed and they are noticed again immediately before every mitosis [6].

ARTIFICIAL OOCYTE ACTIVATION

Any deviation in the above-mentioned crucial biochemical processes and molecules involved, such as PLCζ, PIP_2, or IP_3, would automatically cause a reduction or loss of intracellular Ca^{2+}, in particular the absence of calcium oscillations. Such a scenario could also be the result of oocyte-related problems, *e.g.* in case the ooplasma is immature (which in turn would result in maldistribution of sER) or in case IP3-receptors are not functional. However, irrespective of the fact whether cytosolic calcium deficiency is caused by sperm- or oocyte-derived problems it will have a profound impact on cell physiology, and as a consequence, the egg remains unfertilized. As long as calcium is the underlying cause of the dilemma any AOA technique would assist oocyte activation and fertilization by literally requisitioning extracellular Ca^{2+} from the culture medium to compensate for the observed loss. It has to be kept in mind that the vast majority of such artificial methodologies do not generate physiological oscillations of Ca^{2+} but rather produce one single calcium peak. Despite this obvious unphysiological condition oocyte activation can be triggered artificially. This has to do with the fact that it is the amplitude of the very first calcium peak that drives further downstream events (*e.g.*, Ca^{2+}/calmodulin-dependent protein kinase II performance). Once started these processes remain active even in the absence of calcium due to the autophosphorylation of enzymes. It is generally accepted that the mammalian oocyte is relatively susceptible to a wide range of Ca^{2+} signals as long as it is guaranteed that a certain threshold of intracellular calcium is passed [7]. To put it differently, human eggs have a surprisingly high degree of tolerance for changes in ooplasmic calcium levels.

It is nevertheless of importance to distinguish between sperm- or oocyte-borne complications. This would not only allow to tailor the least invasive technique but also guarantee optimal patient counseling in case heterologous oocyte or sperm donation would be the only remaining treatment alternative. The most common method to identify the "root of all evil" and to distinguish between an impaired oocyte-activating potential of the sperm and the failure of the ovum to respond to an internalized sperm is a heterologous ICSI approach known as the mouse oocyte activation test (MOAT) [8]. Based on the potential of a given sperm to activate murine oocytes (as assessed as the rate of 2-cell formation) patients are allocated to a low ($\leq$20% mouse egg activation), intermediate (21-84%) or high activation group ($\geq$85%). The latter cohort is the only one that is composed of men having normal oocyte fertilization capacity. In other words, if the sperm is supposed to be normal the oocyte must be the critical factor in terms of fertilization deficiency. Recently, a more informative test of higher sensitivity, such as mouse/human oocyte Ca^{2+} analysis (M-OCA and H-OCA, respectively), has been introduced [9].

Methods of AOA

Apart from AOA techniques involving injection of calcium (*e.g.* $CaCl_2$) during ICSI, all approaches aim for culture medium and/or intracellular storages as the source of calcium.

Every penetration of the outer shell of the oocyte during ICSI leads to permeabilization of the membranes which in turn results in a micromanipulation-driven calcium entry from the outside to the ooplasm. The effect of the Ca^{2+}influx can further be enhanced if a more invasive intracellular manipulation is performed during ICSI. Such modified ICSI techniques additionally recruit calcium ions from intracellular storages which finally initiates oocyte activation [10, 11].

Alternatively, embryologists may take advantage of the fact that electrical fields generate micropores in the oocyte membranes which again leads to an influx of

Ca^{2+} ions from the culture medium [12, 13]. In principle, both direct or alternate current, as well as a combination of both, are feasible options. Of note, energy and the number of required piezoelectrical stimuli were found to be variable. Nevertheless, the missing of the required technical equipment in most IVF labs is a substantial limitation of the applicability of electrical AOA.

In terms of AOA, chemical activation leads the way and is the most commonly used technique amongst embryologists worldwide. A vast number of compounds could be used to artificially increase calcium concentration within the egg. These chemical approaches include unusual substances like 6-DMAP, puromycin, thimerosal, or phorbol esters, to name but a few. They all have in common that their use is limited to some species, that they are less effective, and that they may cause cytoskeletal damage such as spindle disruption. Amongst these exotic agents used for AOA strontium chloride has a special status since it is thought to provoke multiple Ca^{2+} peaks in mouse oocytes and may do so in humans as well, although this is still subject to controversy [14]. Although the mechanism by which SrCl2 induces Ca^{2+} oscillations is not fully understood it is very likely that strontium ions act *via* calcium channels since both ions are closely related alkaline earth metals.

However, the most extensively studied activating substances in use for AOA is a group of mobile ion carriers, so-called ionophores, which coalesce with divalent cations (*e.g.*, a charge of 2+) thus assisting them to cross the membranes. Due to its physiological importance calcium is the key factor in this process. In everyday practice, two such ionophores are relevant, iono- and calcimycin the latter which is also referred to as A23187.

Vanden Meerschaut and colleagues [15] impressively highlighted the main problem with AOA which is the lack of standardization of the said procedure. Not only that either iono- or calcimycin can be used, there are also major differences in the number of stimuli (single or double), the dosage (mostly 5- 10μM), and the timing of ionophore application (10-30 min). This inconsistency makes it difficult to draw meaningful conclusions from different IVF studies available [16, 17].

Recently, a ready-to-use A23187 (GM508 CultActive, Gynemed, Germany) was introduced which can be considered a step towards the standardization of AOA. Using CultActive in diverse patient cohorts 96 healthy births were reported [6, 17 - 20]. Application of the ready-to-use compound is very easy and self-explaining. The activation agent has a shelf life of six months from production and should be stored in the refrigerator (2 to 8°C). Once a bottle is opened, it should not be used for longer than one week. Before performing AOA an appropriate 4-well culture dish has to be prepared and pre-equilibrated (CO_2 atmosphere). The said dish should be composed of one well containing a 50µl CultActive drop and three wells with a washing medium. Usage of mineral oil is obligatory. It is of utmost importance to agitate the bottle containing GM508 CultActive before the preparation of the dish to avoid an ionophore gradient. The preparation steps can be done the day before AOA is planned. On the day of AOA, ICSI is performed as usual and injected oocytes should be transferred to CultActive within 30 minutes after the end of the injection procedure. The concept should be incubated for 15 minutes in the ionophore upon which they are sequentially transferred to the wells containing the washing medium. The last step then is to transfer the oocytes to the routinely used culture dishes and media for further in vitro culture.

Indications for AOA

To guarantee proper counseling of the patients the correct indication has to be made in advance. The classical indication for the application of ionophores is complete activation failure (0% fertilization, no transfer) in a previous ICSI cycle. A German-Austrian group of scientists [17] reported that even if previous ICSI was associated with a low fertilization rate the outcome of ICSI could be increased with AOA in the following treatment cycle. The respective cut-off value was found to be 30%; in other words, if less than a third of MII eggs were fertilized, a switch to AOA is recommended in the next treatment cycle.

A significant proportion of patients who faced fertilization problems (0-30%) suffer from male infertility so "severe male factor sterility" is another indication that would require consideration of AOA. Indeed, there is growing evidence that using A23187 patients with azoospermia [19], cryptozoospermia [19], Kartagener syndrome [20], or frozen sperm of poor quality [21] can yield acceptable rates of fertilization and pregnancy. While there is little evidence that ionophore treatment is justified for isolated teratozoospermia, its usage in globozoospermic patients is a prerequisite for successful fertilization and pregnancy. It should be highlighted that for globozoospermia, ionomycin was found to be most effective [22].

As mentioned above, calcium peaks reappear immediately before every mitosis and a problem with available Ca^{2+} at these critical stages could lead to mitotic delay or arrest. Consequently, several authors used A23187 to boost mitosis in cases of developmental arrest, developmental delay, and reduced blastocyst formation [6, 20]. Although ionophores worked well in such scenarios it should be clarified that it is no proper AOA approach since this term is reserved for restoring fertilization.

Safety of AOA

Embryologists are still waiting for more prospective studies with a larger number of patients to confirm the reassuring observations so far and to further substantiate the safety of AOA. Meanwhile, an increasing number of publications are available which all indicate that neonatal and neurodevelopmental outcomes (up to the age of 10), as well as language skills, are within the expected ranges [17, 18, 23]. So far only one malformation has been published after the application of A23187 for AOA [17]. Since the observed anal atresia is a late-onset malformation in gastrointestinal tract formation, a causal relation with ionophore application (applied before fusion of the gametes) is very unlikely.

In addition to the collection of data on children born after AOA, embryologists try to increase their knowledge of the potential physiological effects of this artificial method. Concerning this, it is noteworthy that in terms of gene expression of AOA (using $SrCl_2$) was closer to conventional IVF than to ICSI alone [24]. Even in a 10-fold higher concentration as used in routine AOA (*e.g.*, 100μM) A23187 did not cause a widespread increase in chromosome segregation errors. Normal embryo quality and, particularly, comparable morphokinetic behavior [25] further suggest the acceptance of AOA in daily IVF work.

These reassuring findings should not hide the fact that AOA is not physiological, that is not beneficial for all patients, and as such that it has to be considered experimental and should only be applied with a proper indication.

CONCLUSION AND FUTURE PERSPECTIVES

Despite every effort taken, there will be situations in which the paternal or maternal Ca^{2+} balance will not be the cause of fertilization problems or in which the artificial Ca^{2+} transient will not be sufficient to pass the threshold that is required to initiate AOA. In such a scenario, all current AOA methods will be prone to failure. As an alternative solution, one could try to amplify the "trigger" for the Ca^{2+} machinery by increasing the number of stimuli, thus aiming for the additive capacity of Ca^{2+} signals. If more invasive techniques are a feasible option after appropriate ethical and legal consideration, the active injection of $CaCl_2$ (0.1 mol/L) preceding an ionomycin treatment would be a feasible approach [8, 15].

The ultimate and most "physiological" AOA approach would be the utilization of recombinant PLCζ [26]. Since oocytes from other species might differ in response to human recombinant PLCζ (as compared to human eggs) finding the right dosage of the enzyme for clinical use remains a challenge.

Usually, AOA should rather not be used in the first stimulation cycle, except for globozoospermia or PLCζ mutations. In case of doubt that poor sperm quality might affect fertilization, a sibling oocyte model should be chosen. Although it has been stated that nowadays AOA is considered safe and ready for clinical application [20], the doctrine "primum non-nocere" should always be kept in mind.

CONSENT FOR PUBLICATION

Not applicable

CONFLICT OF INTEREST

The author acts as consultant for Gynemed.

ACKNOWLEDGEMENT

Declared none.

REFERENCES

[1]   ESHRE Special Interest Group of Embryology and Alpha Scientists in Reproductive Medicine.. The Vienna consensus: report of an expert meeting on the development of ART laboratory performance indicators. Reprod Biomed Online 2017; 35(5): 494-510.
        [http://dx.doi.org/10.1016/j.rbmo.2017.06.015] [PMID: 28784335]

[2]   Palermo G, Joris H, Devroey P, Van Steirteghem AC. Pregnancies after intracytoplasmic injection of single spermatozoon into an oocyte. Lancet 1992; 340(8810): 17-8. b
        [http://dx.doi.org/10.1016/0140-6736(92)92425-F] [PMID: 1351601]

[3]   Ebner T, Filicori M, Tews G, Parmegiani L. A plea for a more physiological ICSI. Andrologia 2012; 44 (Suppl. 1): 2-19.
        [http://dx.doi.org/10.1111/j.1439-0272.2011.01266.x] [PMID: 22211911]

[4]   Coticchio G, Mignini Renzini M, Novara PV, et al. Focused time-lapse analysis reveals novel aspects of human fertilization and suggests new parameters of embryo viability. Hum Reprod 2018; 33(1): 23-31.
        [http://dx.doi.org/10.1093/humrep/dex344] [PMID: 29149327]

[5]   Ferrer-Buitrago M, Bonte D, De Sutter P, Leybaert L, Heindryckx B. Single Ca^{2+} transients *vs* oscillatory Ca^{2+} signaling for assisted oocyte activation: limitations and benefits. Reproduction 2018; 155(2): R105-19.
        [http://dx.doi.org/10.1530/REP-17-0098] [PMID: 29122969]

[6]   Ebner T, Oppelt P, Wöber M, et al. Treatment with Ca2+ ionophore improves embryo development and outcome in cases with previous developmental problems: a prospective multicenter study. Hum Reprod 2015; 30(1): 97-102.
        [http://dx.doi.org/10.1093/humrep/deu285] [PMID: 25376461]

[7]   Ebner T, Montag M. Artificial oocyte activation: evidence for clinical readiness. Reprod Biomed Online 2016; 32(3): 271-3.
        [http://dx.doi.org/10.1016/j.rbmo.2015.12.004] [PMID: 26776820]

[8]   Heindryckx B, De Gheselle S, Gerris J, Dhont M, De Sutter P. Efficiency of assisted oocyte activation as a solution for failed intracytoplasmic sperm injection. Reprod Biomed Online 2008; 17(5): 662-8.

[http://dx.doi.org/10.1016/S1472-6483(10)60313-6] [PMID: 18983750]

[9]   Ferrer-Buitrago M, Dhaenens L, Lu Y, *et al.* Human oocyte calcium analysis predicts the response to assisted oocyte activation in patients experiencing fertilization failure after ICSI. Hum Reprod 2018; 33(3): 416-25.
[http://dx.doi.org/10.1093/humrep/dex376] [PMID: 29329390]

[10]   Tesarik J, Rienzi L, Ubaldi F, Mendoza C, Greco E. Use of a modified intracytoplasmic sperm injection technique to overcome sperm-borne and oocyte-borne oocyte activation failures. Fertil Steril 2002; 78(3): 619-24.
[http://dx.doi.org/10.1016/S0015-0282(02)03291-0] [PMID: 12215343]

[11]   Ebner T, Moser M, Sommergruber M, Jesacher K, Tews G. Complete oocyte activation failure after ICSI can be overcome by a modified injection technique. Hum Reprod 2004; 19(8): 1837-41.
[http://dx.doi.org/10.1093/humrep/deh325] [PMID: 15166130]

[12]   Mansour R, Fahmy I, Tawab NA, *et al.* Electrical activation of oocytes after intracytoplasmic sperm injection: a controlled randomized study. Fertil Steril 2009; 91(1): 133-9.
[http://dx.doi.org/10.1016/j.fertnstert.2007.08.017] [PMID: 18829025]

[13]   Baltaci V, Ayvaz OU, Unsal E, *et al.* The effectiveness of intracytoplasmic sperm injection combined with piezoelectric stimulation in infertile couples with total fertilization failure. Fertil Steril 2010; 94(3): 900-4.
[http://dx.doi.org/10.1016/j.fertnstert.2009.03.107] [PMID: 19464000]

[14]   Kyono K, Kumagai S, Nishinaka C, *et al.* Birth and follow-up of babies born following ICSI using SrCl2 oocyte activation. Reprod Biomed Online 2008; 17(1): 53-8.
[http://dx.doi.org/10.1016/S1472-6483(10)60293-3] [PMID: 18616891]

[15]   Vanden Meerschaut F, Nikiforaki D, Heindryckx B, De Sutter P. Assisted oocyte activation following ICSI fertilization failure. Reprod Biomed Online 2014; 28(5): 560-71.
[http://dx.doi.org/10.1016/j.rbmo.2014.01.008] [PMID: 24656559]

[16]   Vanden Meerschaut F, Nikiforaki D, De Gheselle S, *et al.* Assisted oocyte activation is not beneficial for all patients with a suspected oocyte-related activation deficiency. Hum Reprod 2012; 27(7): 1977-84.
[http://dx.doi.org/10.1093/humrep/des097] [PMID: 22493027]

[17]   Ebner T, Montag M, Montag M, *et al.* Live birth after artificial oocyte activation using a ready-to-use ionophore: a prospective multicentre study. Reprod Biomed Online 2015; 30(4): 359-65.
[http://dx.doi.org/10.1016/j.rbmo.2014.11.012] [PMID: 25596904]

[18]   Ebner T, Köster M, Shebl O, *et al.* Application of a ready-to-use calcium ionophore increases rates of fertilization and pregnancy in severe male factor infertility. Fertil Steril 2012; 98(6): 1432-7.
[http://dx.doi.org/10.1016/j.fertnstert.2012.07.1134] [PMID: 22921909]

[19]   Ebner T, Maurer M, Oppelt P, *et al.* Healthy twin live-birth after ionophore treatment in a case of theophylline-resistant Kartagener syndrome. J Assist Reprod Genet 2015; 32(6): 873-7.
[http://dx.doi.org/10.1007/s10815-015-0486-2] [PMID: 25956262]

[20]   Darwish E, Magdi Y. A preliminary report of successful cleavage after calcium ionophore activation at ICSI in cases with previous arrest at the pronuclear stage. Reprod Biomed Online 2015; 31(6): 799-804.
[http://dx.doi.org/10.1016/j.rbmo.2015.08.012] [PMID: 26507280]

[21]   Karabulut S, Aksünger Ö, Ata C, Sağıroglu Y, Keskin İ. Artificial oocyte activation with calcium ionophore for frozen sperm cycles. Syst Biol Reprod Med 2018; 64(5): 381-8.
[http://dx.doi.org/10.1080/19396368.2018.1452311] [PMID: 29620419]

[22]   Nikiforaki D, Vanden Meerschaut F, de Roo C, *et al.* Effect of two assisted oocyte activation protocols used to overcome fertilization failure on the activation potential and calcium releasing pattern. Fertil Steril 2016; 105(3): 798-806.e2.

[http://dx.doi.org/10.1016/j.fertnstert.2015.11.007] [PMID: 26632207]

[23]   Vanden Meerschaut F, D'Haeseleer E, Gysels H, *et al.* Neonatal and neurodevelopmental outcome of children aged 3-10 years born following assisted oocyte activation. Reprod Biomed Online 2014; 28(1): 54-63.
[http://dx.doi.org/10.1016/j.rbmo.2013.07.013] [PMID: 24125944]

[24]   Bridges PJ, Jeoung M, Kim H, *et al.* Methodology matters: IVF *versus* ICSI and embryonic gene expression. Reprod Biomed Online 2011; 23(2): 234-44.
[http://dx.doi.org/10.1016/j.rbmo.2011.04.007] [PMID: 21665548]

[25]   Martínez M, Durban M, Santaló J, Rodríguez A, Vassena R. Assisted oocyte activation effects on the morphokinetic pattern of derived embryos. J Assist Reprod Genet 2021; 38(2): 531-7.
[http://dx.doi.org/10.1007/s10815-020-02025-9] [PMID: 33405007]

[26]   Yoon SY, Eum JH, Lee JE, *et al.* Recombinant human phospholipase C zeta 1 induces intracellular calcium oscillations and oocyte activation in mouse and human oocytes. Hum Reprod 2012; 27(6): 1768-80.
[http://dx.doi.org/10.1093/humrep/des092] [PMID: 22456923]

CHAPTER 7

Advances in the Human Preimplantation Embryonic Culture System

Manar M. Hozyen[1,*] and **Islam M. Saadeldin**[2,3,4,*]

[1] *Ganin Fertility Center, Cairo, Egypt*

[2] *Research Institute of Veterinary Medicine, Chungnam National University, Daejeon, Republic of Korea*

[3] *College of Veterinary Medicine, Chungnam National University; 34134 Daejeon, Republic of Korea*

[4] *Department of Physiology, Faculty of Veterinary Medicine, Zagazig University, 44519 Zagazig, Egypt*

Abstract: The embryonic culture system is the main part of *in-vitro* embryonic development. The culture system consists of chemical factors such as the chemical composition of the culture media and physical factors such as temperature, pH, oil overlay and, incubation systems. The optimal culture conditions are needed to support embryonic development *in-vitro*, increase pregnancy, implantation rates, and decrease the level of pregnancy loss. Any suboptimality in culture conditions may impair embryonic development and affect subsequent viability. In this chapter, we will focus on the current culture system in the field of assisted reproductive technology (ART), passing by current culture media formulations, deficiencies, culture devices used, and the areas that need further investigations.

Keywords: Culture dishes, Culture media, Culture system, Embryo culture, Oil overlay, pH.

THE EMBRYONIC CULTURE SYSTEM: A CHEMICAL PERSPECTIVE

Culture system is the surrounding environment of the gametes and embryos growing *in vitro* based basically on the normal composition of the female reproductive tract, in which embryos would normally be exposed [1, 2]. Mimicking nature could support preimplantation embryonic development *in vitro*, which might result in higher post-implantation outcomes (pregnancy, implantation, and live birth rates).

---

* **Corresponding authors Manar Mohamed Hozyen and Islam M. Saadeldin:** Ganin fertility center, Egypt; E-mail: manarnrc@gmail.com and Research Institute of Veterinary Medicine, Chungnam National University, 34134 Daejeon, South Korea; Tel: 00821024817666; E-mails: islamms@cnu.ac.kr and islamsaad82@gmail.com

Culture Media Formulation

Many commercial media are developed for assisted reproductive technologies (ART), to support embryonic development [1, 3, 4]. Basically, the chemical composition of *in-vitro* fertilization media is composed of:

1. Water, which is the major component, about 99% of the media [5].
2. Ions such as potassium, which has an important role in sperm capacitation, chloride ions, which affect mRNA or protein synthesis, and calcium as an essential ion for the compaction stage of the embryos [6].
3. Carbohydrates and other components such as amino acids, vitamins, nucleic acid precursors, chelators, antioxidants, antibiotics, buffer system, protein/macromolecules, hormones, and growth factors [1, 4, 7].

Each component of the media has a specific role to complete the orchestra. Different commercial culture media have different compositions or nutrient concentrations [3, 4]. The exact media formulations and their concentrations are usually not disclosed by the manufacturers due to commercial competition [3]. Up to date, the optimum medium formulation is still unknown and needs more investigations [3]. Hence, each IVF laboratory should choose and validate the media that match their needs.

Back to Nature and Let the Embryo Choose Theories

The development of the embryonic culture media passed with many evolutionary steps [6]. Starting with: The introduction of the culturing idea, *in vitro* first successful fertilization of human gametes in simple media by Robert Edwards, and further modifications to the current complex human culture media forms [6, 8]. Now, the design of culture media emerged from two distinct main theories. The first one is a "back to nature" or "sequential culture media", which developed to mimic the oviduct and uterine component as the natural environment surrounding the embryos. This approach involves refreshing the embryos from one media, which supports pre-compaction embryo development to another that supports post-compaction development regarding the physiology and embryonic need [6, 9]. Although the beneficial use of the sequential media; raises the concern about the stress that evolved from moving the embryos from one media to another [9]. To eliminate this concern, the other approach has been emerged as "let the embryo choose" or "single-step media." In this approach embryo, itself will choose its requirements from the surrounding media. As the media contains the concentration of nutrients and adequate substrates required for embryonic development from fertilization to the blastocyst stage with no need for media change [6, 9]. Both single-step and sequential media are commercially available and have their unique concentrations of energy substrates to support preimplantation embryonic development as well as clinical outcomes (Fig. **1**) [10]. Many studies compare those two types of media, their effect on the embryological outcomes, clinical outcomes, aneuploidy, or even the live birth rate [11 - 15]. However, it was difficult to identify which media is superior due to many cofounders that may affect the results [10, 16]. Emerging from those main approaches, many protocols can be used for embryo culture such as (I) interrupted culture system with the sequential culture media by refreshing the embryos in new media on day 3, (II) daily refreshment system with sequential culture media, (III) uninterrupted culture system with no refreshments using single-step media, (IIII) interrupted culture system either by refreshing the embryos on day 3 or with daily refreshment using single-step media.

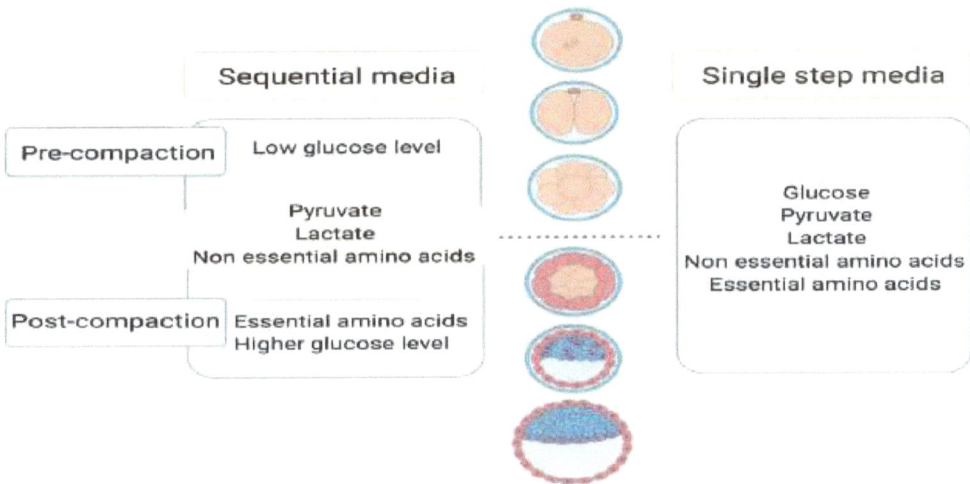

Fig. (1). Main energy substrates of a single step and sequential media for pre-compacting and post-compacting embryos.

Oxygen

The gas phase, especially oxygen is considered one of the most important factors for embryonic development [1, 17]. Oxygen has an important role in the energy production process, although the high concentration may lead to the formation of the reactive oxygen species, which had a toxic effect on gametes and embryos [18]. Its concentration within the uterus and oviduct is approximately 2%, 5-8%, respectively which differs from the oxygen concentration of the air 20% (Fig. **2**) [17, 19]. In 1971, Wes Whitten found no blastocyst formation for mouse embryos cultured under 20% O_2, whereas 100% blastocyst formed under 5% O_2 [17]. He reported that excessive oxygen concentration may harm the surface membrane of the embryonic cells [17]. Also, in the case of human embryonic cells, different O_2

concentrations might have different effects on embryonic development, success, and utilization rates of the *in-vitro* fertilization (IVF) cycles [18]. Culturing the human embryonic cells under atmospheric O_2 or gas mixture containing reduced O_2 level 5% could produce viable embryos [18, 20]. Many studies and published data have been conducted to determine the optimum O_2 level for ART outcomes, but the results are still conflicting [18 - 22]. To minimize this conflict, Bontekoe *et al.* 2013 conducted a systematic review of 7 randomized controlled trials with 2422 participants comparing reproductive outcomes of IVF/ICSI out from the embryo culture under the atmospheric O_2 or low O_2 concentration 5% conditions [19]. They concluded that there are improvements under reduced oxygen levels 5% in the clinical outcomes (clinical pregnancy, ongoing pregnancy, and live birth rates) with no evidence of increased adverse effects (miscarriage, multiple pregnancy, or congenital abnormalities) [19]. Another study compared the ultra-low oxygen concentration of 2% *vs.* 5% and they did not find a beneficial effect on ART outcomes using extremely low O_2 as 2% [22]. To date, no studies have compared the off-spring health resulting from atmospheric O_2 *vs.* low O_2 concentration [16].

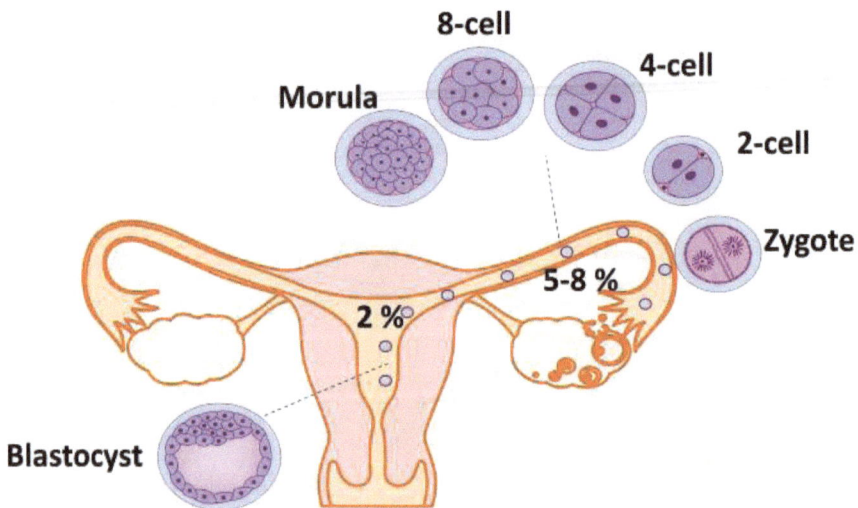

Fig. (2). Different oxygen concentrations within the uterus and oviduct, the higher oxygen level in the oviduct 5-8% than uterus 2%.

Macromolecules and Growth Factors Supplementation

Culture media supplementation with macromolecules such as protein or growth factors (GFs) is not an obligatory component of the IVF media [1, 23]. The use of protein-free media may result in embryonic development till the blastocyst stage, hatching, or even achieving pregnancy [23, 24]. However, the consensus recommends the protein as an important component that should be present in all embryo culture media [16]. Macromolecules supplementation may affect the ART outcomes in several ways to modulate the embryonic interaction with the chemical environment (Fig. **3**) [1, 23, 25]. It has many advantages that could improve the reproductive outcomes of IVF/ICSI cycles. The most common protein in the culture media is albumin, as it is in the oviduct in abundant levels [23, 26]. Albumin is supplemented to the culture media as, human serum albumin, purified albumin products or, recombinant albumin [23].

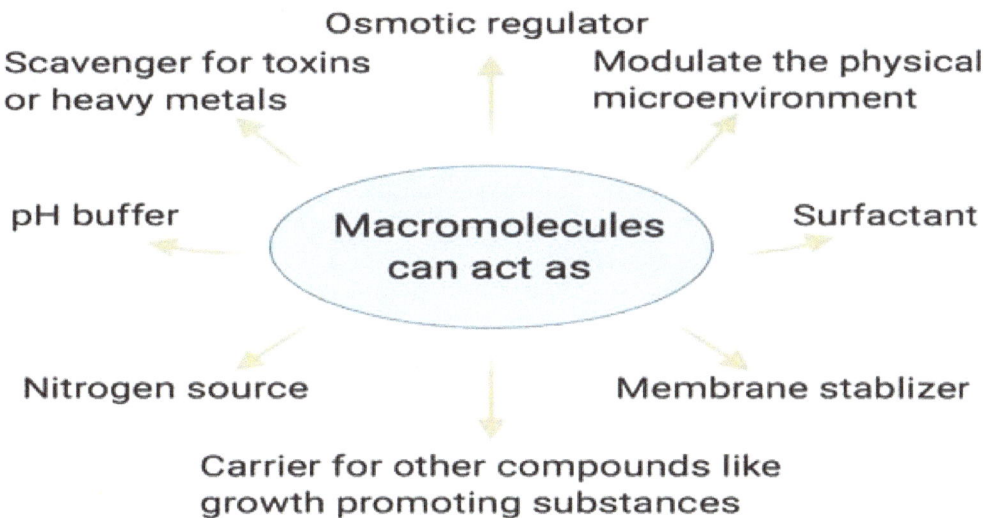

Osmotic regulator

Scavenger for toxins or heavy metals

Modulate the physical microenvironment

pH buffer

Macromolecules can act as

Surfactant

Nitrogen source

Membrane stablizer

Carrier for other compounds like growth promoting substances

Fig. (3). The putative roles of macromolecules supplementation in embryonic culture media.

Naturally, the embryos also are exposed to waves of different cytokines and growth factors such as the insulin-like growth factor, fibroblast growth factors, transforming growth factor, leukemia inhibitory factor, vascular endothelial growth factor, granulocyte colony-stimulating factors, and others [27 - 29]. The presence or absence of those cytokines naturally or when supplemented with IVF media could affect the preimplantation embryonic development (cleavage, blastocyst rates, and quality), and the subsequent reproductive outcomes especially the implantation rate [7, 24, 28]. In addition, they may consider as anti-apoptotic factors, and assist in repairing mechanisms of the abnormal cells [7, 30]. However, the supplementation of GFs in culture media if not well regulated, might have adverse effects on the reproductive outcomes [7]. The actual role of each GF is still under investigation [28]. More research is needed to determine the safety of using GFs, and to clear the view in such a promising area.

THE EMBRYONIC CULTURE SYSTEM: A PHYSICAL PERSPECTIVE

Not only does the chemical composition of the culture media affect embryonic development, but also the physical conditions surrounding the embryos [17]. During embryo culture, the stability of pH, osmolality, and temperature is crucial, especially outside the incubator. Also, the manipulation of gametes and embryos could result in mechanical stress or instability of some physical conditions [17]. The stressors could negatively impact the gametes, embryos, and their reproductive outcomes. From this point, we should pay attention to the physical compartment in addition to the chemical conditions of the culture system.

Temperature

Temperature stability is one of the critical factors in the embryonic culture [16]. The common temperature used during embryonic culture is around 37°C based on human core body temperature [31]. However, the optimal temperature is still under investigation [32]. Temperature variations could negatively impact the meiotic spindles of oocytes, and embryonic metabolism resulting in impairment of development [25]. Even transit cooling leads to dissemble of meiotic spindles in metaphase I (MI) and metaphase II (MII), hence reducing the fertilization rate, delaying embryonic development, and lowering the clinical pregnancy rate (Fig. **4**) [11, 33]. Up to date, we have limited data regarding the effect of different temperatures on ART outcomes [34]. Some studies found higher embryonic development using a lower temperature (36.5-36.9°C) [35, 36], while others have reported no differences even in 37.1°C [31, 32, 37]. Controlling the temperature of the environment around the gametes or embryos may be regulated by (I) the setup of laboratory incubators, (II) usage of oil overlay (as we will discuss later), (III) heated stage during the manipulation of the gametes or embryos, (IIII) limiting the manipulation time outside the incubators, and (IIIII) avoid multiple door opening of the incubators [25].

pH and Osmolality

The success of the embryonic culture system depends on "how the culture conditions could minimize the stress on gametes or embryos and optimize the *in vitro* conditions" [16, 38]. pH and osmolality are crucial parameters that need to be regulated, to reduce stress resulting from the culture conditions [38, 39]. Embryos have limited ability to regulate the intracellular pH of the cytoplasm (pHi), oocytes lack any mechanisms and they are susceptible to any fluctuations in extracellular pH of the culture media (pHe), especially in the case of denuded oocytes [38]. In practical work, we cannot control the pHi, however, we can regulate the pHe of the culture media by controlling the carbon dioxide level of culture incubators [38]. The acceptable pH range for embryo culture media is

~7.2-7.4 depending on the physiology of the uterus and fallopian tube, which can generally be obtained from 5-6% CO_2 [38, 40]. Media pH is measured in more than one simple way, pH meter is the most common way, blood gas analyzer as a recent accurate choice, and continual pH measurement by a new technology using fluoroscopy [16, 41, 42].

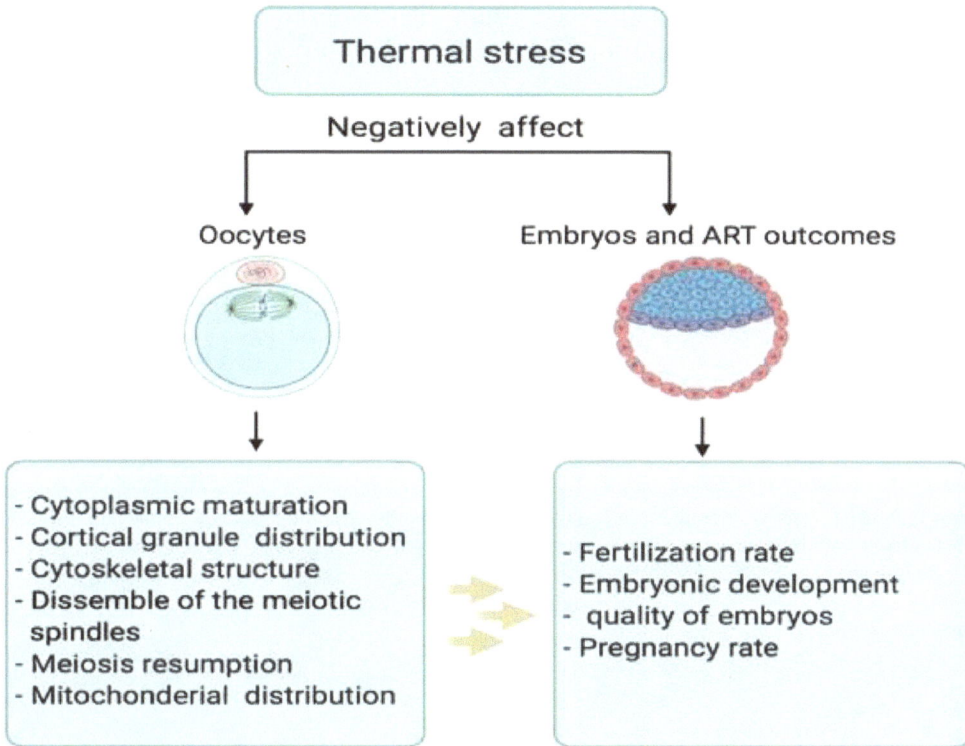

Fig. (4). Negative impact of thermal stress on oocytes, embryos, or clinical assisted reproductive technologies (ART) outcomes.

On the other hand, osmolality is the osmotic pressure of a solution [1], proper embryo culture media osmolality ~260□290mOsm [39]. More than 300mOsm can affect the cellular volume, impair embryo development *in vitro*, impact the chromosome dynamics, and may give a higher aneuploidy rate [11]. Osmolality could be measured by many types of osmometers to insure the proper limits. pH, osmolality, and temperature are interlinked parameters, higher temperature increases the evaporation rate of the media, leads to increment in osmolality level, and hence increase the pH level (Fig. **5**) [43]. Any fluctuation of those parameters less or above the proper limits can negatively affect oocyte cytoskeleton, mitochondrial localization, cellular metabolism, embryonic development, aneuploidy rate, and subsequent reproductive outcomes (Fig. **5**) [38, 40, 43]. Controlling and minimization pH and osmolality variations can be obtained by (I) daily regular checks for a CO_2 level of the incubators [16], (II) regular pH checks by an accurate device [16], (III) the use of media with a strong buffer system has good pK_a values and optimum buffering capacity [38], (IV) maintain a constant temperature [43], (V) the use of oil overlay to maintain temperature control [44], (VI) and control the evaporation rate of the media that may occur due to, the type of incubator (humidified/dry) [39], culture dish material and configuration [45], the method of media preparation (oil overlay/oil underlay) (Fig. **6**) [39], the type and amount of oil overlay [44].

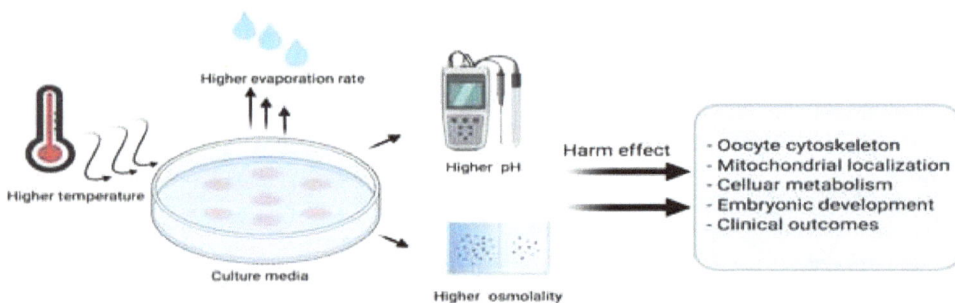

Fig. (5). The interlink between temperature, osmolality, and pH. higher temperature increases the evaporation rate of the media, leads to increment in osmolality level, and increase the pH. And, the negative impact of higher pH and osmolality level on the ART reproductive outcomes.

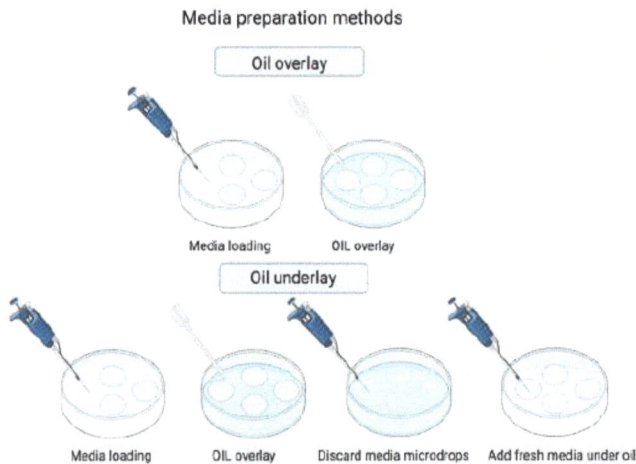

Fig. (6). Different methods of media preparation, the standard oil overlay method, and the oil underlay method to minimize the osmolality change by loading media drops under oil.

Oil Overlay

IVF culture media are susceptible to excessive evaporation because it is usually used in small volumes [39]. Oil overlay is widely used to minimize that evaporation and hence, maintain stable pH and osmolality [46]. In addition, it reduces the temperature variations of the culture media surrounding the gametes or embryos, acting as a chelating barrier to the microbial contamination, and prevent it from passing to the media [16, 17, 46] Mineral or paraffin oils as derivatives of petroleum production are typically used in culture media overlay [47]. The difference between mineral and paraffin oils is due to the polycarbonic lipid tail as mineral oil contains more unsaturated bonds than paraffin oil [47]. There is another type of oil used to overlay the culture media called synthetic oil, which undergoes chemical treatment to improve its structure. Despite the benefits of using oil overlay; there are many concerns about its negative impact on embryonic development and reproductive outcomes [46]. Those concerns may be due to (I) the probability of crude oil toxic remnants, which might contaminate the culture oil. (II) the sensitivity to the photooxidation and peroxidation reactions resulting from the unsaturated bonds, especially in the mineral oil [17, 48, 49]. The types of oils used to overlay the culture media may also differ in their chemical composition or physical characteristics such as density and viscosity which might affect the pH/osmolality and temperature stability of the embryonic culture media [46, 50]. Many studies have reported the differences between the types of oil, but there are limited data that compared the effect of different culture oils overlay and its effect on the reproductive outcomes [44, 46]. One prospective randomized trial by Sifer *et al* showed the effect of four types of oil on human embryonic development [51]. They reported that the embryo quality might be affected by the type of oil used [51]. As they find a better quality of day 3 embryos cultured in media overlaid with paraffin oil (PO) than mineral oil (MO) [51]. However, Sifer's study is applied to human embryos; it lacks the siblings' oocytes design and the other embryological parameters as the blastulation rate [46, 51]. Cristina *et al* performed another study on the animal model and they showed significantly higher developmental rates in terms of cleavage and blastulation rates in PO groups rather than MO [48]. Soraya *et al* in their prospective randomized study divided sibling oocytes between culture media overlaid with PO and MO [46]. They did not find any significant differences in the embryos till day 3 either in quality or the utilization rate [46]. On the other hand, Otsuki *et al* concluded in their study that mineral oil could result in a significant reduction in the fertilization, cleavage, and blastocyst formation rates [49]. This area is still open for more well-designed studies to optimize embryonic culture conditions.

Light

During the procedures of IVF or ICSI, gametes and embryos are subjected to different light sources such as aa: sunlight, microscopic light, time-lapse cameras, or the light emitted from laminar flow, which naturally not happened [17, 52, 53]. Exposure of gametes or embryos to light during retrieval, injection, manipulation, or at the time of embryo transfer may hurt the reproductive outcomes depending on light wavelength [52 - 54]. Many studies have suggested that visible light may activate stress genes, cause DNA damage through the ionization process, and may exert reactive oxygen species (ROS), which affect mitochondrial functions and cause cellular damage [52, 53, 55]. In addition, light affects the embryonic culture media and quality of oil through the photooxidation proces, consequently inducing sperm/oocyte photooxidation altering the membrane structure inhibiting the fertilization,n and reducing the subsequent outcomes even the implantation rate (Fig. 7) [17, 55, 56]. Although, there is no strong evidence about the harmful effect of light on human gametes, embryos, and their reproductive outcomes; we can take mouse embryos as a model due to the similarities between the two species [56]. Until the appearance of new data on the human model, we should take caution to reduce the harmful effect of light on the gametes, embryos, or media by multiple tips [52, 55, 56]. (I) Use protective light filters might be a helpful solution, as reported in a recent study by Bódis *et al* [52]. (II) The inclusion of antioxidants in culture media may reduce the negative impact of ROS. (III) The use of short-wavelength light inside IVF laboratories, as it has less harmful effects than a longer one. (IV) Reduce the use of direct light and avoid fluorescent light. (V) IVF laboratories should be placed in areas far from direct sunlight. (VI) Embryologists should be careful when using hood light, or microscopic light.

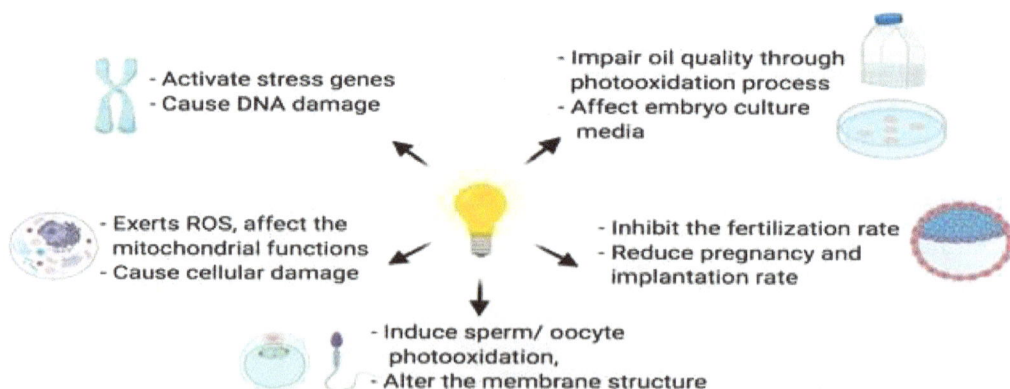

- Activate stress genes
- Cause DNA damage

- Impair oil quality through photooxidation process
- Affect embryo culture media

- Exerts ROS, affect the mitochondrial functions
- Cause cellular damage

- Inhibit the fertilization rate
- Reduce pregnancy and implantation rate

- Induce sperm/ oocyte photooxidation,
- Alter the membrane structure

Fig. (7). Visible light effects on the gametes and the subsequent reproductive outcomes.

Single Embryo Culture and Groups Culture

During ART procedures, embryologists determine to culture the embryos either in groups or individually [17]. Single or group cultures are different strategies, each has its pros and cons [57]. Embryos may gain benefits from the production of autocrine or paracrine factors that accumulated during group culture [17, 58]. Embryo-trophic factors were found to be useful for embryo development in many animal species and even in human embryos [57 - 59]. The success of the group culture approach may be affected by many variables [57, 58, 60]: (I) Embryo to volume ratio as several embryos per media drop, (II) spacing between embryos, if the embryo-trophic factors only extended to a certain distance, (III) quality of embryos per drop as good quality embryos may improve the development of other embryos, also bad quality embryos may impact the rest of embryos, (IV) culture media utilization and its by-product, (V) frequency of media refreshment, (VI) volume of oil overlay, (VII) and the type of culture dish used [57, 58, 60]. On the contrary, single culture lacks the benefits of embryo-trophic factors, however, it gives a chance to track each embryo individually for a better selection, and is used in case of embryo biopsy for preimplantation genetic diagnosis or preimplantation genetic sequencing PGD/PGS (Fig. **8**) [17, 57].

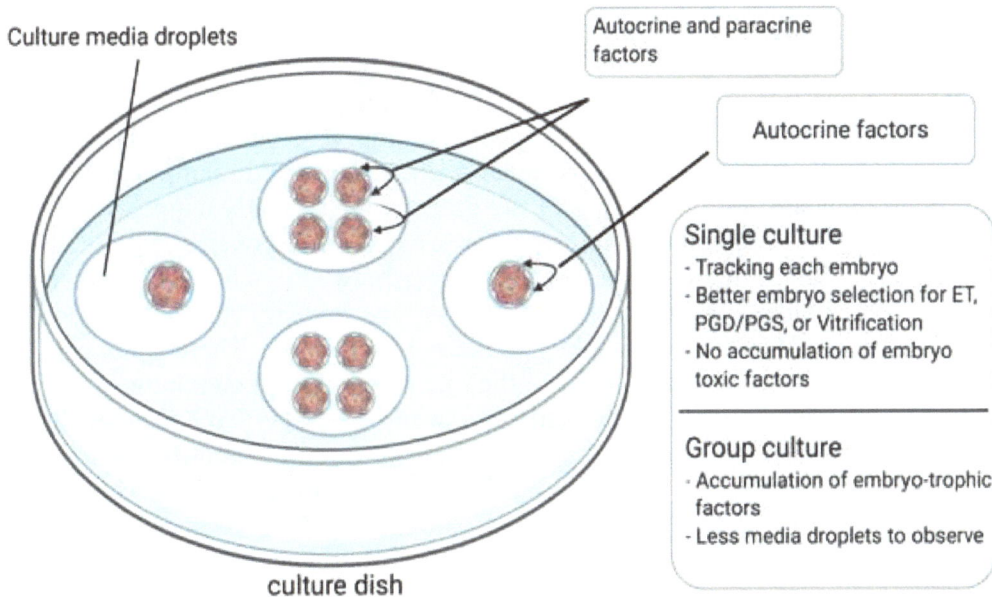

Culture media droplets

Autocrine and paracrine factors

Autocrine factors

Single culture
- Tracking each embryo
- Better embryo selection for ET, PGD/PGS, or Vitrification
- No accumulation of embryo toxic factors

Group culture
- Accumulation of embryo-trophic factors
- Less media droplets to observe

culture dish

Fig. (8). The advantages of single and group embryo culture, illustrating the autocrine/paracrine factors. ET: embryo transfer, PGD: preimplantation genetic diagnosis, PGS: preimplantation genetic sequencing.

Plasticware and Culture Dishes

A vast amount of plasticware is used through ART procedures, varied from cups that are used in semen collection, different types of tubes for semen processing, pastures/tips and strippers for oocyte and embryo manipulation, dishes for oocytes collection, embryo culture, and injection [61]. Although their wide usage; few types are embryo tested and are suitable for IVF procedures [62]. A few data reported the effect of plasticware on *in vitro* reproductive outcomes. The most important item of the plasticware used in IVF is the culture dish, which has direct contact with gametes or embryos. IVF culture dish material is varied between glass or plastic polymers usually polystyrene as an inert surface [45]. Polyvinyl chloride-based, or other plastics that are undeveloped especially for IVF, might have toxic materials, the release of such toxic materials in presence of gametes or preimplantation embryos could suppress the cellular development [61]. A four-year survey by Martine Nijs *et al* to test the reprotoxicity of the consumables used in ART finds that thirteen of 36 products were reprotoxic [61]. In addition, he reported the importance of quality control procedures to identify potential reprotoxic products and consumables used in ART procedures [61]. Not only the material that affected the reproductive outcomes but also the design of the dish whether it is two-dimensional (2D) or three-dimensional dish (3D) (Fig. **9**) [45]. Three-dimensional dishes are the advanced upgrading of two-dimensional conventional dishes. Although, the high cost of 3D dishes compared to conventional dishes; there is more than one benefit from using 3D over 2D dishes: (I) 3D configuration of the dish mimic the physiological tissue nature, (II) the concave well design could help to collect the embryos together, (III) promote rapid identification for the embryos under a microscope and manipulation, (IV) minimizing the shifting or merging of the drops, which usually happens with the conventional dishes, (V) maintain easily tracking for embryos in a single culture with an extra feature that allows media sharing between embryos (in a special design). (VI) help in the reduction of the surface area to volume around the embryos that could maintain a stable osmolality [29, 63 - 65]. There is not enough published data that studied the effect of the internal dish design on human *in vitro* embryonic development. A system called the well of well (WOW) was developed by Vajta *et al* as a modification of the 2D dishes to form a 3D configuration [65, 66] (Fig. **9**). His system showed significant improvement in the embryonic development of the animal and human models [65, 66]. Also, the usage of 3D dishes that are manufacturing, especially for IVF purposes resulted in a higher developmental rate of mouse embryos compared to 2D dishes [67]. To date, there is no adequate data regarding the effect of the plasticware or culture dishes on ART outcomes.

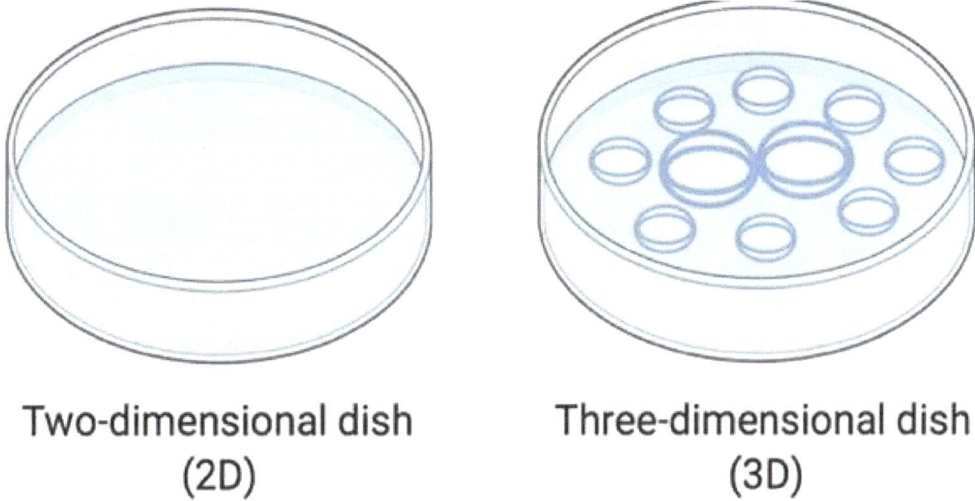

Two-dimensional dish
(2D)   Three-dimensional dish
(3D)

Fig. (9). Illustration for different culture dish configurations.

Incubation Platforms

The development of embryos needs controlled conditions to mimic the natural environment of the fallopian tube and uterus. For this purpose, incubators are used, with the warm temperature and controlled environmental factors (gas concentration, temperature, and humidity) [68, 69]. Any suboptimal conditions of the incubators could negatively impact embryonic development [69]. Different incubator technologies have been emerged over the years, incipiently, incubators were "big boxes" as a refrigerator with one big door [69]. Many culture dishes containing gametes or embryos from different cases could be placed inside. Large gaseous disturbance and temperature instability when the door was opened, and a long time for recovery [69]. Then, small box incubators with smaller volumes have been emerged to reduce the recovery time [68]. Generally, big and small box incubators are humidified to mimic nature and maintain stable osmolality [70]. Big and small box incubators are still in use till the current days in many IVF laboratories. The other type is the benchtop, as an incubator with small separate chambers each has its separate door. Benchtop incubators resulted in more stable conditions than big boxes [68, 69]. Opining one chamber will not affect the conditions of the other chambers [68]. That led to a short time for recovery of co_2 or nitrogen gas in small volume benchtop incubators 0.31-0.5 L compared to big box 150-200 L [68]. In addition, increase the safety of the cases as each case can be placed in a separate chamber (Fig. 10). Going up to more advanced technology in the time-lapse incubators (Fig. **10**), which has built-in cameras inside to take photos for every single embryo every 10-12 minutes [71]. Time-lapse (TL) technology offers better assessment for the embryos without disturbing the surrounding environment for up to 5-6 days with the uninterrupted culture system

[69, 71]. This technology not only offers the morphological assessment of the embryo, but also the kinetics of development. Giving a better chance of embryo selection for embryo transfer, preimplantation genetic testing (PGT), or vitrification [69]. However, it is considered a very expensive culturing platform [71]. Often, benchtop and TL incubators are dry without a humidification source, nowadays, many incubators have two versions (humid and dry). Humidity in the incubators may induce microorganism growth, which may reduce embryonic development [16]. Non-humid incubators overcome this problem, while it may induce osmolality raise even with oil overlay [16]. That also impacts the development and quality of embryos [16]. The published data failed to determine the superiority of a particular incubator over the other due to the multiple variables and confounders [69].

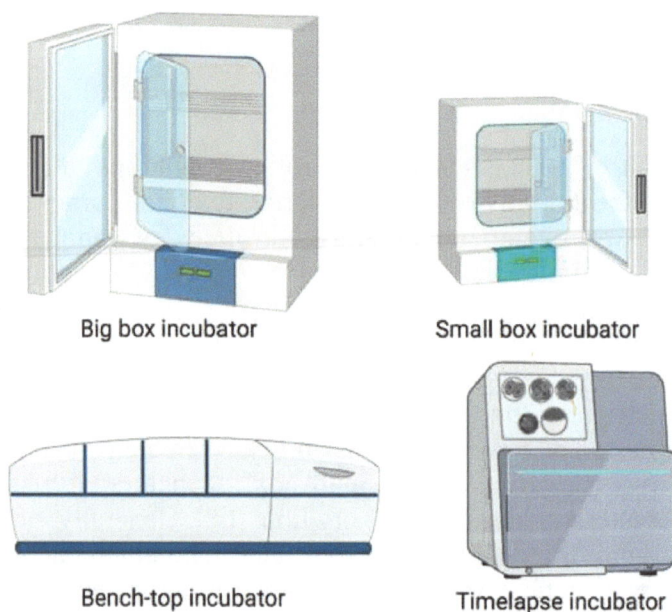

Big box incubator   Small box incubator

Bench-top incubator   Timelapse incubator

Fig. (10). Different incubators used in embryo culture.

Static and Dynamic Nature

The traditional culture system has a static nature, which uses inert dishes with different configurations (as we will discuss later) or test tubes as containers of gametes/embryos with the culture media [29, 45]. A static culture approach involves no active movement of the media around the embryos [29]. The embryonic development in that approach depends on the volume of the media, embryo density, and dish configuration [29, 45]. The natural embryonic environment within the uterus and oviduct is a dynamic nature involving fluid exchange, hormonal changes in addition to mechanical stimulants [29]. To optimize the culture conditions by imitating nature, studies have been conducted to determine the active media movement effect on embryos during culture days as a term of "dynamic culture system" [45, 72] (Fig. **11**). No culture system is considered completely static, due to the mechanical manipulation during media refreshment, or moving dishes during routine embryo checks [29, 45]. However, the dynamic system here refers to the culture devices that control media movement [29]. *In vitro* dynamic embryo culture can be achieved in many simple and complex ways: shaking, rotation, tilting, vibration, or by a controlled fluid flow of the media by specific devices (Fig. **11**) [45, 72]. All those methods may assist in replenishing media nutrients, disturbing the local constituents, or help in removing harmful byproducts [29, 45]. A dynamic culture system is an exciting approach that may improve embryonic development, embryonic quality and achieve higher pregnancy rates in ART protocols [72 - 74]. However, it has limited commercial availability with low implementation in clinical use [74]. This approach needs more research to optimize the cultural conditions [74].

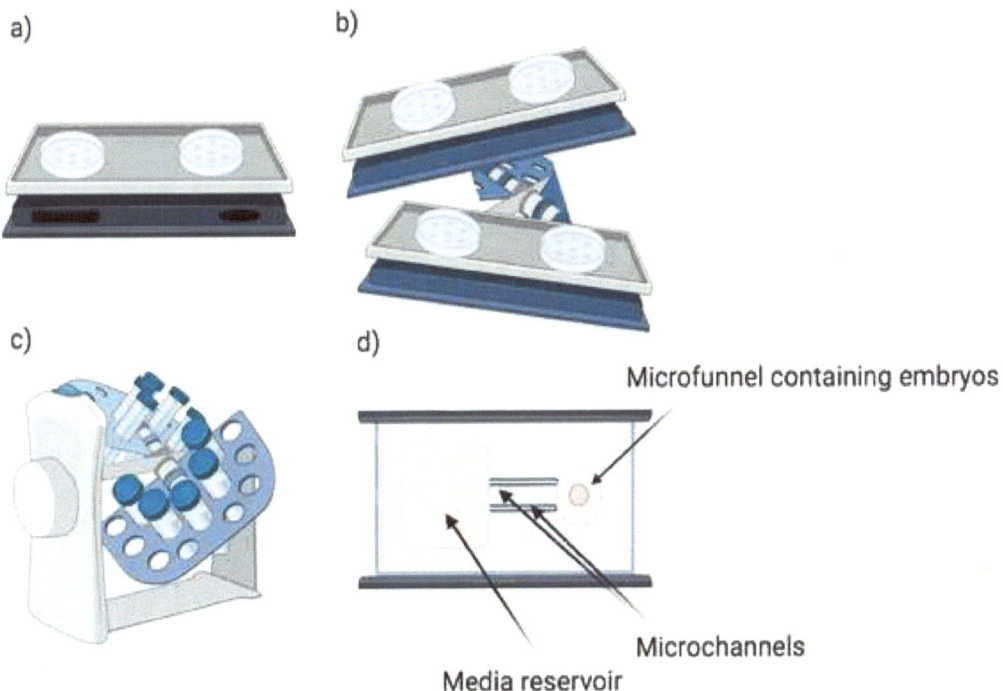

Fig. (11). Some of the dynamic culture system devices used. a) Vibration and rotation, b) Tilting movement for dishes, c) Tilting movement for tubes, and d) Controlled fluid flow by dynamic micro funnel chip that regulates media flow to embryos through microchannels.

Deficiencies in the Current Culture Systems

Our ultimate goal is to optimize the culture system, to raise reproductive outcomes through the ART journey. Despite, the large data on the chemical composition of the culture media, its effect on the *in vitro* embryonic development, and subsequent reproductive outcomes; there are limited studies on the physical factors of the culture system. The key for culture system optimization is not only the sole dependence of the chemical composition of the culture media, the addition of antioxidants or growth factors to the media, or even the physical compartment of the culture system. It is about every small factor introduced to the culture system in combination with the other factors. The current literature lacks research that studies the combinations of different culture factors on the ART outcomes.

Further Investigations

There is always a chance to improve; more well-designed studies that extended to off-spring health are needed to optimize the embryonic culture system. Exact media compositions and their concentrations need to be disclosed for accurate evaluation. More data will be better on some promising areas such as embryo density per drop, dynamic culture, GFs addition, or antioxidants in addition to monitoring for any single factor added to the media as GFs and antioxidants to determine the safety. Paying attention to the combinations between different factors that affect the culture system.

CONCLUSION

Every single factor in the embryo culture system could affect the reproductive results. The optimal culture system could support embryonic development, increase the pregnancy rate, and lead to healthy offspring. Suboptimal conditions impair the outcomes in different embryological and clinical levels. Considering the fine details for every single procedure, new elements added to the media, advanced device, or technique could lead to better management for each case, better embryo quality, higher developmental rates, and higher success rates for ART cases.

CONSENT FOR PUBLICATION

Not applicable

CONFLICT OF INTEREST

The authors declare no conflict of interest, financial or otherwise.

ACKNOWLEDGEMENT

The authors would like to thank their corresponding families for the support and patience.

REFERENCES

[1]   Quinn P. Culture media, solution, and systems in human ART [Internet]. Cambridge University Press 2014; 278.
        [http://dx.doi.org/10.1017/CBO9781139059053]

[2]   Leese HJ. Human embryo culture: back to nature. J Assist Reprod Genet 1998; 15(8): 466-8.
        [http://dx.doi.org/10.1023/A:1022526219202] [PMID: 9785193]

[3]   Tarahomi M, Vaz FM, van Straalen JP, *et al.* The composition of human preimplantation embryo culture media and their stability during storage and culture. Hum Reprod 2019; 34(8): 1450-61.
        [http://dx.doi.org/10.1093/humrep/dez102] [PMID: 31348827]

[4]   Morbeck DE, Krisher RL, Herrick JR, Baumann NA, Matern D, Moyer T. Composition of commercial media used for human embryo culture. Fertil Steril 2014; 102(3): 759-766.e9.
        [http://dx.doi.org/10.1016/j.fertnstert.2014.05.043] [PMID: 24998366]

[5]   Gardner D. *In vitro* fertilization: a practical approach. 2007.http://evilevil.pixub.com/o/*in-vitro-fertilization-a-practical-approach-by-david-k-gardner*.pdf

[6]   Gruber I, Klein M. Embryo culture media for human IVF: which possibilities exist? J Turk Ger Gynecol Assoc 2011; 12(2): 110-7.
        [http://dx.doi.org/10.5152/jtgga.2011.25] [PMID: 24591972]

[7]   Chronopoulou E, Harper JC. IVF culture media: past, present and future. Hum Reprod Update 2015; 21(1): 39-55.
        [http://dx.doi.org/10.1093/humupd/dmu040] [PMID: 25035437]

[8]   Edwards RG. Test-tube babies, 1981. 1981; 293(September): 253–6.

[9]   Fertiliz KJ, Vitro I, Keskintepe L. Journal of Fertilization : *In vitro* - IVF-Worldwide, Reproductive Medicine, Genetics & Stem Cell Biology Human Embryo Culture Dilemma Continues : " Back to Nature " or. Let the Embryo Choose 2012; 2(2): 7491.

[10]   Swain JE. Controversies in ART: considerations and risks for uninterrupted embryo culture. Reprod Biomed Online 2019; 39(1): 19-26.
        [http://dx.doi.org/10.1016/j.rbmo.2019.02.009] [PMID: 31109893]

[11]   Swain JE. Controversies in ART: can the IVF laboratory influence preimplantation embryo aneuploidy? Reprod Biomed Online 2019; 39(4): 599-607.
        [http://dx.doi.org/10.1016/j.rbmo.2019.06.009] [PMID: 31492521]

[12]   Dieamant F, Petersen CG, Mauri AL, *et al.* Single versus sequential culture medium: which is better at improving ongoing pregnancy rates? A systematic review and meta-analysis. JBRA Assist Reprod 2017; 21(3): 240-6.
        [http://dx.doi.org/10.5935/1518-0557.20170045] [PMID: 28837034]

[13]   Sfontouris IA, Martins WP, Nastri CO, *et al.* Blastocyst culture using single versus sequential media in clinical IVF: a systematic review and meta-analysis of randomized controlled trials. J Assist Reprod Genet 2016; 33(10): 1261-72.
        [http://dx.doi.org/10.1007/s10815-016-0774-5] [PMID: 27491772]

[14]   Reed ML, Hamic A, Thompson DJ, Caperton CL. Continuous uninterrupted single medium culture without medium renewal versus sequential media culture: a sibling embryo study. Fertil Steril 2009; 92(5): 1783-6.
        [http://dx.doi.org/10.1016/j.fertnstert.2009.05.008] [PMID: 19523617]

[15]   Stimpfel M, Bacer-Kermavner L, Jancar N, Vrtacnik-Bokal E. The influence of the type of embryo culture media on the outcome of IVF/ICSI cycles. Taiwan J Obstet Gynecol 2020; 59(6): 848-54.
[http://dx.doi.org/10.1016/j.tjog.2020.08.001] [PMID: 33218400]

[16]   Consensus Group C. 'There is only one thing that is truly important in an IVF laboratory: everything' Cairo Consensus Guidelines on IVF Culture Conditions. Reprod Biomed Online 2020; 40(1): 33-60.
[http://dx.doi.org/10.1016/j.rbmo.2019.10.003] [PMID: 31836437]

[17]   Wale PL, Gardner DK. The effects of chemical and physical factors on mammalian embryo culture and their importance for the practice of assisted human reproduction. Hum Reprod Update 2016; 22(1): 2-22.
[http://dx.doi.org/10.1093/humupd/dmv034] [PMID: 26207016]

[18]   Van Montfoort APA, Arts EGJM, Wijnandts L, *et al.* Reduced oxygen concentration during human IVF culture improves embryo utilization and cumulative pregnancy rates per cycle. Hum Reprod Open 2020; 2020(1): hoz036.
[http://dx.doi.org/10.1093/hropen/hoz036] [PMID: 31989044]

[19]   Mantikou E, Bontekoe S, van Wely M, Seshadri S, Repping S, Mastenbroek S. Low oxygen concentrations for embryo culture in assisted reproductive technologies. Hum Reprod Update 2013; 19(3): 209.
[http://dx.doi.org/10.1093/humupd/dms055] [PMID: 23377864]

[20]   Dumoulin JCM, Meijers CJJ, Bras M, Coonen E, Geraedts JPM, Evers JLH. Effect of oxygen concentration on human *in-vitro* fertilization and embryo culture. Hum Reprod 1999; 14(2): 465-9.
[http://dx.doi.org/10.1093/humrep/14.2.465] [PMID: 10099995]

[21]   Kovačič B, Vlaisavljević V. Influence of atmospheric versus reduced oxygen concentration on development of human blastocysts *in vitro*: a prospective study on sibling oocytes. Reprod Biomed Online 2008; 17(2): 229-36.
[http://dx.doi.org/10.1016/S1472-6483(10)60199-X] [PMID: 18681997]

[22]   Nakagawa K, Shirai A, Nishi Y, *et al.* A study of the effect of an extremely low oxygen concentration on the development of human embryos in assisted reproductive technology. Reprod Med Biol 2010; 9(3): 163-8.
[http://dx.doi.org/10.1007/s12522-010-0052-7] [PMID: 29657553]

[23]   Blake D, Svalander P, Jin M, Silversand C, Hamberger L. Protein supplementation of human IVF culture media. J Assist Reprod Genet 2002; 19(3): 137-43.
[http://dx.doi.org/10.1023/A:1014788821965] [PMID: 12005309]

[24]   Meintjes M. 2012.

[25]   Swain JE, Carrell D, Cobo A, Meseguer M, Rubio C, Smith GD. Optimizing the culture environment and embryo manipulation to help maintain embryo developmental potential. Fertil Steril 2016; 105(3): 571-87.
[http://dx.doi.org/10.1016/j.fertnstert.2016.01.035] [PMID: 26851765]

[26]   Pérez-Cerezales S, Ramos-Ibeas P, Acuña OS, *et al.* The oviduct: from sperm selection to the epigenetic landscape of the embryo. Biol Reprod 2018; 98(3): 262-76.
[http://dx.doi.org/10.1093/biolre/iox173] [PMID: 29228115]

[27]   Richter KS. The importance of growth factors for preimplantation embryo development and *in-vitro* culture. Curr Opin Obstet Gynecol 2008; 20(3): 292-304.
[http://dx.doi.org/10.1097/GCO.0b013e3282fe743b] [PMID: 18460945]

[28]   Raheem KA. Cytokines, growth factors and macromolecules as mediators of implantation in mammalian species. Int J Vet Sci Med 2017; 6(sup1) (Suppl.): S6-S14.
[http://dx.doi.org/10.1016/j.ijvsm.2017.12.001] [PMID: 30761315]

[29]   Swain JE, Smith GD. Advances in embryo culture platforms: novel approaches to improve preimplantation embryo development through modifications of the microenvironment. Hum Reprod

Update 2011; 17(4): 541-57.
[http://dx.doi.org/10.1093/humupd/dmr006] [PMID: 21454356]

[30]   Hardy K, Spanos S. Growth factor expression and function in the human and mouse preimplantation embryo. J Endocrinol 2002; 172(2): 221-36.
[http://dx.doi.org/10.1677/joe.0.1720221] [PMID: 11834440]

[31]   Neelke M, Ronny J, Samuel SR, Herman T, Hilde VV, Greta V. The effect of different temperature conditions on human embryos*in vitro*: two sibling studies. Reprod Biomed Online 2019; 38(4): 508-15.
[http://dx.doi.org/10.1016/j.rbmo.2018.11.020] [PMID: 30782420]

[32]   Hong KH, Lee H, Forman EJ, Upham KM, Scott RT Jr. Examining the temperature of embryo culture in *in vitro* fertilization: a randomized controlled trial comparing traditional core temperature (37°C) to a more physiologic, cooler temperature (36°C). Fertil Steril 2014; 102(3): 767-73.
[http://dx.doi.org/10.1016/j.fertnstert.2014.06.009] [PMID: 25044079]

[33]   Sun XF, Zhang WH, Chen XJ, Xiao GH, Mai WY, Wang WH. Spindle dynamics in living mouse oocytes during meiotic maturation, ageing, cooling and overheating: a study by polarized light microscopy. Zygote 2004; 12(3): 241-9.
[http://dx.doi.org/10.1017/S0967199404002850] [PMID: 15521714]

[34]   Baak NA, Cantineau AEP, Farquhar C, Brison DR. Temperature of embryo culture for assisted reproduction. Cochrane Database Syst Rev 2019; 9(9): CD012192.
[http://dx.doi.org/10.1002/14651858.CD012192.pub2] [PMID: 31529804]

[35]   Hunter RHF. Temperature gradients in female reproductive tissues. Reprod Biomed Online 2012; 24(4): 377-80.
[http://dx.doi.org/10.1016/j.rbmo.2011.12.007] [PMID: 22369989]

[36]   Higdon HL III, Blackhurst DW, Boone WR. Incubator management in an assisted reproductive technology laboratory. Fertil Steril 2008; 89(3): 703-10.
[http://dx.doi.org/10.1016/j.fertnstert.2007.03.040] [PMID: 17524397]

[37]   Fawzy M, Emad M, Gad MA, *et al.* Comparing 36.5°C with 37°C for human embryo culture: a prospective randomized controlled trial. Reprod Biomed Online 2018; 36(6): 620-6.
[http://dx.doi.org/10.1016/j.rbmo.2018.03.011] [PMID: 29625822]

[38]   Swain JE. Optimizing the culture environment in the IVF laboratory: impact of pH and buffer capacity on gamete and embryo quality. Reprod Biomed Online 2010; 21(1): 6-16.
[http://dx.doi.org/10.1016/j.rbmo.2010.03.012] [PMID: 20570214]

[39]   Swain JE, Cabrera L, Xu X, Smith GD. Microdrop preparation factors influence culture-media osmolality, which can impair mouse embryo preimplantation development. Reprod Biomed Online 2012; 24(2): 142-7.
[http://dx.doi.org/10.1016/j.rbmo.2011.10.008] [PMID: 22197605]

[40]   Swain JE. Is there an optimal pH for culture media used in clinical IVF? Hum Reprod Update 2012; 18(3): 333-9.
[http://dx.doi.org/10.1093/humupd/dmr053] [PMID: 22311857]

[41]   Diaz de Pool JDN, Van Den Berg SAA, Pilgram GSK, Ballieux BEPB, Van Der Westerlaken LAJ. Validation of the blood gas analyzer for pH measurements in IVF culture medium: Prevent suboptimal culture conditions. PLoS One 2018; 13(11): e0206707.
[http://dx.doi.org/10.1371/journal.pone.0206707] [PMID: 30418977]

[42]   Swain JE, Wilding M. Practical pH for the IVF Laboratory. J Reprod Stem Cell Biotechnol 2012; 3(2): 62-76.
[http://dx.doi.org/10.1177/205891581200300205]

[43]   Simopoulou M, Sfakianoudis K, Rapani A, Giannelou P, Anifandis G, Bolaris S, *et al.* Considerations on Embryo Culture Conditions. 2018.

[44]   Swain JE. Different mineral oils used for embryo culture microdrop overlay differentially impact media evaporation. Fertil Steril 2018; 109(3): e53. [Internet].
[http://dx.doi.org/10.1016/j.fertnstert.2018.02.101]

[45]   Smith GD, Takayama S, Swain JE. Rethinking *in vitro* embryo culture: new developments in culture platforms and potential to improve assisted reproductive technologies. Biol Reprod 2012; 86(3): 62.
[http://dx.doi.org/10.1095/biolreprod.111.095778] [PMID: 21998170]

[46]   Labied S, Jouan C, Wenders F, *et al.* Comparison between paraffin and mineral oil covering on early human embryo culture: a prospective randomized study. Syst Biol Reprod Med 2019; 65(1): 81-6.
[http://dx.doi.org/10.1080/19396368.2018.1492645] [PMID: 29985652]

[47]   Speight JG. Handbook of Industrial Hydrocarbon Processes. 2020.

[48]   Martinez CA, Nohalez A, Parrilla I, *et al.* The overlaying oil type influences *in vitro* embryo production: differences in composition and compound transfer into incubation medium between oils. Sci Rep 2017; 7(1): 10505.
[http://dx.doi.org/10.1038/s41598-017-10989-5] [PMID: 28874873]

[49]   Otsuki J, Nagai Y, Chiba K. Peroxidation of mineral oil used in droplet culture is detrimental to fertilization and embryo development. Fertil Steril 2007; 88(3): 741-3.
[http://dx.doi.org/10.1016/j.fertnstert.2006.11.144] [PMID: 17292894]

[50]   Yumoto K, Iwata K, Sugishima M, *et al.* Unstable osmolality of microdrops cultured in non-humidified incubators. J Assist Reprod Genet 2019; 36(8): 1571-7.
[http://dx.doi.org/10.1007/s10815-019-01515-9] [PMID: 31267335]

[51]   Sifer C, Pont JC, Porcher R, Martin-Pont B, Benzacken B, Wolf JP. A prospective randomized study to compare four different mineral oils used to culture human embryos in IVF/ICSI treatments. Eur J Obstet Gynecol Reprod Biol 2009; 147(1): 52-6.
[http://dx.doi.org/10.1016/j.ejogrb.2009.06.023] [PMID: 19615809]

[52]   Bódis J, Gödöny K, Várnagy Á, *et al.* How to reduce the potential harmful effects of light on blastocyst development during IVF. Med Princ Pract 2020; 29(6): 558-64.
[http://dx.doi.org/10.1159/000509016] [PMID: 32474560]

[53]   Ottosen LDM, Hindkjaer J, Ingerslev J. Light exposure of the ovum and preimplantation embryo during ART procedures. J Assist Reprod Genet 2007; 24(2-3): 99-103.
[http://dx.doi.org/10.1007/s10815-006-9081-x] [PMID: 17216346]

[54]   Li R, Pedersen KS, Liu Y, *et al.* Effect of red light on the development and quality of mammalian embryos. J Assist Reprod Genet 2014; 31(7): 795-801.
[http://dx.doi.org/10.1007/s10815-014-0247-7] [PMID: 24854483]

[55]   Pomeroy KO, Reed ML. The Effect of Light on Embryos and Embryo Culture. J Reprod Stem Cell Biotechnol 2012; 3(2): 46-54.
[http://dx.doi.org/10.1177/205891581200300203]

[56]   Bognar Z, Csabai TJ, Pallinger E, *et al.* The effect of light exposure on the cleavage rate and implantation capacity of preimplantation murine embryos. J Reprod Immunol 2019; 132(132): 21-8.
[http://dx.doi.org/10.1016/j.jri.2019.02.003] [PMID: 30852462]

[57]   Reed ML, Woodward BJ, Swain JE. Single or Group Culture of Mammalian Embryos: The Verdict of the Literature. J Reprod Stem Cell Biotechnol 2011; 2(2): 77-87.
[http://dx.doi.org/10.1177/205891581100200203]

[58]   Ebner T, Shebl O, Moser M, Mayer RB, Arzt W, Tews G. Group culture of human zygotes is superior to individual culture in terms of blastulation, implantation and life birth. Reprod Biomed Online 2010; 21(6): 762-8.
[http://dx.doi.org/10.1016/j.rbmo.2010.06.038] [PMID: 21051291]

[59]   Watanabe H. Correlation between the number of cultured human embryos and embryo development in

the wow culture dish system. Fertil Steril 2017; 108(3): e158. [Internet].
[http://dx.doi.org/10.1016/j.fertnstert.2017.07.476]

[60]   Ebner T, Moser M, Sommergruber M, Shebl O, Tews G. Incomplete denudation of oocytes prior to ICSI enhances embryo quality and blastocyst development. Hum Reprod 2006; 21(11): 2972-7.
[http://dx.doi.org/10.1093/humrep/del272] [PMID: 16835216]

[61]   Nijs M, Franssen K, Cox A, Wissmann D, Ruis H, Ombelet W. Reprotoxicity of intrauterine insemination and *in vitro* fertilization-embryo transfer disposables and products: a 4-year survey. Fertil Steril 2009; 92(2): 527-35.
[http://dx.doi.org/10.1016/j.fertnstert.2008.07.011] [PMID: 18937937]

[62]   Vajta G, Rienzi L, Cobo A, Yovich J. Embryo culture: can we perform better than nature? Reprod Biomed Online 2010; 20(4): 453-69.
[http://dx.doi.org/10.1016/j.rbmo.2009.12.018] [PMID: 20202911]

[63]   Tung YC, Hsiao AY, Allen SG, Torisawa YS, Ho M, Takayama S. High-throughput 3D spheroid culture and drug testing using a 384 hanging drop array. Analyst (Lond) 2011; 136(3): 473-8.
[http://dx.doi.org/10.1039/C0AN00609B] [PMID: 20967331]

[64]   Mullen SF. Humidity and osmolality : Can we avoid media osmolality shifts in a dry environment? 2019.

[65]   Vajta G, Korösi T, Du Y, *et al.* The Well-of-the-Well system: an efficient approach to improve embryo development. Reprod Biomed Online 2008; 17(1): 73-81.
[http://dx.doi.org/10.1016/S1472-6483(10)60296-9] [PMID: 18616894]

[66]   Vajta G, Peura TT, Holm P, *et al.* New method for culture of zona-included or zona-free embryos: the Well of the Well (WOW) system. Mol Reprod Dev 2000; 55(3): 256-64.
[http://dx.doi.org/10.1002/(SICI)1098-2795(200003)55:3<256::AID-MRD3>3.0.CO;2-7]   [PMID: 10657044]

[67]   Rieger D. T. Tschimmel JC and MC. Comparison of GPS and standard dishes for embryo culture: set-up and observation times, and embryo development. Proceeding 14th World Congr IVF 3 rd World Congr IVM. 1202.

[68]   Swain JE. Decisions for the IVF laboratory: comparative analysis of embryo culture incubators. Reprod Biomed Online 2014; 28(5): 535-47.
[http://dx.doi.org/10.1016/j.rbmo.2014.01.004] [PMID: 24656561]

[69]   Swain JE. A Comparison of Embryo Culture Incubators for the IVF Laboratory. 2020.
[http://dx.doi.org/10.1007/978-981-15-2377-9_77]

[70]   Fawzy M, AbdelRahman MY, Zidan MH, *et al.* Humid versus dry incubator: a prospective, randomized, controlled trial. Fertil Steril 2017; 108(2): 277-83.
[http://dx.doi.org/10.1016/j.fertnstert.2017.05.036] [PMID: 28651960]

[71]   Armstrong S, Bhide P, Jordan V, Pacey A, Farquhar C. Time-lapse systems for embryo incubation and assessment in assisted reproduction. Cochrane Database Syst Rev 2018; 5(5): CD011320.
[http://dx.doi.org/10.1002/14651858.CD011320.pub3] [PMID: 29800485]

[72]   Monteiro A, Smith GD. Chapter 20 Culture Systems: Fluid Dynamic Embryo Culture Systems (Micro fl uidics). 912: 355–65.

[73]   Heo YS, Cabrera LM, Bormann CL, Shah CT, Takayama S, Smith GD. Dynamic microfunnel culture enhances mouse embryo development and pregnancy rates. Hum Reprod 2010; 25(3): 613-22.
[http://dx.doi.org/10.1093/humrep/dep449] [PMID: 20047936]

[74]   Hur YS, Ryu EK, Yoon SH, Lim KS, Lee WD, Lim JH. Comparison of static culture, micro-vibration culture, and micro-vibration culture with co-culture in poor ovarian responders. Clin Exp Reprod Med 2016; 43(3): 146-51.
[http://dx.doi.org/10.5653/cerm.2016.43.3.146] [PMID: 27689036]

CHAPTER 8

Assisted Hatching

Jayesh Parasharam Shinde[1,*]

[1] *ART Fertility Clinic PVT LTD, B 404 Aishwaryam Comfort, Akrudi, Pune-411019, Maharashtra, India*

Abstract: The selected Spermatozoa when it reaches the ovulated Cumulus Oocyte Complex after ovulation, dispersion of the granulosa cells and corona radiata cells occur. The Spermatozoa then must cross the Zona Pellucida (ZP), fuse with the oolemma, and then subsequently fertilize the oocyte. Embryologist Karl Ernst von Baer coined the term 'Zona Pellucida' from Greek work Zone which means belt or girdle and Latin work Pellucida which means transparent or shining. This extracellular matrix is about 13-15 um thick and surrounds all the mammalian eggs and pre-implantation embryos. Zona Pellucida structure is made up of carbohydrates, specific proteins, glycoproteins, hyaluronic acid, heparin, collagen, and fibrous proteins. Human Zona Pellucida contains 4 glycosylated proteins namely ZP1, ZP2, ZP3, and ZP4. ZP plays an important role in helping oocytes to transport essential nutrients and helps in avoiding polyspermy by hardening after fertilization. The embryos must break open the protective ZP layer to the implant, the process is called hatching. It is said that in Assisted reproductive treatment (ART) factors such as the non-availability of enzymes from the endometrium which helps in hatching, extended culture, vitrification may lead to failure in the hatching of embryos from ZP. It was postulated that micromanipulation of ZP to create an opening will help the embryos to hatch and thus implant and will lead to an increase in Implantation rates (IR). This process was later called Assisted Hatching (AH). Various methods were discovered for Assisted hatching such as mechanical ZP AH, zona digestion using enzymes, and laser-Assisted hatching. This chapter will focus on the advantages and disadvantages of each method of AH and their applications in ART along with the impact of AH on clinical outcomes. The use of any method of AH should be chosen carefully to avoid damage to the embryo which will defy the whole purpose of application of AH. In any case, laser-assisted hatching is widely used for Pre- Implantation Genetic Testing (PGT) of the embryos as it is very safe if applied properly, convenient, easy to use, and faster compared to other methods of AH. Each laboratory should identify the correct time and stage at which application of AH is considered based on whether it is helping to improve clinical rates or not.

Keywords: Assisted hatching, Chemical hatching, Laser-assisted hatching, Mechanical hatching, Zona Drilling, Zona Pellucida.

---

[*] **Corresponding author Jayesh Parasharam Shinde:** ART Fertility Clinic PVT LTD, B 404 Aishwaryam Comfort, Akrudi, Pune- 411019. Maharashtra, India; Tel: +91 7875891475;
E-mail: jayesh.shinde45@yahoo.com

INTRODUCTION

The mammalian female gamete-Oocyte is surrounded by cumulus and corona cells and is called Cumulus Oocyte Complex (COC). This COC is released during ovulation. When the selected spermatozoa, after passing through various barriers of the female reproductive tract, reaches and penetrates the ovulated COC the dispersion of tightly bound granulosa cells and corona radiata cells occurs. The spermatozoa then have to cross the Zona Pellucida (ZP) to fuse with the Oolemma and then subsequently fertilize the oocyte [1]. In 1827, Embryologist Karl Ernst von Baer adopted the word 'Zona Pellucida' from Greek work Zone- meaning belt or girdle and Latin word pellucida meaning transparent or shining while describing the Human egg. This thick extracellular matrix (ECM) which is about 13-15 um in diameter, surrounds all the mammalian eggs and preimplantation embryos [1, 2].

The structure of ZP is made up of carbohydrates, ZP-specific proteins, and Glycoproteins. Zp also contains hyaluronic acid, heparin, chondroitin (proteoglycans) and collagen, laminins, elastins, and fibronectins (fibrous proteins). Most mammalian eggs have 3 glycosylated proteins, human ZP contains 4 glycosylated proteins, namely ZP1, ZP2, ZP3, and ZP4. All these ZP1-4 have a polypeptide chain which includes the ZP domain (ZPD). The ZPD has 2 subdomains ZP-N and ZP-C connected by the linker region, 8 conserved cysteine residues, and ~270 amino acids (aa) [1, 2].

The main events during fertilization are sperm-COC recognition, penetration of ZP, and sperm oocyte fusion. The interaction between ZP and sperm is species-specific by receptor-ligand interaction leading to activation of sperm. Molecules like murine tyrosine kinase and galactosyltransferase are believed to be the sperm's receptors for ZP. Glycoprotein ZP3 acts as a ligand for receptors present on the sperm surface. This binding of sperm results in exocytotic event releasing proteolytic enzyme which will facilitate the penetration of ZP by the sperm, the process is called Acrosome reaction, where ZP2 acts as secondary sperm binding site resulting in gamete interaction *via* contact of ZP with the inner membrane of sperm which has undergone Acrosome reaction [3]. Such a complex interaction and process may hint towards the indirect role of ZP in sperm selection. Post fertilization, exocytosis of cortical granule content in perivitelline space (PVS) happens and comes in contact with the oolema and ZP. This exocytosis and cortical reaction lead to the blocking of multiple sperms entering the oocytes and thus avoiding polyspermy [4].

Zona hardening happens post-fertilization and is helpful to avoid polyspermy, protect the preimplantation embryo during early embryo development, and blastomere integrity and passage of embryo in the female reproductive tract before implantation. It is also argued that ZP protects the embryos from various uterine factors and immune responses. Zona hardening is also observed in mouse and human embryos cultured *in vitro* [1, 5, 6].

For the embryo to implant, it must hatch out of its protective layer-ZP. Based on time-lapse observations it was believed that repeated expansions and contractions of the blastocyst help the embryo exert pressure on the ZP to assist its hatching and implant. The volume of the blastocyst increases thereby stretching the epithelium of trophoblast cells and exerting increasing hydrostatic pressure leading to thinning of ZP. The presence of specialized plump cells called 'Zona Breakers' are also found to be present near the opening of the ZP [7]. Now there are sufficient studies in the database which suggest that various enzymes, proteolysins such as trypsin, proteases, and uterolysins, which are secreted by either the blastocyst and/or endometrium are responsible for the hatching of the embryo [7, 8, 16].

ASSISTED HATCHING AND ART

There are many reasons which may lead to the failure of implantation of a viable embryo. Sometimes the embryo may be slow-growing or its inherent quality may not be good to implant. Endometrium may not provide adequate uterine factors to assist implantation or might not be receptive enough. Another reason might be due to the inability of the embryo to hatch owing to circumstances related to laboratory and clinical conditions [11]. Implantation of the embryo after hatching is based on a series of events that need to be completed in a coordinated manner. Failure in completing any one of these numerous steps might stop the embryo from implanting. Patients' ability to produce enough uterolysins may be diminished or the ability of such uterolysins to carry out their function may also be influenced due to assisted reproductive treatment cycles they are undertaking [13].

Improvement of culture conditions led to extended cultures in many IVF labs around the world to select good quality blastocyst and transfer fewer embryos. This might lead to compromise the hatching process as enzymes that are thought to be released by endometrium are present to assist the hatching in natural conception. Extended culture might also lead to disrupting the timing and mechanism of hatching as Zona hardening is observed [1, 7, 9, 13]. *In vitro* culture embryos are found to be slower in development or poor in quality or many of them may not reach the blastocyst stage or hatch when compared to the development of embryos in natural conditions [10]. FSH levels, pre-ovulatory oestradiol, cause of infertility, and advanced maternal age also have an impact on the ZP mostly on its thickness and evenness [9, 10, 14]. In some cases, the enzymatic lysins, which digests the ZP, produced by the trophectoderm cells may have diminished production affecting the hatching of the embryo [12]. It is also reported that the uterolysin action in some patients or some cycles may be affected due to assisted reproductive treatment is ineffective to assist the embryo to hatch [13]. It was suggested that a patient's age may also have an impact on the production of lysins thereby affecting the implantation potential of the embryos [17].

There has been a good improvement seen in the survival rate of embryos in Cleavage and Blastocyst stage embryos since the popularity and use of Vitrification have been seen. Many studies have concluded that the Overall Live birth rates (LBR) are improved when the embryos are not transferred in the same cycle when oocytes are retrieved, the rates are much better when the embryos are transferred in the subsequent cycle when the endometrium is ready [15]. Even though the morphology of the vitrified and warmed embryos has resumed mitosis and metabolism, it is believed that the cryopreservation of embryos may affect the ZP. It is suggested that ZP may become hard and/or thick after cryopreservation of the embryos. This hardening and/or thickening of the ZP may lead to implantation failure as the embryo may not hatch owing to changes imparted on zona [14].

ZP thickness measurements taken during fertilization and embryo morphology assessments have been used to predict the success of the IVF cycles. Assisted hatching was proposed for patients because it was observed that the implantation rates (IR) correlate with the ZP character and thickness (thinner ZP had better IR) and that prolonged culture of embryos in the sub-optimal condition *in vitro* may have an impact on the thickness and/or hardening of ZP affecting about 75% embryos and stop them from hatching [18]. An interesting study was carried out by Cohen *et al.* where his group compared the impact of Zona thickness and Zona thickness variation measured during culture and its impact on the IR were compared retrospectively. It was observed that Zona thickness variation was able to predict the IVF outcomes better than Zona thickness alone. Gradual decline in Zona thickness and Zona thickness variation was observed with an increase in age, more effect was seen in patients older than 35 [18].

ASSISTED HATCHING

It was postulated that opening made in ZP may help the embryo to hatch and thus implant, based on observations of an increase in IR in embryos subjected to partial zona dissection [16]. The randomized control trial was carried out by Cohen where a patient group named poor prognosis was selected based on embryo assessments. These patients' embryos were selected for micro-manipulation of zona based on blastomere number, fragmentation in embryo, Thickness of zona, and maternal age. The embryos of such patients were subjected to a mechanical breach of the ZP using a micro-manipulator.

This micro-manipulation of zona was later labeled as 'Assisted Hatching' (AH) which involves breaching the zona in the selected region away from embryo cells or thinning of the ZP depending upon the requirement [8, 19].

Methods for Assisted Hatching

In the literature, many methods are available for AH and many labs around the world have tried and compared which one works better for them. It is important to note, before selecting the suitable AH method in your lab, that the method causes the least or no damage to the embryo while it is micro manipulated outside the incubator. Usual fluctuations seen in pH and temperature which happens when the embryos are handled outside the incubator should be avoided when embryos are subjected to the selected AH method.

As we know these changes can cause detrimental effects on the embryo, affects its implantation, and thus defy the whole purpose of doing AH in cases where it is done to improve the IR or address issues with ZP. AH must always be carried out at 37oC with the use of nonbuffered (culture media) or buffered media. In case culture media is used the procedure has to be quick enough to avoid the changes in pH and temperature. It is recommended to thoroughly wash the embryos in different micro drops of culture or buffered media after subjecting them to any of the AH methods. For embryos that are transferred after the AH method, embryos should be washed thoroughly and should wait for some time before the Embryo transfer (ET) is done. The whole idea is to make sure that the micromanipulation of the embryo should not affect its structural and implantation potential. Following are the different methods for AH with their advantages and disadvantages.

Mechanical Zona Pellucida Assisted Hatching

This method was first introduced by Cohen *et al.* in which the ZP of early cleavage embryos was artificially opened before the transfer to increase the implantation potential [1]. This method is also called 'Partial Zona Dissection' (PZD). In this method, the selected embryos were held steadily with help of suction pressure using the holding pipette. The sharp microneedle is then inserted in the ZP where the Perivitelline space (PVS) is large and then the ZP is pierced tangentially such that the micropipette is pierced out from the opposite side. For understanding purpose imagine the ZP is pierced between 5 o'clock and 6 o'clock on one side and then inserting it tangentially in large PVS is taken out from ZP region between 6 o'clock and 7 o'clock all the while making sure that micropipette never disturbs blastomeres. The suction from the holding pipette is then released and the trapped ZP is rubbed against the bottom of the holding pipette. The ZP was then removed by scrubbing to the holding pipette leaving a cut in the previously intact ZP [6].

The second method called 'Controlled Zona Dissection' (CZD)was also introduced, when the bigger hole was needed to be created. This was a slight modification of the PZD. In CZD, the place to insert the micropipette is selected (large PVS) and embryos are rotated in such a way that this large PVS region between 5 o'clock and 6 o'clock. The micropipette is then inserted and then this micropipette tip is further inserted in the holding pipette while tangentially piercing the embryo. The 'caught Zona' is then rubbed against the bottom of the dish to cut the ZP.

Another method in which the PZD is carried out twice in the same region, where the second slit is made below the previous slit. This process is called 'Three-dimensional Partial Zona Dissection (3D-PZD).

The above two processes-CZD and 3D-PZD aim to make a bigger opening compared to PZD.

The advantages of these methods (PZD, CZD, 3D-PZD) are that they are faster and cheaper as compared to Laser-Assisted hatching (LAH). Also, it does not involve the use of enzymes like pronace or Acid tyrode and heat to create a hole in ZP thus less harmful compared to LAH and Acid Tyrode (AT)assisted hatching thus not affecting further embryo development and improving implantation [22] [23].

The main disadvantages of these methods are that the openings created can never be uniform, may be very small thus affecting the cells of blastocyst while hatching, and needs a lot of practice to create a proper opening without damaging the embryo.

Zona Digestion using Enzyme Pronase (EZD)

It was believed that culturing the embryos in improved culture media was unable to support the physiological process of hatching. Proteolytic enzyme 'Pronase' was thus used to either soften or remove the zona completely. In the mouse model, it was reported that protease enzyme similar to Pronase is released by the endometrium or the blastocyst to assist the hatching process and thus implant. *In vitro* culture of embryos may lack these enzymes and Zona hardening may impair the hatching process. Also, Zona softening or removal will increase the chances of embryo dialogue with the endometrium thus improving its chances to attach to the endometrium and better cell to cell interactions. This may help in transferring fewer embryos and reduction of the embryonic losses. First, the live birth of a normal baby after such enzymatic treatment and zona removed blastocyst was reported by Fong *et al.* [24] [25].

This method of assisted hatching is considered safe if the concentration and duration of the enzyme exposure are controlled as confirmed by studying the ultrastructure of pronase treated and non-treated naturally hatched blastocysts in co-culture. By doing so, studies have suggested improvement in the implantation rates as compared to the unhatched blastocysts however no prospective randomized control trial is conducted to prove this on the larger sample size.

As mentioned previously right concentration of Pronase and duration of exposure is important to carry out the hatching process safely. A study conducted by Fong *et al.* compared the effects of variable exposure time of Pronase on the blastocysts. The embryos were exposed to Pronase for 1.5 min, 2 min, and 5 min and later checked with Light Microscopy (LM) and Transmission Electron Microscopy (TEM). The group found that the exposure of 10 IU Pronase at 37oC will dissolve the ZP completely in the experiment group with no differences seen in the blastocyst which were naturally hatched when observed under the LM and TEM. Subtle changes were observed in tonofibrils when the embryos were exposed to Pronase for 2 mins in TEM, and no changes were seen under LM when compared with the control group. In the third group where the embryos were exposed to pronase for 5 mins, Trophectoderm cells started showing blebbing under LM, the blastocele was collapsed and Trophectoderm cells started showing blebbing towards the blastocele and vacuolization increased when observed under TEM.

Safety of this method of AH can be established as the Fong *et al.* group reported no pregnancy losses up to 22 weeks of gestation and no congenital anomalies have been reported thus signifying that enzymatic treatment of ZP will not interfere with implantation of the embryo and its further growth [24] [25].

Zona Drilling by Use of Acid Tyrode's Solution (ZD)

It is said that the ZP is bi-layered, the inner layer is more resilient and compact as compared to the outer layer, which is less compact, comparatively thin, and can be digested. Acid Tyrode's solution was first reported to be used to remove abnormal or thick Zona in 1992. In 1992 and 1993, partial drilling and circumferential thinning were reported using Acid Tyrode's solution on mouse embryos to assist hatching. Since then many studies have reported the use of Acid Tyrode's solution to assist hatching of the embryos in a selected group of patients to increase the pregnancy rates and thus implantation rates [26] [27].

The Acid Tyrode's solution (ATS) can be prepared in-house with the help of existing established protocols and is also available commercially (SIGMA or *Vitro* life). The pH of the ATS is usually around 2.5+/-0.3. In this method of AH, the important thing to be controlled here is precise manipulation and controlled and limited exposure of the ATS to the embryos. There are two methods by which the embryos can be exposed to ATS.

In the first method also known as Partial Thinning, the embryo is held in place with the use of a Holding pipette set on a micromanipulator. Then using a mouth-controlled suction system is used to expel the ATS gently at the 3 o'clock end of the embryo. The embryo is held in such a way that the large PVS or extracellular fragment is usually exposed to ATS when released from a micro pipette in a controlled manner. When the ZP is dissolved the suction is stopped and the embryo is quickly transferred to another culture media drop where it is washed thoroughly to get rid of the ATS.

In the Circumferential Zona Thinning method, the embryos are incubated for about 30 seconds in approximately 1 ml of ATS drops covered with mineral oil. At the end of 30 seconds, the embryos are removed and checked under the inverted microscope to check if the outer layer of ZP is dissolved or not. The embryos are again incubated in ATS drop for about 10-15 seconds if the Zona was still intact even after the first round of exposure in ATS. An important point to be remembered here is to wash off the ATS thoroughly to avoid adverse effects on the development of the embryos.

Precise and proper depth of Zona dissolution can be obtained in the Partial Thinning method but needs a special mouth suction control system and expertise to carry out the hatching. Circumferential thinning, on the other hand, is relatively easy, fast, and does not need special equipment. However, both these methods of ZD are not as precise as Laser-assisted hatching. Overexposure to ATS can lead to cytoplasmic degeneration because of acidification and changes in the intracellular pH [28].

Even though improved IR and clinical pregnancy rates (CPR) using partial thinning were reported, there were no differences in the rate were seen when circumferential thinning was applied *vs* the control group. The main disadvantage of ZD using ATS is that the method is very inconsistent in the hatching of the zona if circumferential thinning is taken into consideration. Also, the negative effect on embryo development, trauma to oocytes, low rates of embryo cleavage, and higher abnormal cleavages resulting in a lower number of embryos available were reported by Garrisi *et al.* [29].

Laser-assisted Hatching (LAH)

Tadiran and Palankar were the firsts to report the use of laser for zona drilling in 1991 [31]. The use of laser beam causes thermolysis after its exposure to the Zona leaving a hole in the targeted area. In the initial stages, an argon fluoride excimer laser beam in the deep ultraviolet (UV) region at 193 nm of the spectrum was used. Then this laser beam was bombarded on the ZP after passing through a glass capillary or pipette with positive air pressure. Another option was to use Infrared (IR) wavelength using quartz fiber. The number of pulses and emitted energy per pulse were controlled to manipulate the depth of the drilled hole. The size of the pipette tip was used to control the diameter of the drilled hole. This method of LAH was called the contact mode as the pipette or the optical fiber carrying the UV or IR rays would be touching the ZP. The hole appears in the zona as the laser energy is absorbed and leads to photoablation. This contract method was later considered to be harmful to the embryos as it was associated with the damage of DNA and cellular macromolecules because of absorption of UV/IR light leading to mutagenic effects. Also needing sterile micropipettes and optical fibers which also constricted the size of the opening, light diffraction at the tip, and requirement of complex setup restricted their use later [30] [31] [32] [33].

The non-contact mode later became famous in which 1,480 nm, diode, IR were used. These were considered safe as they had a longer wavelength of the IR range which is not absorbed by the DNA and hence associated with the mutagenic effects [33] [32] [31]. The ZP drilling is easily carried out using a special objective through which laser pulse is applied without actually being in contact with the embryo.

The size of the hole drilled can be easily controlled by increasing or decreasing the exposure time of the laser to the embryo. It is important here to understand that the hole created is a furrow in the ZP and would look like a tunnel (extending from ZP touching the bottom of the dish, passing tangentially upwards as the laser moves upwards) when imagined in a 3D and not just a mere circular hole which appears on the ZP under the microscope. This method became famous as it was considered safe for embryos, was comparatively easy to apply with minimum training required to master the technique.

LAH can be applied to the ZP starting from outside and moving inwards towards the Oolemma while giving multiple shots. It is recommended to apply the laser pulses in the region where PVS is large thus avoiding exposing them to blastomeres to avoid damage. LAH needs to be performed on the ICSI machine set up on the anti-vibration table as the performance will be affected by vibrations. Also, the embryos should be held in a stable position using a holding pipette to accurately control the size of the ZP openings.

Although the LAH is safe, one should always remember that the temperature of media in the areas near the hole drilled in ZP would be increased to about 75oC when a 2 ms laser pulse is used [34]. Applying the laser shots should be done quickly and the embryos should be shifted immediately to avoid the adverse effects of an increase in temperature due to LAH [31] [32] [33].

Contact method is almost obsolete nowadays as non-contact methods are very easy to operate but the laser equipment is comparatively expensive to install and maintain. As AH is only popular when Preimplantation genetic testing (PGT) is required, centers in developing countries and underdeveloped countries may find it difficult to manage the expenses.

There are various ways in which the LAH may be applied to ZP to carry out Complete hatching, Partial hatching, or Zona thining which can be described as follows:

A-Complete hatching-This involves complete manipulation of the ZP in the selected region where the ZP is drilled starting from outside to inside ultimately also opening the Oolema. It is important to take into consideration that complete hatching of the ZP can affect the natural hatching process as naturally, the intracellular pressure increases during the expansion of the embryo leading to hatching. Complete hatching might lead to dropping in this intracellular pressure. Efflux of the fluid located inside the blastocele, collapsing the embryo was also reported after complete hatching of the ZP.

B-Partial hatching-In this method the ZP is not drilled completely and some part of ZP on the inner layer is left as it without breaking the oolema. Thus this type of LAH will make sure that the embryo is not harmed by the application of the AH and as opposed to complete hatching the risk of premature zona breaking due to expansion will not happen as the intracellular pressure will not be affected. There are not many studies comparing partial hatching *vs* complete hatching but both methods have their pros and cons and the clear advantage of one over the other is still not established. One should take into consideration that in cases of partial hatching the hatching may not be sufficient to assist embryo hatch completely as ZP left behind may be still thick.

C- Zona Thinning (ZT)-This type of hatching involves ablation of the quarter of the ZP, in short carrying out partial hatching but in a much larger portion of the ZP. Blake *et al.* applied ZT for the first time and reported that initiation and completion of hatching are better *in vitro*, compared to untreated embryos. The same group also reported that the embryos hatched from the same sites where ZT was done and not from the other site to support their initial findings.

Only one study suggested that partial hatching and ZT are better compared to complete hatching. There was a drawback supporting this claim as complete hatching was done only by applying 2 laser shots as the opening created would have been smaller for the embryo to hatch completely. Santhananthan *et al.* also reported better outcomes with ZT as it may increase the chance that artificially ablated area corresponds to the natural site of hatching [33].

Applications of AH

To Improve the Implantation

In IVF cycles where the exogenous gonadotropins are administered during ovarian stimulation, it is believed that the implantation window is 1-2 days earlier compared to natural cycles. As it is known that proper synchronization between the endometrium and hatching of the embryo is required for implantation, AH may play an important role and assist the embryo to implant. Also as mentioned previously, the prolonged culture of embryos in IVF lab and/or cryopreservation may lead to Zona hardening and AH will be used to improve the implantation of the embryo. ZP is known to assist the embryo in the transport of certain metabolites and Growth factors because of its porous nature. Too thick ZP or uneven thickness of ZP or hardening of the Zona may fail the transport of these important metabolites and factors leading to implantation failure. Creating an artificial opening using AH will act as a channel to improve the exchange of metabolites, growth factors, and messages between the endometrium and the embryo, thus improving the synchronization between the two in addition to facilitating the hatching process mechanically and improving the efficacy of the implantation [20] [21].

Although ZP can be thinned or drilled at any stage during the culture of the embryos, applying AH on cleavage stage embryos seems to be a popular approach. Many studies have reported the application of AH on cleavage stage embryos compared to the blastocyst stage embryos. This can be because the previously extended culture was not very common and the majority of the embryo were transferred or frozen on either day2 or day3. It can be also because the PVS is larger on the day1/day2/day3 stage of embryo development compared to the morula stage or blastocyst stage. Thus it becomes easier to apply the AH without affecting the blastomeres and/or embryos negatively. On the other hand, applying AH on blastocyst seems to be more logical as the embryo hatches out just before implantation *in vivo*. Because the ZP is very thin and the PVS is smaller compared to cleavage stage embryos, application of AH at the blastocyst stage may be dangerous and affect the blastomeres. Technically AH can be applied safely on cleavage stage embryos but theoretically, it would be much more physiological to apply the AH on blastocyst stage.

LAH became more popular as it was found to be safer, easier, and fast compared to all other AH methods mentioned above [33].

Laser-Assisted ICSI

Reinzi *et al.* reported that the use of LAH in oocytes with poor elasticity of ZP and oolema can be beneficial to avoid the degeneration of the poor quality oocytes. The patients were selected based on the quality of oocytes retrieved during the previous cycle. The Laser shots were applied to ZP to make a small hole to facilitate easy injecting of sperm inside the oocyte while performing ICSI. Prospective study on Laser-assisted ICSI where 5-6 um hole was created on ZP leaving the inside layer intact, concluded that the survival rate of oocytes post-ICSI and embryo development rate was better in patients where degeneration rate of >20% was reported in the previous cycle. Moser *et al.* modified the process a little where Laser shots were applied in such a way that half of ZP thickness was left behind. This modification was done keeping in mind the increased risk of embryo trapping which might cause embryo degeneration or monozygotic twinning. This group reported that the modification of technique suggested by Reinzi *et al.* worked better in terms of a lower rate of leakage of oocytes and more hatching blastocysts [33].

Removal of Fragments and Necrotic Blastomeres

Various groups have used LAH as the size of the opening created can be easily controlled. They have tried to remove the fragments from the cleavage stage embryos as the fragments are present in proximity and by removing them the spatial relationship can be improved. Others have reported removing the blastomeres which have not survived the cryopreservation process. The degenerated blastomeres may release metabolites that are harmful to the surrounding blastomeres. As it is well known that having a full viable embryo will have better outcomes than partial degenerated ones, Reinzi *et al.* reported that improvement in the overall development and increase in IR by removing the degenerated fragment. Removing fragments and degenerated blastomeres will be easy if they are close to each other. It would be rather difficult to remove them if they are not placed near each other without affecting other viable blastomeres and the risk of damage would be high and the skills of the embryologists will play an important role [33].

Cryopreservation

Mostly LAH is used in two ways in the cryopreservation of embryos, first, laser shots have been tried to collapse the blastocyst so that the chances of survival of the embryo post vitrification are better. Many groups also tried using a micropipette or LAH which is more popular, to create a hole in ZP and oolema to take out the fluid in the blastocyst as it is believed that ice crystals may form during vitrification if the fluid is inside thus affecting the survival [35]. Many studies have reported having increased survival rates by collapsing the blastocyst before vitrification. On the other side, many have tried vitrifying the blastocyst without collapsing and the survival rates were similar as compared to collapsed blastocyst. The data available to compare the outcomes of collapsing the blastocyst *vs* not collapsing are not sufficient enough to consider removing fluid from the blastocoel as an option.

The second use of AH in cryopreservation was to create a hole in ZP or thinning of ZP to assist the embryo to hatch. Although there are studies that reported the application of PZD, EZD, AT, and LAH to create an opening or thinning the ZP after cryopreservation, LAH is the most widely and popular technique of choice of AH to be applied post cryopreservation. As mentioned earlier Zona hardening may be seen in embryos after culturing them *in vitro* and/or cryopreservation. For embryos to hatch, thinning and elasticity of ZP play an important role. It is also said that opening produced will help in the exchange of metabolites and growth factors between the endometrium and the embryo and thus help in implantation [36]. There is no conclusive evidence that supports the AH of cryopreserved embryos improves the PR and thus IR. If embryos who have undergone Zona thinning or drilling in a selected group of patients do not improve the outcomes if compared to the control group [37] [38].

Hemizona Assay

This particular application is very rarely applied in Assisted reproduction techniques and is particularly difficult to perform as it requires highly skilled personnel. This assay is used to test the ability and competence of spermatozoa of binding and thus predict the fertilization potential and help to understand the difference between the zona binding capacity of spermatozoa obtained from fertile *vs* infertile male patients. In the initial days, the ZP was cut in half using micro blades or bisected manually and then these matching counterparts would act as an internal control to perform the assay. Nowadays diode laser is used to cut the ZP and perform the assay as it does not alter the binding characteristics of the sperm. The sliced ZP of the gamete is rotated and held in such a way that mild suction can be applied using another specialized holding pipette without damaging the ZP or the gamete. Using LAH for performing this assay one has to be careful as it involves multiple laser shots which may end up damaging the ZP and/or the gamete [33] [39] [40].

Preimplantation Genetic Testing (PGT)

Opening a hole in the ZP and taking the biopsy of the genetic material for testing the chromosomal content and/or checking the embryos for single gene disorder has been the most useful application of any method of AH. Opening of the ZP with the help of ATS and mechanical opening of the zona with the use of sterile glass pipettes was very common in its initial days. Using ATS brings along issues like openings of uncertain sizes and acidification of cells which may hamper the process of taking out the required number of cells or blastomeres from the embryo and hampering the embryo development and/or leases of cells. Mechanical opening of the zona also was difficult as the hole size would be inconsistent and may not be sufficient enough to take the cells out. As described in the respective sections of both the techniques in the discussion, technical expertise is a prerequisite and inexperienced embryologists can struggle to perform the procedures consistently.

The majority of the IVF labs have now started using LAH, except for a handful of labs still using the mechanical method to create an opening, as the method of choice to create an opening in the ZP and complete the biopsy of the cells. The use of LAH has made the process fairly stable, safe, and consistent as the opening created in the ZP can be easily controlled by varying the laser pulse exposure. As changing the dishes is also not required post-hatching the zone chances of contamination are also reduced which can affect the amplification of the genetic material and thus affect the result of the test. Zona openings of the desired size can be made with the respective laser shot to carry out the process rapidly decreasing the exposure of the embryos outside the incubator thus decreasing the metabolic stress on the embryos.

The important thing to be noted here is that the implantation potential of the embryos is reported to be similar when comparing the use of ATS *vs* LAH to hatch the ZP. The advantage of LAH was in terms of the survival of the blastomeres [33].

LAH can be used to perform Polar body (PB) biopsies, biopsies of day3, and blastocyst stage embryos. For all the biopsies of the mature oocyte, cleavage stage embryos, and blastocysts, the laser shots can be given on the same day as planned for biopsies or in cases of blastocyst-stage biopsy, LAH can be done on day 3 and embryos can be kept in culture. The trophectoderm cells will start hatching out which will be easily held in position using the blunt biopsy pipette and by applying the laser shots on the cells junctions of the trophectoderm layer the biopsy can be taken. Some embryologists still choose to apply the laser shots on the blastocyst stage and perform the embryo biopsy depending upon the protocol of their labs. In cases of blastocyst biopsies the selection of the day to apply the laser shots, *i.e* day 3 or blastocyst stage, should be based on taking into consideration the pros and cons of both approaches.

With the use of PGT- A where 'A' stands for Aneuploidy screening, the PR and IR can be improved in the advanced maternal age group as it is well known that with an increase in maternal age the aneuploidy rates increase. Also, the miscarriage rates, time to pregnancy, and multiple pregnancy rates can be decreased as a single embryo with normal chromosomal content can be transferred. The advent of Next Generation Sequencing (NGS) can also share information about the mosaicism state of the embryo and thus help the patient to make informed decisions to decide about the further course of the mosaic embryo depending upon their need while taking into account the strategy used for sampling the genetic material, test outcomes and frequency of obtaining mosaic embryos as it will create a state of conflict when discarding of the viable embryo is considered [41, 42, 43]. Transfer of a single Chromosomaly normal embryo is proven to be more beneficial when compared to the untested embryos [42].

Polar Body biopsy is used to screen the chromosomal content of oocytes. ATS, mechanical zona opening, and LAH have been used to create a hole from which the first polar body of the mature oocyte is analyzed [33]. In the same way, embryos can be tested for PGT-M that is for monogenic disorders, and PGT-SR for detecting disorders related to structural rearrangements in the chromosomes in the embryos before embryo transfer.

Assisted Hatching and Clinical Outcomes

As mentioned earlier in various topics, the application of AH is considered to be helping in the improvement of the clinical rates that is ongoing pregnancy rates (OPR) and ultimately the live birth rate (LBR). AH has been applied to various groups of patients selected on Age, Previous failures, and transfer of sub-optimal embryos compared to control groups where slight improvement is seen in the CPR. There are also studies concluding that the method of AH has similar outcomes in the selected patient groups and a prospective randomized trial concluded that application of LAH will improve the CPR in frozen embryo transfers [14] [36]. However, not many studies were properly designed and the sample size is not suitable enough to encourage the use of AH in all patient groups to have added benefits as observed by Minh Tam Le *et al.* A systemic review and meta-analysis of randomized controlled trials with 5507 participants concluded that in cases of frozen-thawed embryos or patients with previous failures, the multiple PR and clinical pregnancy was increased. When the data was analyzed for unselected patients using fresh embryos or patients with advanced maternal age, no increase in CPR is seen [44] [39] [21]. Cochrane review also concluded that there is no statistical difference in the CPR of AH and the control group. When subgroup analysis was done, AH was found to be useful for patients with previous failure or poor prognosis patients and the statistical difference was seen in CPR when,

1- complete removal of ZP was carried out by an application on AH.

2- Patients receiving fresh embryos compared to patients receiving fresh or frozen embryos for transfer [45]. In a study involving 760 fresh embryo transfers and 200 frozen embryo transfers which were randomly assigned to AH and no AH group, no statistically significant difference was found in the CPR in both the groups [14].

The main aim of assisted reproductive treatment is healthy childbirth. As most AH techniques have been considered to be safe the question remains whether the application of AH will improve live birth or not. As per the systemic review published by Edmond Edi-Osagie *et al.*, better trials are required to be conducted to justify the application of AH to improve the LBR. The review did mention improvement seen in the clinical pregnancy in advanced maternal age patients and previous failures [10]. Carney *et al.* in their published review stated their disappointment due to the unavailability of a central database and availability of the data as only 9 out of 31 studies reported the LBR. Studies were considered of good quality only when robust randomization methods were used. Subgroup analysis does not provide conclusive evidence that application of AH improved the LBR [45] Even a prospective, randomized, double-blinded study in patients with advanced maternal age concluded that application of AH in women > or = to 36 years has no significant difference in IR and ongoing PR in hatched *vs* control group [16] and more studies are required to prove the advantage of the application of AH on women of older age in frozen embryo transfers [15].

Assisted Hatching and Monozygotic (MZ) Twinning

Monozygotic twinning occurs when a single fertilized oocyte splits into two embryos that are identical genetically. It occurs in approximately 0.42% of all deliveries and factors leading to such events are still unclear [46]. In general scenarios also multiple gestations are associated with an increase in infant mortality and morbidity. The perinatal mortality risk is comparatively 2 - 3 times higher in MZ twins than Dizygotic twins [47]. The study with 35,503 cycles (IVF-ET) and 11,247 pregnancies, after adjusting the variables for such as several embryos transferred, patients age, *etc*, concluded that MZ twinning risk is associated with AH and cautioned that such a potential risk should be not be outweighed for use of AH if the benefits are still unclear. In multiple births, infants are at risk of having low birth weight, congenital malformations, fetal and infant death, long-term morbidity, and disability. Applying AH should be then done with extreme caution [47]. In the Cochrane review mentioned before, the rate of MZ twinning was 0.8% in the AH group when compared to 0% in the control group with the data available in 6 out of 31 RCTs selected for review [45].

However, the risk of MZ twinning is not associated with any particular type of method of AH, and further studies are required to justify the application of assisted hatching in a selected or unselected group of patients [47]. The breach of the zona and size of the opening created might influence the rate of MZ twinning as the hole might be very small to trap the blastocyst, causing harm and thus splitting the ICM. These small openings, created unintentionally during AH, in the majority of cases are not detectable in later stages of development. Such conditions may worsen the hatching of the embryo even more [33]. No studies, at least to the author's knowledge, have mentioned the impact of the size of the opening in ZP and the risk associated with the MZ twinning.

On the other side, the risk of MZ twinning was associated with the extended culture of embryos in fresh and frozen cycles and the number of embryos transferred, ICSI and application of AH does not have any influence on it when 3,464 pregnancies were analyzed [48]. MZ twinning was also seen after transfer of zona-free blastocysts somewhere suggesting that application of AH is not the sole reason and factors such as ovarian stimulation, delay in fertilization, embryo development or implantation, changes in culture conditions and mechanical disruption of early also may play an important role [49].

CONCLUSION

As mentioned before, there are various methods to create an opening or thinning of the ZP. Each of the mentioned methods has its advantages and disadvantages and needs the embryos to be taken out of the incubator for a short period. If the selected method of AH is not performed properly or embryos are kept outside for too long, it may cause damage to the blastomeres/embryos leading to permanent damage and thus failure in implantation or loss of the embryos. This will defy the whole purpose of the application of AH as the whole purpose of it was to assist the embryo to hatch so that it implants and then help the couple achieve a healthy baby. The time selected for application of AH to the embryos should be based on individual lab protocols and should avoid having any negative influence on the implantation of the embryos.

LAH is an extremely popular, safe, fast method of choice to either create an opening or thinning of the ZP. Most of the labs have started offering PGT services to their patients where LAH helps in extracting the genetic material which needs to be of good quality and viability to analyze its chromosomes. Many authors have reported an increase in the PR, IR, and hence the LBR with PGT thus helping patients to achieve their dream of parenthood.

As per the guidelines given by the practice committee of the ASRM and practice committee of the Society for Assisted Reproductive Technology, application of AH should be done extremely carefully as there is no satisfactory data and evidence that AH improves the LBR in patients [50]. As mentioned before, AH may help some groups of patients but charging patients the extra amount of money would be unfair if there is no proven benefit from it. Proper studies and maintaining the proper database are required to justify the application of AH safely and thus improve the take-home baby rate.

CONSENT FOR PUBLICATION

Not applicable

CONFLICT OF INTEREST

The author declares no conflict of interest, financial or otherwise.

ACKNOWLEDGEMENT

Declared none.

REFERENCES

[1]   A. De Vos, A. Van Steirteghem, Zona Hardening, Zona drilling and Assisted Hatching: New Achievements in Assisted Reproduction, Cells Tissues Organs 2000; 166: 220-227.

[2]   Litscher ES, Wassarman PM. Zona pellucida genes and proteins and human fertility. Trends Dev Biol 2020; 13: 21-33.
[PMID: 33335361]

[3]   Marcel Van Duin. Jan E.M. Polman, Ingeborg T. M. De Breet, Karin Van Ginneken, Hans Bunschoten, Arijan Grootenhuis, James Brindle and John Aitken, Recombinant Human Zona Pellucida Protein ZP3 Produced by Chinese Hamster Ovary Cells Induces the Human Sperm Acrosome reaction and Promotes Sper-Egg Fusion. Biol Reprod 1994; 51: 607-17.
[http://dx.doi.org/10.1095/biolreprod51.4.607]

[4]   Wolf DP. The Mammalian egg's block to Polyspermy, Fertilisation and Embryonic Development *In Vitro*, 183-197. 5-Massimo de Felici, Dr. Gregorio Siracusa, Spontaneous hardening of the zona pellucida of mouse oocytes during *in vitro* culture. Mol Reprod Dev 6: 107-13.

[6]   Balaban B, Urman B, Alatas C, Mercan R, Mumcu A, Isiklar A. A comparison of four different techniques of assisted hatching. Hum Reprod 2002; 17(5): 1239-43.
[http://dx.doi.org/10.1093/humrep/17.5.1239] [PMID: 11980745]

[7]   Sathananthan H, Menezes J, Gunasheela S. Mechanics of human blastocyst hatching *in vitro*. Reprod Biomed Online 2003; 7(2): 228-34.
[http://dx.doi.org/10.1016/S1472-6483(10)61757-9] [PMID: 14567898]

[8]   Cohen J, Alikani M, Trowbridge J, Rosenwaks Z. Implantation enhancement by selective assisted hatching using zona drilling of human embryos with poor prognosis. Hum Reprod 1992; 7(5): 685-91.
[http://dx.doi.org/10.1093/oxfordjournals.humrep.a137720] [PMID: 1639990]

[9]   Montag M, van der Ven H. Laser-assisted hatching in assisted reproduction. Croat Med J 1999; 40(3): 398-403.
[PMID: 10411968]

[10]   Edi-Osagie E, Hooper L, Seif MW. The impact of assisted hatching on live birth rates and outcomes of assisted conception: a systematic review. Hum Reprod 2003; 18(9): 1828-35.
[http://dx.doi.org/10.1093/humrep/deg334] [PMID: 12923134]

[11]   Blake DA, Forsberg AS, Johansson BR, Wikland M. Laser zona pellucida thinning--an alternative approach to assisted hatching. Hum Reprod 2001; 16(9): 1959-64.
[http://dx.doi.org/10.1093/humrep/16.9.1959] [PMID: 11527905]

[12]   Antinori S, Selman HA, Caffa B, Panci C, Dani GL, Versaci C. Zona opening of human embryos using a non-contact UV laser for assisted hatching in patients with poor prognosis of pregnancy. Hum Reprod 1996; 11(11): 2488-92.
[http://dx.doi.org/10.1093/oxfordjournals.humrep.a019145] [PMID: 8981141]

[13]   Mandelbaum J. The effects of assisted hatching on the hatching process and implantation. Hum Reprod 1996; 11 (Suppl. 1): 43-50.
[http://dx.doi.org/10.1093/humrep/11.suppl_5.43] [PMID: 8968769]

[14]   Ge HS, Zhou W, Zhang W, Lin J-J. Impact of assisted hatching on fresh and frozen-thawed embryo transfer cycles: a prospective, randomized study. Reprod Biomed Online 2008; 16(4): 589-96.
[http://dx.doi.org/10.1016/S1472-6483(10)60466-X] [PMID: 18413070]

[15]   Jennifer F. Knudtson, Courtney M. Failor, Jonathan A. Gelfond, Martin W. Goros, Teincheng Arthur Chang, Robert S. Schenken and Randal D. Robinson, Assisted hatching and live births in first cycle frozen embryo transfers. Fertil Steril 2007; 108(4): 628-34.

[16]   Susan E. Lanzendorf, Fariba Nehchiri, Jacob F. Mayer, Sergio Oehninger and Suheil J. Muasher, A prospective, randomised, double-blind study for the evaluation of assisted hatching in patients with advanced maternal age. Hum Reprod 1998; 13(2): 409-13.
[http://dx.doi.org/10.1093/humrep/13.2.409] [PMID: 9557848]

[17]   Schiewe MC, Hazeleger NL, Sclimenti C, Balmaceda JP. Physiological characterization of blastocyst hatching mechanisms by use of a mouse antihatching model. Fertil Steril 1995; 63(2): 288-94.
[http://dx.doi.org/10.1016/S0015-0282(16)57357-9] [PMID: 7843433]

[18]   Gabrielsen A, Bhatnager PR, Petersen K, Lindenberg S. Influence of zona pellucida thickness of human embryos on clinical pregnancy outcome following *in vitro* fertilization treatment. J Assist Reprod Genet 2000; 17(6): 323-8.
[http://dx.doi.org/10.1023/A:1009453011321] [PMID: 11042829]

[19]   Cohen J, Elsner C, Kort H, *et al.* Impairment of the hatching process following IVF in the human and improvement of implantation by assisting hatching using micromanipulation. Hum Reprod 1990; 5(1): 7-13.
[http://dx.doi.org/10.1093/oxfordjournals.humrep.a137044] [PMID: 2324246]

[20]   Liu HC, Cohen J, Alikani M, Noyes N, Rosenwaks Z. Assisted hatching facilitates earlier implantation. Fertil Steril 1993; 60(5): 871-5.
[http://dx.doi.org/10.1016/S0015-0282(16)56289-X] [PMID: 8224273]

[21]   Wellington P. Martins, Isa A. Rocha, Rui A. Ferriani and Carolina O. Nastri, Assisted hatching of human embryos: a systematic review and meta-analysis of randomised controlled trials. Hum Reprod

2011; 17(4): 438-53.

[22]   Henry E. Malter, Jacques Cohen, Partial Zona Dissection of the human oocytes: a non traumatic method using micro manipulation to assist zona pellucida penetration. Fertil Steril 1989; 51(1): 139-48.
[http://dx.doi.org/10.1016/S0015-0282(16)60443-0] [PMID: 2910709]

[23]   Fong CY, Bongso A, Ng SC, Anandakumar C, Trounson A, Ratnam S. Ongoing normal pregnancy after transfer of zona-free blastocysts: implications for embryo transfer in the human. Hum Reprod 1997; 12(3): 557-60.
[http://dx.doi.org/10.1093/humrep/12.3.557] [PMID: 9130759]

[24]   Chui- Yee Fong, Ariff Bongso, Soon- Chye Ng, Jothi Kuman, Alan Trounson and Shan Ratnam, Blastocyst transfer after enzymatic treatment of the zona pellucida: improving *in-vitro* fertilization and understanding implantation. Hum Reprod 1998; 13(10): 2926-32.
[http://dx.doi.org/10.1093/humrep/13.10.2926]

[25]   Fong CY, Bongso A, Sathananthan H, Ho J, Ng SC. Ultrastructural observations of enzymatically treated human blastocysts: zona-free blastocyst transfer and rescue of blastocysts with hatching difficulties. Hum Reprod 2001; 16(3): 540-6.
[http://dx.doi.org/10.1093/humrep/16.3.540] [PMID: 11228226]

[26]   Yano K, Yano C, Kubo T, Ohashi I, Maeda N, Fukaya T. Chemical zona pellucida thinning with acidified Tyrode's solution: comparison between partial and circumferential techniques. J Assist Reprod Genet 2007; 24(10): 471-5.
[http://dx.doi.org/10.1007/s10815-007-9131-z] [PMID: 17701000]

[27]   Bider D, Livshits A, Yonish M, Yemini Z, Mashiach S, Dor J. Assisted hatching by zona drilling of human embryos in women of advanced age. Hum Reprod 1997; 12(2): 317-20.
[http://dx.doi.org/10.1093/humrep/12.2.317] [PMID: 9070719]

[28]   Depypere HT, Leybaert L. Intracellular pH changes during zona drilling. Fertil Steril 1994; 61(2): 319-23.
[http://dx.doi.org/10.1016/S0015-0282(16)56525-X] [PMID: 8299790]

[29]   Garrisi GJ, Talansky BE, Grunfeld L, Sapira V, Navot D, Gordon JW. Clinical evaluation of three approaches to micromanipulation-assisted fertilization. Fertil Steril 1990; 54(4): 671-7.
[http://dx.doi.org/10.1016/S0015-0282(16)53828-X] [PMID: 2209888]

[30]   Palanker D, Ohad S, Lewis A, *et al.* Technique for cellular microsurgery using the 193-nm excimer laser. Lasers Surg Med 1991; 11(6): 580-6.
[http://dx.doi.org/10.1002/lsm.1900110613] [PMID: 1753851]

[31]   Mohamad Eid Hammadeh. Constanze Fischer-Hammadeh, Khaled Refaat Ali, Assisted hatching in assisted reproduction: a state of the art. J Assist Reprod Genet 2011; 28: 119-28.
[http://dx.doi.org/10.1007/s10815-010-9495-3]

[32]   Mantoudis E, Podsiadly BT, Gorgy A, Venkat G, Craft IL. A comparison between quarter, partial and total laser assisted hatching in selected infertility patients. Hum Reprod 2001; 16(10): 2182-6.
[http://dx.doi.org/10.1093/humrep/16.10.2182] [PMID: 11574513]

[33]   Ebner T, Moser M, Tews G. Possible applications of a non-contact 1.48 microm wavelength diode laser in assisted reproduction technologies. Hum Reprod Update 2005; 11(4): 425-35.
[http://dx.doi.org/10.1093/humupd/dmi009] [PMID: 15817523]

[34]   Zakharchenko EO, Zalessky AD, Osychenko AA, *et al.* Effect of laser optoperforation of the zona pellucida on mouse embryo development *in vitro*. Biochemistry (Mosc) 2015; 80(6): 769-75.
[http://dx.doi.org/10.1134/S0006297915060127] [PMID: 26531022]

[35]   Darwish E, Magdi Y. Artificial shrinkage of blastocoel using a laser pulse prior to vitrification improves clinical outcome. J Assist Reprod Genet 2016; 33(4): 467-71.
[http://dx.doi.org/10.1007/s10815-016-0662-z] [PMID: 26843389]

[36]   Wan Cai-Yun, Song Cheng, Diao Liang-Hui, *et al.* Laser assisted hatching improved clinical outcomes of vitrified0warmed blastocyst developed from low-grade cleavage-stage embryos: a prospective randomized study, Reproductive Biomedecine online 2014: 28: 582-589.

[37]   Petersen CG, Mauri AL, Baruffi RL, *et al.* Laser-assisted hatching of cryopreserved-thawed embryos by thinning one quarter of the zona. Reprod Biomed Online 2006; 13(5): 668-75.
        [http://dx.doi.org/10.1016/S1472-6483(10)60657-8] [PMID: 17169177]

[38]   Le MT, Nguyen TTA, Nguyen TTT, *et al.* Thinning and drilling laser-assisted hatching in thawed embryo transfer: A randomized controlled trial. Clin Exp Reprod Med 2018; 45(3): 129-34.
        [http://dx.doi.org/10.5653/cerm.2018.45.3.129] [PMID: 30202743]

[39]   Oehninger S, Morshedi M, Franken D. The hemizona assay for assessment of sperm function. Methods Mol Biol 2013; 927(91): 91-102.
        [http://dx.doi.org/10.1007/978-1-62703-038-0_9] [PMID: 22992906]

[40]   Mayenco-Aguirre A M, Perez Cortes A B. Preliminary results of hemizona assay as a fertility test for canine spermatozo, Theriogenology 1998: 15: 50: 2: 195-204.

[41]   Maxwell SM, Grifo JA. Should every embryo undergo preimplantation genetic testing for aneuploidy? A review of the modern approach to *in vitro* fertilization. Best Pract Res Clin Obstet Gynaecol 2018; 53: 38-47.
        [http://dx.doi.org/10.1016/j.bpobgyn.2018.07.005] [PMID: 30146380]

[42]   Anderson R E, Whitney JB, Schiewe M C. Clinical benefits of preimplantation genetic testing for aneuploidy (PGT-A) for all *in vitro* fertilisation treatment cycles, European Journal of Medical Genetics; 2020: 63: 2: 103731.

[43]   Sciorio R, Dattilo M. PGT-A preimplantation genetic testing for aneuploidies and embryo selection in routine ART cycles: Time to step back? Clin Genet 2020; 98(2): 107-15.
        [http://dx.doi.org/10.1111/cge.13732] [PMID: 32141057]

[44]   Ghannadi A, Kazerooni M, Jamalzadeh F, Amiri S, Rostami P, Absalan F. The effects of laser assisted hatching on pregnancy rates, Iranian Journal of Reproductive Medecine; 2011: 9: 2: 95-98 45-Carney SK, Das S, Blake D, Farquhar C, Seif MM and Nelson L, Assisted hatching on assisted conception (*in vitro* fertilisation (IVF) and intracytoplasmic sperm injection (ICSI)). Cochrane Database Syst Rev 2012; 12: CD001894. [Review].

[46]   Sills ES, Moomjy M, Zaninovic N, *et al.* Human zona pellucida micromanipulation and monozygotic twinning frequency after IVF. Hum Reprod 2000; 15(4): 890-5.
        [http://dx.doi.org/10.1093/humrep/15.4.890] [PMID: 10739838]

[47]   Laura A. Scheive, Susan F. Meikle, Herbert B. Peterson, Gary Jeng, Nancy M. Burnett and Lynne S. Wilcox, Does assisted hatching pose a risk for monozygotic twinning in pregnancies conceived through *in vitro* fertilisation? Fertil Steril 2000; 74(2): 288-94.
        [http://dx.doi.org/10.1016/S0015-0282(00)00602-6] [PMID: 10927046]

[48]   Liu H, Liu J, Chen S, Kang X, Du H, Li L. Elevated incidence of monozygotic twinning is associated with extended embryo culture, but not with zona pellucida manipulation or freeze-thaw procedure. Fertil Steril 2018; 109(6): 1044-50.
        [http://dx.doi.org/10.1016/j.fertnstert.2018.01.040] [PMID: 29871792]

[49]   Aston KI, Peterson CM, Carrell DT. Monozygotic twinning associated with assisted reproductive technologies: a review. Reproduction 2008; 136(4): 377-86.
        [http://dx.doi.org/10.1530/REP-08-0206] [PMID: 18577552]

[50]   Role of assisted hatching in *in vitro* fertilization: a guideline. Fertil Steril 2014; 102(2): 348-51.
        [http://dx.doi.org/10.1016/j.fertnstert.2014.05.034] [PMID: 24951365]

CHAPTER 9

Oocytes and Embryos Cryopreservation

Mohamed Fadel El Mohr[1,2,*] and **Islam M. Saadeldin**[3,4,5,*]

[1] *Dr. Faris Medical Center for Infertility and Human Reproduction, Heliopolis, Cairo, Egypt*

[2] *Dar El Om for Infertility and Human Reproduction, Mokatam, Cairo, Egypt*

[3] *Research Institute of Veterinary Medicine, Chungnam National University, Daejeon, Republic of Korea*

[4] *College of Veterinary Medicine, Chungnam National University; 34134 Daejeon, Republic of Korea*

[5] *Department of Physiology, Faculty of Veterinary Medicine, Zagazig University, 44519 Zagazig, Egypt*

Abstract: The availability of cryopreservation technology has extended the scope of human-assisted conception treatment and made it more convenient for patients. Freezing of oocytes and ovarian tissue is now taking place in research institutions and should hopefully become widely available for clinical use in the future if the women have to be away at a critical time during the treatment; the frozen oocyte was thawed and used when required. Good quality embryos frequently remain after the transfer of the required number into the woman; these can be frozen and used at a later date, if required, instead of the couple going through another cycle of in vitro fertilization (IVF) treatment. The conduct of cryopreservation in humans is based on information originally derived from animal work, where sperm cryopreservation, for example, has been carried out for several decades for animal breeding. This chapter will consider the principles of cryopreservation followed by a general description of the practical steps. The clinical applications of cryopreservation will also be described. A final section will deal with frozen embryo replacement.

Keywords: Oocyte and embryos cryopreservation, Slow freezing, Thawing embryos, Vitrification.

HISTORY OF OOCYTE AND EMBRYOS CRYOPRESERVATION

Early basic scientific advancements in measurements of temperature and therefore the chemistry of solutions and gases are certainly sentinel events for cryobiology. It's been suggested that original versions of the tools to live temperature were

* **Corresponding authors Mohamed Fadel El Mohr and Islam M. Saadeldin:** Dr. Faris Medical Center for Infertility and Human Reproduction, Heliopolis, Cairo, Egypt; Tel: +201006311052; E-mail: fadell20@gmail.com and Research Institute of Veterinary Medicine, Chungnam National University, 34134 Daejeon, South Korea; Tel: 00821024817666; E-mails: islamms@cnu.ac.kr and islamsaad82@gmail.com

Islam M. Saadeldin (Ed.)

made by Galileo within the early seventeenth century. The primary accurate means of measuring temperature were developed in the early 1700s by the German physicist Gabriel Fahrenheit through the application of mercury in glass.

Since then, modifications of instruments to assess temperature became significantly more accurate and easier to use. Equally important were early advancements made within the nineteenth century involving understandings of liquefaction of gases and the potential use of such refrigerants to chill and store specimens at extremely low temperatures. When one traces the history of mammalian gamete cryopreservation, numerous accounts reference the start of low-temperature biology to 1866, when an Italian military physician Mantegazza documented the observation that human spermatozoa became immotile when cooled in the ice [1, 2].

Investigations in the late 1940s and early 1950s by Chang [3, 4] on low-temperature storage of rabbit oocytes, zygotes, and embryos paved the way for studies on the cryopreservation of female gametes and embryos. Subsequent experiments by Sherman and Lin [5 - 8] demonstrated that mouse oocytes could even be cooled in glycerol, stored, and subsequently fertilized in recipients; furthermore, resulting in embryos supported pregnancies.

In the 1960s and early 1970s, a merging of basic/theoretical cryobiology and practical studies ultimately gave rise to the increased success of embryo cryopreservation. Classical basic science investigations by Mazur [9 - 11] formed the inspiration for understanding cell-specific optimal cooling and warming rates which today remain a pivotal key to successful mammalian gamete and embryo cryopreservation. It had been the combined strengths of Mazur, Leibo, and Whittingham that resulted in successful cryopreservation of mouse embryos.

The cryoprotective agent used by these researchers was 1.5 M dimethylsulfoxide (DMSO), which was combined with a slow cooling rate (0.3C/min to 80C) and deposited in liquid nitrogen [12]. Slow cooling rates supported dehydration during cooling and prevented intracellular ice formation, so this technique was used. In addition, it was discovered that the cryoprotective agent (here DMSO) should be added and removed in a stepwise manner to avoid osmotic shock or injury. In the early 1980s, the same cryopreservation methods were used to successfully create the first human pregnancies after freezing and thawing [13, 14].

Testart *et al.* pioneered the use of 1, 2-propanediol (PROH) as a permeating cryoprotectant for pronuclear-stage zygotes [15]. Furthermore, as an osmotic buffer, these researchers used sucrose in the cryo-media as a non-permeable cryoprotectant. Slow cooling rates were used until the temperature reached 30°C, after which the samples were submerged in liquid nitrogen and the warming rate was rapid.

Cryopreserving pronuclear- and cleavage-stage mammalian embryos has become a popular practice in assisted reproductive technology (ART) labs throughout the United States and around the world. As experience with cryopreservation of different cell types grew, it became clear that as cell size grows, so does the complexity of cryopreservation [11].

This principle is particularly important in the cryopreservation of mammalian oocytes and embryos slow-rate freezing and vitrification are the two approaches currently used to cryopreserve mammalian oocytes and embryos [16].

Effects on oocyte and embryonic cellular functions may impair abilities to develop normally following the cryopreservation process, regardless of the cryopreservation method used. Oocyte and/or embryo "cryo-damage" is the collective term for these damaged cellular events although the effects of cryopreservation on documented and/or theoretically specific cellular structures and functions, as well as subsequent effects on oocyte and embryonic developmental competence, have been previously reviewed [17].

Cells are subjected to a variety of stresses during cryopreservation, including mechanical, thermal, and chemical stresses [11, 18], which can disrupt cell function and cause cell death. Oocytes, in general, are more vulnerable to cryo-damage than later embryonic stages moreover slow-rate freezing attempts to control biophysical properties of freezing, such as cooling and warming rates, in conjunction with cryoprotectants to minimize adverse cellular events. This approach allows cells to be cooled to extremely low temperatures while minimizing the production of intracellular ice crystals and attempting to avoid the negative effects of increased solute concentrations and osmotic stress [19]. As a result, extracellular ice formation drives cellular dehydration into an equilibrium mechanism when slow-rate freezing is used.

In recent years the new technique of vitrification came to avoid the mistakes of slow freezing, the word "vitrification" comes from the Latin word vitreous, which means glassy or similar to glass. Vitrification is a non-equilibrium cryopreservation method that was originally designed for the cryopreservation of mammalian sperm [20] and embryos [21]. This technique used a special device known as straw-like Fig. (**1**) and is divided into three regions 1st leaf which put embryos or oocytes on its 2nd plastic mark to write the name and id of the couple and 3rd the coverslip.

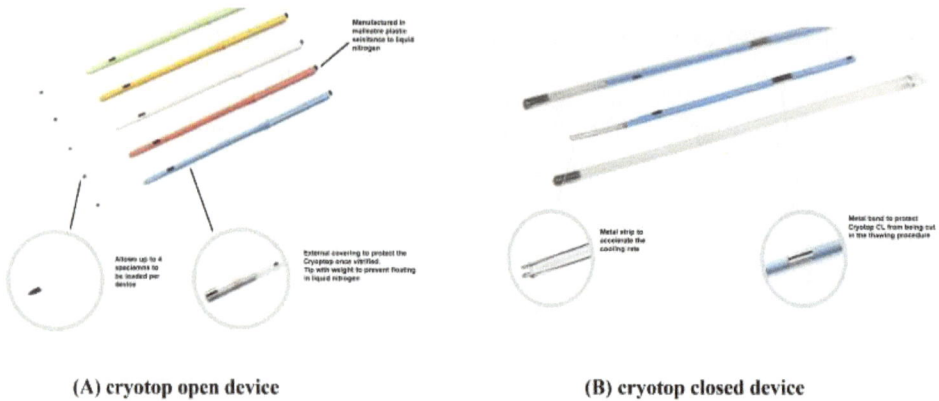

(A) cryotop open device   (B) cryotop closed device

Fig. (1). Vitrification device (Straw Kitazato) open and closed system.

As a result, the vitrified solids retain the initial liquid's natural molecular and ionic distributions and can be called extremely viscous, supercooled liquid [22]. In this technique, oocytes or embryos are dehydrated by brief exposure to a concentrated solution of cryoprotectant before plunging the samples directly into liquid nitrogen; Oocytes or embryos are dehydrated in this method by exposing them to a concentrated cryoprotectant solution for a brief period before being frozen in liquid nitrogen. As shown in Fig. (2).

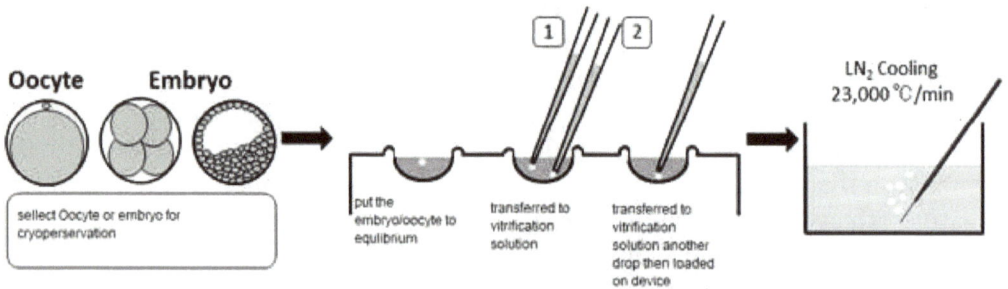

Fig. (2). Dehydration by brief exposure to a concentrated solution of cryoprotectant before plunging the samples directly into liquid nitrogen.

Many clinical, rodent and domestic animal production laboratories are currently focusing on the use of vitrification for both oocytes and embryos. Human oocytes [23] have been vitrified, as have pronuclear zygotes [24], cleavage-stage embryos [25], and blastocysts [26 - 28]. There are excellent reviews of vitrification's past, use, and potential benefits available [29, 30].

The Mechanism of Cryopreservation

Slow Freezing

Since its adoption in 1985, the protocol for embryo cryopreservation was standardized 20 years ago and has been widely used with minor changes, yielding positive results [31]. The slow freezing technique has been chosen because it allows the embryos to cool at a very slow rate after being treated with cryoprotectants at a low concentration (1.0–1.5 M), minimizing toxic and osmotic damage by using a software program as shown in Fig. (**3**).

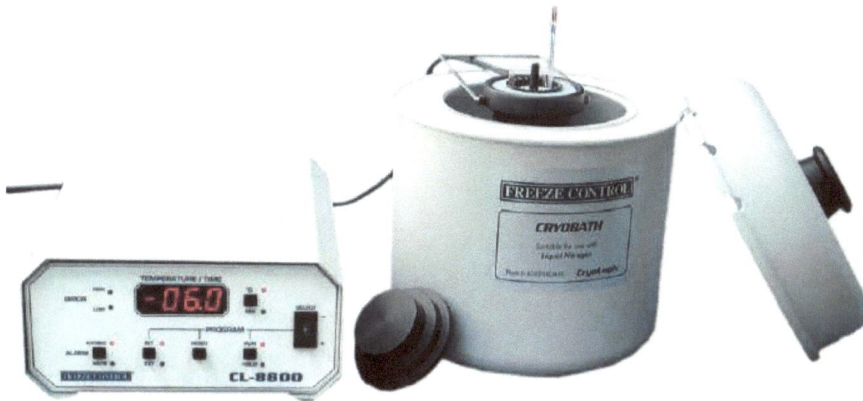

Fig. (3). The FREEZE CONTROL CL863 programmable system produced by Cryologic Pty Ltd, Mt Waverly, Australia

The dehydration of the cells and the absorption of cryoprotectant agents (CPAs) into the cells occur very slowly in this procedure over a long period. The technique causes the extracellular and intracellular fluids to equilibrate at the end, which is why this process is also known as "equilibrium freezing." The position of embryos or oocytes in straw is shown in Fig. (**4**).

The extracellular liquid crystallizes after the process is completed, and intracellular ice formation cannot be completely prevented, which is one of the strategy's major flaws. Since these molecules are removed from the inner structure of the ice crystals, extracellular ice formation results in a significant increase in the concentration of ions, macromolecules, and other components in the remaining fluid. As a consequence, the resulting hyperosmolarity can cause serious osmotic damage. Slow freezing exposes the cells to chilling damage, which occurs between +15°C and +5°C and affects the lipids of the cell membrane and the microtubules of the meiotic spindle, causing hardening of the zona pellucid [32]. These effects have mostly been identified in the context of slow oocyte freezing. The modification of the membrane's lipid composition causes permanent harm [33], while the restoration of the meiotic spindle has been recorded in humans [34].

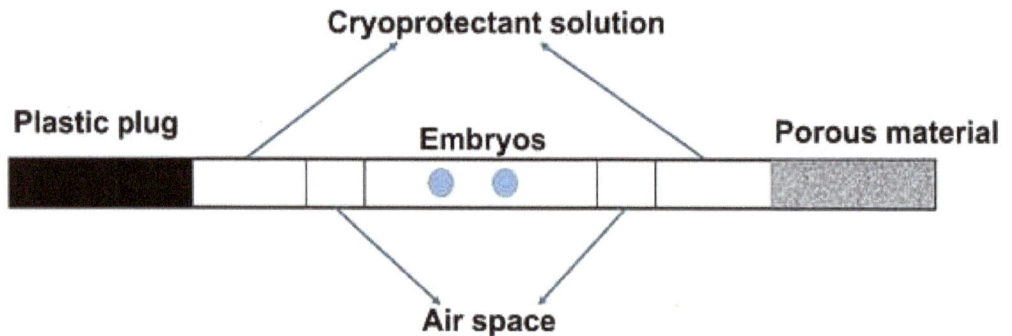

Fig. (4). Straw containing two cells for cryopreservation. Normally, gentle suction is applied to the end of the straw that is plugged loosely with a porous material. The opposite end of the straw is dipped into the cryoprotectant solution and used in aspirating a small volume of fluid before being raised and aspirating some air. The straw is then dipped into the cryoprotectant solution again to aspirate the cells together with some of the solution.

The accumulation of small crystals during thawing, resulting in larger crystals, is the mechanism by which intracellular ice formation causes damage. This is caused by high activation energy. Mechanical effects are the primary cause of damage from these massive crystals, which disrupt cell structures. The negative consequences of partial crystallization in cells and tissues differ depending on the cell type and the number of cells. The lysis of several cells in a tissue, for example, may be balanced by those that remain, resulting in a functioning tissue. That is true for spermatozoa, ovarian cortex, and even embryos; however, it is not true for oocytes, which are "all or nothing" in terms of survival. One of the causes of partial blastomere loss found in a high proportion of embryos after slow freezing is intracellular ice formation. The effect of partial lysis on the implantation potential of frozen-thawed embryos is well known [35 - 37]. Survival after slow freezing was previously described as 50% percent or more blastomeres surviving the process.

Because of the connection between lower implantation and partial blastomere failure, it has been proposed that rather than evaluating survival, the efficiency of cryopreservation programmers should be measured by the proportion of completely intact and partially lysed embryos recovered after the freezing/thawing phase [38]. Owing to the direct conversion of a liquid to a vitreous solid during vitrification, ice formation is avoided in both the intracellular and extracellular spaces.

Rapid Freezing (Vitrification)

The combination of high cryopreservative concentrations and rapid cooling causes substantial dehydration of the cell and partial intracellular diffusion of CPA, resulting in a lack of equilibrium between the intracellular and extracellular compartments, which is why this procedure is often known as "non-equilibrium freezing."If the concentration of CPA rises, so does the glass transition. A vitreous solid is formed as a result. Other factors, in addition to cryoprotectant concentration and cooling rate, play a significant role in achieving effective vitrification.

The complete absence of crystallization achieved by vitrification results in an exceptionally high proportion of completely intact embryos (95%) [39], which may explain the similar implantation rates observed when vitrified/warmed embryos were compared to fresh embryos of the same morphological group [40]. This discovery, along with the higher survival rates for embryos cryopreserved at both the cleavage and blastocyst levels, suggests that vitrification has an advantage over slow freezing [41] and explains the emerging trend in many IVF clinics to turn to vitrification.

Vitrification achieves a delicate balance between several variables. As previously mentioned, ice formation is prevented during vitrification by exposing the oocytes to high CPA concentrations while cooling at very high rates as a result, increasing the cooling rate and viscosity of the solution, as well as decreasing the volume containing the cells, would increase the likelihood of vitrification [42]. As shown in Fig. (**5**), the wrong and correct loading of embryos and oocytes on vitrification device.

High warming rates are also linked to success because they prevent recrystallization during the warming process [43]. In recent years, vitrification technology has advanced, resulting in a variety of devices, protocols, and methodologies that meet these requirements to varying degrees. As a result, open and closed systems are possible, with the former requiring liquid nitrogen contact and the latter not. Different cryoprotectant combinations, as well as osmotic agents and protein substitutes, have been identified [32, 44]. Chilling injury is effectively avoided during vitrification because the oocytes are simply plunged into liquid nitrogen, passing from room temperature to 196°C without passing through the temperature range where the chilling injury occurs (+15°C and 5°C).

Loading point of the oocyte/embryo

Cryotop®

Cryotop® CL

For Cryotop®, place the oocyte/embryo by
the black mark of the strip.

For Cryotop® CL, place the oocyte/embryo on
the thin part of the transparent strip.

Removal of the excess VS on the strip

After putting oocyte/embryo on the Cryotop®/Cryotop® CL strip, the excess VS should be removed by using pipette.

Place the pipette in the VS drop.

Slide the pipette horizontally.

Aspirate the excess VS, and minimize
the VS drop.

Good example

Bad example

Planar droplet.

The volume of VS2 is too much.

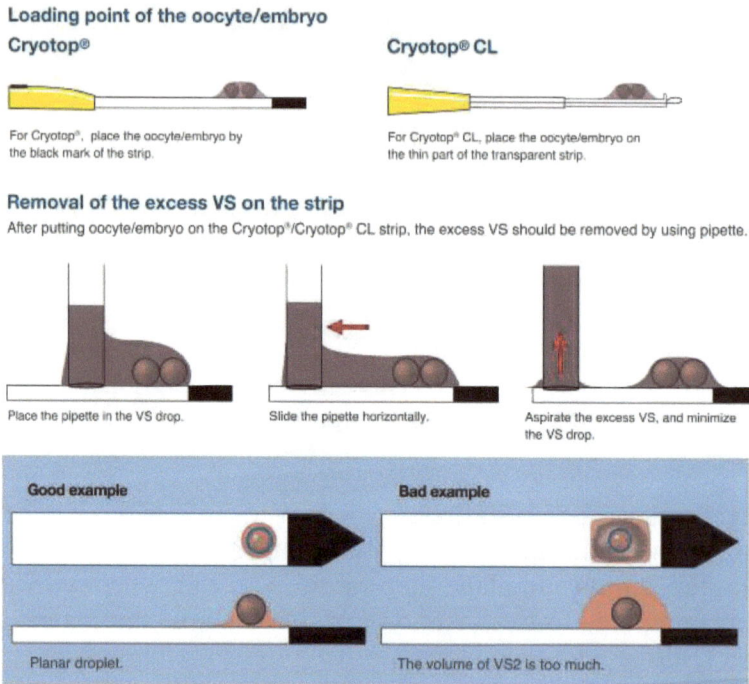

Fig. (5). diagram illustrated the position of loading embryos on the device and the differences between the good and bad positions.

Depolymerization of the meiotic spindle does not occur before or during vitrification because there is no chilling injury. CPAs have a stabilizing effect on tubulin fibers, according to early research, which has been confirmed by noninvasive studies [45]. Depolymerization of cryoprotectants occurs during the dilution of cryoprotectants, according to these reports. The inner cryoprotectant is replaced by water during rehydration at room temperature in these circumstances, which explains why depolymerization occurs at this stage. It has also been shown that the complete restoration of the spindle occurs depending on the incubation time post warming [45, 46] and happens faster-following vitrification than after slow freezing [45].

These findings have allayed initial concerns about the possibility of unbalanced gametes being generated after cryopreservation due to the disassembly of the meiotic spindle [47].

The results of comparable aneuploidy rates between embryos formed from vitrified and fresh oocytes from infertile patients [48], as well as the study of perinatal outcomes of babies born from vitrified oocytes, support the restoration of a functional meiotic spindle [49].

Risk of Vitrification

The hypothetical risk of cross-contamination associated with the direct interaction with liquid nitrogen needed for the vitrification phase and during the storage period has raised concerns about open systems. Although there have been no reports of cross-contamination following a cryo-transfer in ART, several authors have expressed concern, and some country regulations have been developed to address this problem. The presence of viral sequences in various samples such as follicular fluid, culture media, and liquid nitrogen used for vitrification and storage in seropositive patients for human immune deficiency virus (HIV), hepatitis C virus (HCV), and hepatitis B virus (HBV) undergoing IVF cycles was not detected in any of the samples examined, according to a report [50]. This is compelling evidence for open vitrification systems having low to no cross-contamination risks, particularly while working with samples from seropositive women who had a positive blood viral load in some cases. Even so, certain precautions can be taken to make the tool safer, such as using liquid nitrogen to sterilize it [51, 52], storing it in vapor-phase systems [53], or storing it in a closed unit even after open vitrification [52]. Needless to mention, using powerful closed devices eliminates the risk of cross-contamination while producing effective results for both embryos and oocytes [54 - 58]. Figs. (**6** and **7**) illustrated the open and closed system and how to close the straw and insert it in liquid nitrogen.

Device (Open system)

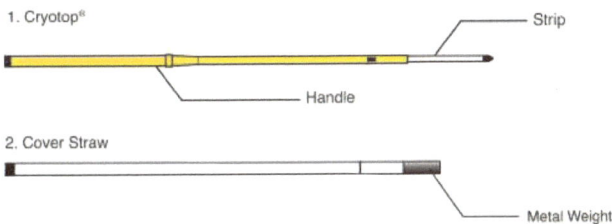

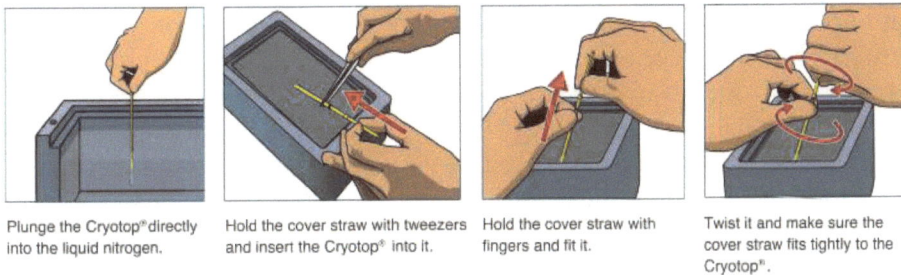

1. Cryotop® — Strip

— Handle

2. Cover Straw

— Metal Weight

Plunge the Cryotop® directly into the liquid nitrogen.

Hold the cover straw with tweezers and insert the Cryotop® into it.

Hold the cover straw with fingers and fit it.

Twist it and make sure the cover straw fits tightly to the Cryotop®.

Fig. (6). plunging the straw (open system) into liquid nitrogen after putting the embryos or oocytes on it.

Device (Closed system)

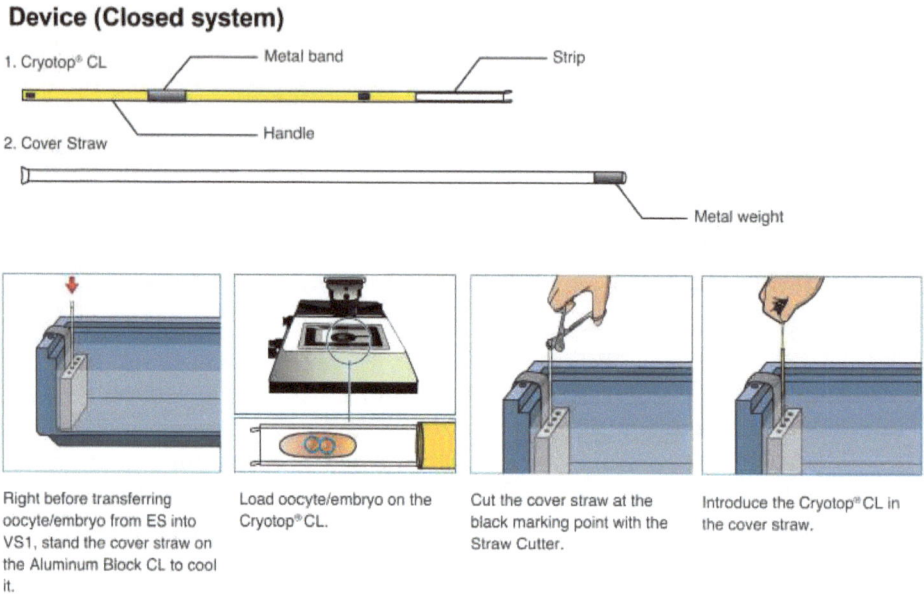

1. Cryotop® CL — Metal band — Strip

2. Cover Straw — Handle

— Metal weight

Right before transferring oocyte/embryo from ES into VS1, stand the cover straw on the Aluminum Block CL to cool it.	Load oocyte/embryo on the Cryotop®CL.	Cut the cover straw at the black marking point with the Straw Cutter.	Introduce the Cryotop®CL in the cover straw.

Fig. (7). plunging the straw (closed system) Insert the straw into the cover straw pre-set at the Aluminum Block Then seal the cover straw.

Despite growing evidence of comparable outcomes obtained with closed versus open devices in cases of oocyte vitrification, major prospective randomized trials demonstrating the safety of closed devices, as opposed to open ones, are still lacking [59].

Oocyte Cryo-preservation

The reasons for cryopreserving human oocytes are well-known diseases and therapies are the most common indicators: i.e. preservation of reproductive integrity in young cancer patients who need pelvic irradiation or chemotherapy; or other surgical operation requiring the removal of ovaries before or after reproductive age [60, 61]. Premature menopause, ovary hyperstimulation syndrome, and inadequate response to ovary stimulation are all issues linked to ovary function. Legal, ethical, social, and practical issues, such as embryo cryopreservation restrictions in several countries, can necessitate oocyte cryopreservation.

Desire to postpone motherhood for a variety of reasons, as well as a scarcity of available sperm following successful oocyte retrieval. Oocyte cryopreservation is also required to compensate for the pronounced age-related decline in fertility seen in women over 35 years, as discussed in detail recently [61].

Women, like most mammalian animals, suffer more and make greater sacrifices for their offspring. Their reproductive capacity, on the other hand, is severely limited in terms of quantity and length. Every 28 days, males produce millions of sperm with a single ejaculation, while only one or two oocytes mature. Men's reproductive ages are almost limitless, while women's reproductive ages are limited to 15–20 years (without additional concerns). Assisted reproductive techniques did not help to close the gap; however, the use of intracytoplasmic sperm injection (ICSI) and active sperm cryopreservation has significantly expanded the gap. Apart from the practical objectives, we have a moral obligation to assist in the development of an effective and secure oocyte cryopreservation process. Regrettably, the job is very difficult. Even though the first pregnancy from a cryopreserved oocyte occurred more than 20 years ago Progress was sluggish until recently [62].

In general, the two most serious issues have been inefficiency and lack of continuity. Oocytes are unique objects: their scale, shape, low cell number, and general fragility account for the majority of cryopreservation challenges. Oocytes are the largest cells in the mammalian body, according to popular belief [63].

In cryobiology, size appears to be a significant factor. Viruses and bacteria can withstand deep freezing without any special treatment (cryoprotectants or managed rate cooling), potentially causing a slew of problems in the embryology field. Freezing fibroblasts and epidermal cells is a simple and effective routine task in tissue culture laboratories that requires no special equipment.

A controlled-rate freezer can be used to effectively conduct sperm cryopreservation. Embryos in the early cleavage stage with cells 10–50 percent the size of oocytes survive slow-rate freezing and retain their developmental competence.

In comparison to fully grown ones, metaphase two (MII), large oocytes, prenatal and primary follicles may be successfully frozen. From the standpoint of cryopreservation, the nearly fully spherical form is not advantageous. Permeable cryoprotectants must be dispersed quickly and evenly in the ooplasm during equilibration and dilution before and after cooling and warming, respectively.

In this regard, a large spherical body is a disadvantage when compared to an elongated form like a fibroblast or a neuron. Since there is no backup at the 1-cell level, it is also a handicapped situation.

Chilling, severe deformation of shape during equilibration and dilution with cryoprotectants and hardening of the zona-pellucida are all known factors that affect their sensitivity. Chilling, which involves damage to lipid droplets, lipid-rich membranes, and microtubules, is perhaps the least known form of cryopreservation injury. The temperature range in which this injury occurs is very broad, ranging from +15°C (in some biological objects +20°C) to 5°C [64]. The lipid damage is permanent, and the oocyte dies as a result.

Human oocytes have low lipid droplet content as compared to other animals. On the other hand, their susceptibility to chilling remains high, owing to membrane damage and microtubule depolymerization, with all the implications that entail, such as chromosome misalignment and aneuploidy [65 - 69]. Chilling damage to membranes appears to be much more severe in matured human oocytes than in later developmental stages (zygotes), possibly explaining the well-known stage-dependent sensitivity [70, 71].

When oocytes are exposed to cryoprotectant solutions, the osmotic effect can cause serious deformation of the shape. Despite the frightening morphological appearance, human oocytes seem to withstand these deformations very well. Cryoprotectants added with care will help reduce the misshape and potential damage. The addition of cytoskeleton relaxants [72], the other suggested technique, may not be necessary for humans. The spherical form, on the other hand, makes only minimal expansion at dilution, and thus the accumulating water can disrupt the cell membrane.

Embryos Cryo-preservation

Embryo cryopreservation has become a critical therapeutic technique in assisted reproductive technology (ART), allowing for the possibility of multiple pregnancies from a single monitored ovarian stimulation (COS) period. In this way, embryo storage is a requirement for improving the overall results. Furthermore, the availability of an effective embryo cryopreservation program enables single embryo transfer to be used as a technique to reduce multiple pregnancy rates. These goals were historically achieved in a reasonably satisfactory manner using the slow freezing technique, but vitrification has undeniably made a major contribution to embryo cryopreservation programs in recent years, significantly enhancing the results and therefore the reach of these programs.

The efficacy of embryo vitrification over slow freezing was verified in a meta-analysis published in 2008 [73]. Before vitrification, the most common method was to freeze embryos at the early cleavage stage rather than the blastocyst stage [74], nowadays, the literature is profuse in reports showing successful outcomes after applying vitrification at all embryo's developmental stages. Table **1** shows results from the most recent publications including data from the last 2 years showing high and similar outcomes between embryo developmental stages and different vitrification devices. Comparable survival, implantation, and delivery rates were shown in a large series of cryo transfers of vitrified embryos from cleavage to blastocyst stage [87].

Table 1. Survival and pregnancy rate after embryo vitrification at cleavage or blastocyst stage using open and/ or closed devices.

Author	Embryo Stage	Survival Rate (%)	Pregnancy Rate (%)	Observations
Liu *et al.* [75]	Cleavage (Day 3)	97.6	38.5 p	-
Panagiotidis *et al.* [76]	Blastocyst	84.1	45.9	Open system
-	-	82.1	42.4	Closed system
Chen *et al.* [77]	Blastocyst	98	47.6	Open system
-	-	95.8	42.2	Closed system
Hashimoto *et al.* [78]	Blastocyst	96.9	46.8	Open system
-	-	97	45.4	Closed system
Kang *et al.* [79]	Blastocyst	96.6	41.8	Single ET
-	-	97.8	48.1	Double ET
Van Landuyt *et al.* [80]	Cleavage (Day 3)	94	–	20.7% IR
Muthukumar *et al.* [81]	Blastocyst	85.5	52.6	Day 5
-	-	79.6	32.6	Day 6
Roy *et al.* [82]	Blastocyst	94.4	58.8	-
Murakami *et al.* [83]	Blastocyst	98.7	51.5	HSA supplement
-	-	98.9	56.0	rHA supplement
Levron *et al.* [84]	Cleavage (Day 2–3)	81.6	20.0	-
Reed *et al.* [85]	Blastocyst	96.3	46.2	Non-biopsied
-	-	97.6	58.2	Biopsied
Iwahata *et al.* [86]	Blastocyst	96.1	–	49.4% IR
-	-	96.5	–	49.7% IR
Total	-	**96.3**	**45.9**	-

Abbreviations: ET, Embryo Transfer; IR, Implantation rate; HAS, Human Serum Albumin; rHA, recombinant Human Albumin.

When vitrified/warmed expanded day-5 blastocysts were moved, Although a few studies [88, 89] found that high survival and delivery rates were achieved.

Surprisingly, no artificial collapse of the blastocele was used; even though other studies have shown that this technique is essential for better performance.

The artificial collapse of the blastocysts is used to minimize the large cavity volume and, as a result, the likelihood of ice forming inside the cavity. When using a closed device for vitrification, a randomized study looked at the convenience of artificially collapsing the blastocysts before vitrification [88].

We can speculate that the need for artificial shrinkage when using a closed device is due to a higher risk of ice forming, which does not seem to be the case with open devices, as the vitrification of completely expanded blastocysts does not affect the results.

When considering biopsied or non-biopsied cleavage stage or blastocyst embryos, different vitrification approaches have also proven to be successful, making vitrification a valuable complement to preimplantation genetic screening (PGS) study [90, 91]. Blastocyst survival after trophectoderm biopsy is 96.3% in our hands, and clinical results after the transfer of chromosomally normal embryos show that vitrification has little effect on their implant ability (45.2% of implantation rate; 62.5% of clinical pregnancy rate; 54.2% of ongoing pregnancy rate), All of these findings demonstrate the value of vitrification as a supplement to PGS program. The availability of vitrification for embryo storage has also allowed IVF segmentation, resulting in the so-called "freeze-all strategy," which has recently received a lot of attention.

Controlled ovarian stimulation produces exceptionally high estrogen and progesterone levels, causing an asynchrony between the embryo developmental stage and the endometrium, which can affect endometrial receptivity in the long run [92].

As a result, delaying embryo transfer to a separate unstimulated period, in which the uterus is not subjected to supra-physiological doses of reproductive hormones, appears to be a viable option, Furthermore, many believe that the reasoning of transferring the embryos in a more physiological, "non activated" endometrium is why frozen embryo transfers yield better obstetric and perinatal outcomes than fresh embryo transfers [93, 94].

Different results, such as the incidence of small for gestational age were significantly lower after frozen embryo transfers compared to fresh embryo transfers in these trials, despite the latter group having a higher incidence of large for gestational age [95].

In addition to better neonatal outcomes, some studies have shown that freezing all embryos for later transfer improves IVF results [92, 96, 97]. Many people have suggested that the technique be used in IVF regularly as a result of this. However, since some of the studies available have been criticized due to significant deficiencies in their design, this evidence should be used with caution. In reality, we have found no evidence of improved IVF outcomes after a freeze-all technique in terms of ongoing pregnancy or live birth rates (36.2% vs 33.8%) in women who have a normal response to ovarian stimulation in our routine practice [98]. There was no proof of the effect of freezing after adjusting for patient age and other factors that could influence outcomes while these results do not endorse a shift in IVF practice to a freeze-all approach, there are some situations where delaying the switch would be preferable, such as patients at risk of ovarian hyperstimulation syndrome [99] or with elevated serum progesterone levels [100], etc. In this case, delaying the transition will be the preferred option. What is clear right now is that further research is required to prove the approach's validity [101].

CONCLUSION

Vitrification is a cryopreservation technique that allows hydrated living cells to be cooled to cryogenic temperatures without the use of ice. Vitrification simplifies and often improves cryopreservation by removing the mechanical injury from ice, the need to find optimal cooling and warming rates, the importance of different optimal cooling and warming rates for cells in mixed cell-type populations, and the need to find a frequently imperfect compromise between solution effects injury and nitrocellulose injury. While this allows for rapid cooling to "outrun" chilling injury, it complicates the osmotic effects of adding and removing cryoprotective agents, as well as increasing the potential of cryoprotectant toxicity during the addition and removal of cryoprotectants. Fortunately, over the last 30 years, a great number of solutions to the latter problem have been identified, and the former problem can usually be removed or sufficiently regulated by paying close attention to technique. Vitrification is thus beginning to realize its potential for enabling superior and convenient cryopreservation of most types of biological systems (including molecules, cells, tissues, organs, and even some whole organisms), and it is even beginning to be recognized as a successful natural strategy for surviving harsh environmental conditions. Many scientists who use vitrification or what they mistakenly believe is vitrification, on the other hand, have only a rudimentary understanding of the basic principles of this relatively new and growing approach to cryopreservation, which limits the practical findings that may be obtained. As a result, a greater understanding could aid in improving current results while also leading the way to new techniques that could be even more effective in the future.

CONSENT FOR PUBLICATION

Not applicable

CONFLICT OF INTEREST

The authors declare no conflict of interest, financial or otherwise.

ACKNOWLEDGEMENT

Declared none.

REFERENCES

[1]   Human Fertilisation and Embryology Authority. Fifth Annual Report. London: Human Fertilisation and Embryology Authority 1996.

[2]   Kovacs GT. What factors are important for successful embryo transfer after in-vitro fertilization? Hum Reprod 1999; 14(3): 590-2.
[http://dx.doi.org/10.1093/humrep/14.3.590] [PMID: 10221680]

[3]   Cohen J. Embryo replacement technology. San Francisco 31 Annual Post Graduate Course, ASRM, 1998.

[4]   Meldrum DR, Chetkowski R, Steingold KA, de Ziegler D, Cedars MI, Hamilton M. Evolution of a highly successful in vitro fertilization-embryo transfer program. Fertil Steril 1987; 48(1): 86-93.
[http://dx.doi.org/10.1016/S0015-0282(16)59295-4] [PMID: 2954865]

[5]   Lin TP, Sherman JK, Willett EL. Survival of unfertilized mouse eggs in media containing glycerol and glycine. J Exp Zool 1957; 134(2): 275-91.
[http://dx.doi.org/10.1002/jez.1401340204] [PMID: 13428954]

[6]   Sherman JK, Lin TP. Effect of glycerol and low temperature on survival of unfertilized mouse eggs. Nature 1958; 181(4611): 785-6.
[http://dx.doi.org/10.1038/181785a0] [PMID: 13517317]

[7]   Sherman JK, Lin TP. Survival of unfertilized mouse eggs during freezing and thawing. Proc Soc Exp Biol Med 1958; 98(4): 902-5.
[http://dx.doi.org/10.3181/00379727-98-24224] [PMID: 13591382]

[8]   Sherman JK, Lin TP. Temperature shock and cold-storage of unfertilized mouse eggs. Fertil Steril 1959; 10(4): 384-96.
[http://dx.doi.org/10.1016/S0015-0282(16)33485-9] [PMID: 13672294]

[9]   Mazur P. Kinetics of water loss from cells at subzero temperatures and the likelihood of intracellular freezing. J Gen Physiol 1963; 47(2): 347-69.
[http://dx.doi.org/10.1085/jgp.47.2.347] [PMID: 14085017]

[10]   Mazur P. Equilibrium, quasi-equilibrium, and nonequilibrium freezing of mammalian embryos. Cell Biophys 1990; 17(1): 53-92.
[http://dx.doi.org/10.1007/BF02989804] [PMID: 1704816]

[11]   Mazur P, Leibo SP, Chu EHY. A two-factor hypothesis of freezing injury. Evidence from Chinese hamster tissue-culture cells. Exp Cell Res 1972; 71(2): 345-55.
[http://dx.doi.org/10.1016/0014-4827(72)90303-5] [PMID: 5045639]

[12]   Whittingham DG, Leibo SP, Mazur P. Survival of mouse embryos frozen to -196 degrees and -269 degrees C. Science 1972; 178(4059): 411-4.
[http://dx.doi.org/10.1126/science.178.4059.411] [PMID: 5077328]

[13]   Trounson A, Mohr L. Human pregnancy following cryopreservation, thawing and transfer of an eight-cell embryo. Nature 1983; 305(5936): 707-9.
       [http://dx.doi.org/10.1038/305707a0] [PMID: 6633637]

[14]   Zeilmaker GH, Alberda AT, van Gent I, Rijkmans CM, Drogendijk AC. Two pregnancies following transfer of intact frozen-thawed embryos. Fertil Steril 1984; 42(2): 293-6.
       [http://dx.doi.org/10.1016/S0015-0282(16)48029-5] [PMID: 6745463]

[15]   Testart J, Lassalle B, Belaisch-Allart J, *et al.* High pregnancy rate after early human embryo freezing. Fertil Steril 1986; 46(2): 268-72.
       [http://dx.doi.org/10.1016/S0015-0282(16)49524-5] [PMID: 3732533]

[16]   Bernard A, Fuller BJ. Cryopreservation of human oocytes: a review of current problems and perspectives. Hum Reprod Update 1996; 2(3): 193-207.
       [http://dx.doi.org/10.1093/humupd/2.3.193] [PMID: 9079413]

[17]   Smith GD, Silva E Silva CA, Silva CA. Developmental consequences of cryopreservation of mammalian oocytes and embryos. Reprod Biomed Online 2004; 9(2): 171-8.
       [http://dx.doi.org/10.1016/S1472-6483(10)62126-8] [PMID: 15333247]

[18]   Meryman HT. Cryoprotective agents. Cryobiology 1971; 8(2): 173-83.
       [http://dx.doi.org/10.1016/0011-2240(71)90024-1] [PMID: 5578883]

[19]   Friedler S, Giudice LC, Lamb EJ. Cryopreservation of embryos and ova. Fertil Steril 1988; 49(5): 743-64.
       [http://dx.doi.org/10.1016/S0015-0282(16)59879-3] [PMID: 3282929]

[20]   Polge C, Smith AU, Parkes AS. Revival of spermatozoa after vitrification and dehydration at low temperatures. Nature 1949; 164(4172): 666.
       [http://dx.doi.org/10.1038/164666a0] [PMID: 18143360]

[21]   Rall WF, Fahy GM. Ice-free cryopreservation of mouse embryos at -196 degrees C by vitrification. Nature 1985; 313(6003): 573-5.
       [http://dx.doi.org/10.1038/313573a0] [PMID: 3969158]

[22]   Rall WF. Factors affecting the survival of mouse embryos cryopreserved by vitrification. Cryobiology 1987; 24(5): 387-402.
       [http://dx.doi.org/10.1016/0011-2240(87)90042-3] [PMID: 3652721]

[23]   Kuwayama M, Vajta G, Kato O, Leibo SP. Highly efficient vitrification method for cryopreservation of human oocytes. Reprod Biomed Online 2005; 11(3): 300-8.
       [http://dx.doi.org/10.1016/S1472-6483(10)60837-1] [PMID: 16176668]

[24]   Isachenko V, Montag M, Isachenko E, *et al.* Aseptic technology of vitrification of human pronuclear oocytes using open-pulled straws. Hum Reprod 2005; 20(2): 492-6.
       [http://dx.doi.org/10.1093/humrep/deh605] [PMID: 15528262]

[25]   El-Danasouri I, Selman H. Successful pregnancies and deliveries after a simple vitrification protocol for day 3 human embryos. Fertil Steril 2001; 76(2): 400-2.
       [http://dx.doi.org/10.1016/S0015-0282(01)01907-0] [PMID: 11476797]

[26]   Lane M, Schoolcraft WB, Gardner DK, Phil D. Vitrification of mouse and human blastocysts using a novel cryoloop container-less technique. Fertil Steril 1999; 72(6): 1073-8.
       [http://dx.doi.org/10.1016/S0015-0282(99)00418-5] [PMID: 10593384]

[27]   Liebermann J, Tucker MJ, Graham JR, Han T, Davis A, Levy MJ. Blastocyst development after vitrification of multipronuclear zygotes using the Flexipet denuding pipette. Reprod Biomed Online 2002; 4(2): 146-50. a
       [http://dx.doi.org/10.1016/S1472-6483(10)61932-3] [PMID: 12470577]

[28]   Mukaida T, Takahashi K, Kasai M. Blastocyst cryopreservation: ultrarapid vitrification using cryoloop technique. Reprod Biomed Online 2003; 6(2): 221-5.

[http://dx.doi.org/10.1016/S1472-6483(10)61713-0] [PMID: 12676003]

[29]   Kuleshova LL, Lopata A. Vitrification can be more favorable than slow cooling. Fertil Steril 2002; 78(3): 449-54.
[http://dx.doi.org/10.1016/S0015-0282(02)03305-8] [PMID: 12215314]

[30]   Liebermann J, Nawroth F, Isachenko V, Isachenko E, Rahimi G, Tucker MJ. Potential importance of vitrification in reproductive medicine. Biol Reprod 2002; 67(6): 1671-80.
[http://dx.doi.org/10.1095/biolreprod.102.006833] [PMID: 12444040]

[31]   Lassalle B, Testart J, Renard JP. Human embryo features that influence the success of cryopreservation with the use of 1,2 propanediol. Fertil Steril 1985; 44(5): 645-51.
[http://dx.doi.org/10.1016/S0015-0282(16)48981-8] [PMID: 4054343]

[32]   Vajta G, Kuwayama M. Improving cryopreservation systems. Theriogenology 2006; 65(1): 236-44.
[http://dx.doi.org/10.1016/j.theriogenology.2005.09.026] [PMID: 16289262]

[33]   Ghetler Y, Yavin S, Shalgi R, Arav A. The effect of chilling on membrane lipid phase transition in human oocytes and zygotes. Hum Reprod 2005; 20(12): 3385-9.
[http://dx.doi.org/10.1093/humrep/dei236] [PMID: 16055458]

[34]   Cobo A, Pérez S, De los Santos MJ, Zulategui J, Domingo J, Remohí J. Effect of different cryopreservation protocols on the metaphase II spindle in human oocytes. Reprod Biomed Online 2008; 17(3): 350-9.
[http://dx.doi.org/10.1016/S1472-6483(10)60218-0] [PMID: 18765005]

[35]   Edgar DH, Bourne H, Speirs AL, McBain JC. A quantitative analysis of the impact of cryopreservation on the implantation potential of human early cleavage stage embryos. Hum Reprod 2000; 15(1): 175-9.
[http://dx.doi.org/10.1093/humrep/15.1.175] [PMID: 10611209]

[36]   Van den Abbeel E, Camus M, Van Waesberghe L, Devroey P, Van Steirteghem AC. Viability of partially damaged human embryos after cryopreservation. Hum Reprod 1997; 12(9): 2006-10.
[http://dx.doi.org/10.1093/humrep/12.9.2006] [PMID: 9363721]

[37]   El-Toukhy T, Khalaf Y, Al-Darazi K, Andritsos V, Taylor A, Braude P. Effect of blastomere loss on the outcome of frozen embryo replacement cycles. Fertil Steril 2003; 79(5): 1106-11.
[http://dx.doi.org/10.1016/S0015-0282(03)00072-4] [PMID: 12738503]

[38]   Edgar DH, Archer J, Bourne H. The application and impact of cryopreservation of early cleavage stage embryos in assisted reproduction. Hum Fertil (Camb) 2005; 8(4): 225-30.
[http://dx.doi.org/10.1080/14647270500054779] [PMID: 16393822]

[39]   Cobo A, de los Santos MJ, Castello D, Gamiz P, Campos P, Remohi J. Outcomes of vitrified early cleavage-stage and blastocyst-stage embryos in a cryopreservation program: evaluation of 3,150 warming cycles. Fertil Steril. [Video-Audio Media]. 2012 Nov; 98(5): 1138–46 e1.

[40]   Zhang S, Tan K, Gong F, Gu Y, Tan Y, Lu C, *et al.* Blastocysts can be rebiopsied for preimplantation genetic diagnosis and screening. Fertil Steril. [Research Support, Non-U.S. Gov't] 2014; 102(6): 1641–5.
[http://dx.doi.org/10.1016/j.fertnstert.2014.09.018]

[41]   Chamayou S, Romano S, Alecci C, *et al.* Oocyte vitrification modifies nucleolar remodeling and zygote kinetics-a sibling study. J Assist Reprod Genet 2015; 32(4): 581-6.
[http://dx.doi.org/10.1007/s10815-015-0446-x] [PMID: 25701142]

[42]   Saragusty J, Arav A. Current progress in oocyte and embryo cryopreservation by slow freezing and vitrification. Reproduction 2011; 141(1): 1-19.
[http://dx.doi.org/10.1530/REP-10-0236] [PMID: 20974741]

[43]   Seki S, Mazur P. The dominance of warming rate over cooling rate in the survival of mouse oocytes subjected to a vitrification procedure. Cryobiology [Research Support, N.I.H., Extramural] 2009; 5 9(1): 75–82.

[http://dx.doi.org/10.1016/j.cryobiol.2009.04.012]

[44]   Stoop D, Silber S, Cobo A. Fertility preservation for age-related fertility decline - authors' reply. Lancet 2015; 385(9967): 507-8.
[http://dx.doi.org/10.1016/S0140-6736(15)60200-8] [PMID: 25705841]

[45]   Martinez M, Rabadan S, Domingo J, Cobo A, Pellicer A, Garcia-Velasco JA. Obstetric outcome after oocyte vitrification and warming for fertility preservation in women with cancer. Reprod Biomed Online [Evaluation Studies Research Support, Non-US Gov't] 2014; 29(6): 722-8.
[http://dx.doi.org/10.1016/j.rbmo.2014.09.002]

[46]   Bianchi V, Coticchio G, Fava L, Flamigni C, Borini A. Meiotic spindle imaging in human oocytes frozen with a slow freezing procedure involving high sucrose concentration. Hum Reprod 2005; 20(4): 1078-83.
[http://dx.doi.org/10.1093/humrep/deh736] [PMID: 15760952]

[47]   Noyes N, Knopman J, Labella P, McCaffrey C, Clark-Williams M, Grifo J. Oocyte cryopreservation outcomes including pre-cryopreservation and post-thaw meiotic spindle evaluation following slow cooling and vitrification of human oocytes. Fertil Steril [Comparative Study Evaluation Studies Research Support, Non-US Gov't] 2010; 94(6): 2078-82.
[http://dx.doi.org/10.1016/j.fertnstert.2010.01.019]

[48]   Forman EJ, Li X, Ferry KM, Scott K, Treff NR, Scott RT Jr. Oocyte vitrification does not increase the risk of embryonic aneuploidy or diminish the implantation potential of blastocysts created after intracytoplasmic sperm injection: A novel, paired randomized controlled trial using DNA fingerprinting. Fertil Steril Randomized Controlled Trial Research Support, Non-US Gov't 2012; 98(3): 644-9.
[http://dx.doi.org/10.1016/j.fertnstert.2012.04.028]

[49]   Cobo A, Serra V, Garrido N, Olmo I, Pellicer A, Remohí J. Obstetric and perinatal outcome of babies born from vitrified oocytes. Fertil Steril 2014; 102(4): 1006-1015.e4.
[http://dx.doi.org/10.1016/j.fertnstert.2014.06.019] [PMID: 25064408]

[50]   Cobo A, Bellver J, de Los Santos MJ, Remohi J. Viral screening of spent culture media and liquid nitrogen samples of oocytes and embryos from hepatitis B, hepatitis C, and human immunodeficiency virus chronically infected women undergoing in vitro fertilization cycles. Fertil Steril 2011.
[PMID: 22035968]

[51]   Cobo A, Castellò D, Weiss B, Vivier C, De la Macorra A, Kramp F.

[52]   Parmegiani L, Accorsi A, Cognigni GE, Bernardi S, Troilo E, Filicori M. Sterilization of liquid nitrogen with ultraviolet irradiation for safe vitrification of human oocytes or embryos. Fertil Steril 2009.
[PMID: 19591992]

[53]   Cobo A, Romero JL, Pérez S, de los Santos MJ, Meseguer M, Remohí J. Storage of human oocytes in the vapor phase of nitrogen. Fertil Steril 2010; 94(5): 1903-7.
[http://dx.doi.org/10.1016/j.fertnstert.2009.10.042] [PMID: 20138272]

[54]   Parmegiani L, Cognigni GE, Bernardi S, *et al.* Efficiency of aseptic open vitrification and hermetical cryostorage of human oocytes. Reprod Biomed Online 2011; 23(4): 505-12.
[http://dx.doi.org/10.1016/j.rbmo.2011.07.003] [PMID: 21843968]

[55]   Van Landuyt L, Stoop D, Verheyen G, Verpoest W, Camus M, Van de Velde H, *et al.* Outcome of closed blastocyst vitrification in relation to blastocyst quality: Evaluation of 759 warming cycles in a singleembryo transfer policy. Hum Reprod [Research Support, Non-US Gov't] 2011; 26(3): 527-34.

[56]   Vanderzwalmen P, Zech NH, Ectors F, *et al.* Blastocyst transfer after aseptic vitrification of zygotes: an approach to overcome an impaired uterine environment. Reprod Biomed Online 2012; 25(6): 591-9.
[http://dx.doi.org/10.1016/j.rbmo.2012.09.004] [PMID: 23069744]

[57]   Papatheodorou A, Vanderzwalmen P, Panagiotidis Y, *et al.* Open versus closed oocyte vitrification system: a prospective randomized sibling-oocyte study. Reprod Biomed Online 2013; 26(6): 595-602.
[http://dx.doi.org/10.1016/j.rbmo.2013.02.014] [PMID: 23602678]

[58]   Stoop D, De Munck N, Jansen E, *et al.* Clinical validation of a closed vitrification system in an oocyte-donation programme. Reprod Biomed Online 2012; 24(2): 180-5.
[http://dx.doi.org/10.1016/j.rbmo.2011.10.015] [PMID: 22222248]

[59]   Cobo A, Meseguer M, Remohí J, Pellicer A. Use of cryo-banked oocytes in an ovum donation programme: a prospective, randomized, controlled, clinical trial. Hum Reprod 2010; 25(9): 2239-46.
[http://dx.doi.org/10.1093/humrep/deq146] [PMID: 20591872]

[60]   Kuwayama M. Highly efficient vitrification for cryopreservation of human oocytes and embryos: the Cryotop method. Theriogenology 2007; 67(1): 73-80.
[http://dx.doi.org/10.1016/j.theriogenology.2006.09.014] [PMID: 17055564]

[61]   Kuwayama M, Cobo A, Vajta G. Vitrification of oocytes: general considerations and the use of the Cryotop method.Vitrification in Assisted Reproduction. London: Informa Healthcare 2007.
[http://dx.doi.org/10.3109/9780203090022.009]

[62]   Chen C. Pregnancy after human oocyte cryopreservation. Lancet 1986; 1(8486): 884-6.
[http://dx.doi.org/10.1016/S0140-6736(86)90989-X] [PMID: 2870356]

[63]   Liebermann J, Tucker MJ. Comparison of vitrification and conventional cryopreservation of day 5 and day 6 blastocysts during clinical application. Fertil Steril 2006; 86(1): 20-6.
[http://dx.doi.org/10.1016/j.fertnstert.2006.01.029] [PMID: 16762345]

[64]   Leibo SP, Martino A, Kobayashi S, Pollard JW. Stagedependent sensitivity of oocytes and embryos to low temperatures. Anim Reprod Sci 1996; 42(1-4): 45-53.
[http://dx.doi.org/10.1016/0378-4320(96)01543-6]

[65]   Magistrini M, Szöllösi D. Effects of cold and of isopropyl-N-phenylcarbamate on the second meiotic spindle of mouse oocytes. Eur J Cell Biol 1980; 22(2): 699-707.
[PMID: 7192628]

[66]   Sathananthan AH, Ng SC, Trounson AO, *et al.* The effects of ultrarapid freezing on meiotic and mitotic spindles of mouse oocytes and embryos. Gamete Res 1988; 21(4): 385-401.
[http://dx.doi.org/10.1002/mrd.1120210407] [PMID: 3220431]

[67]   Pickering SJ, Braude PR, Johnson MH, Cant A, Currie J. Transient cooling to room temperature can cause irreversible disruption of the meiotic spindle in the human oocyte. Fertil Steril 1990; 54(1): 102-8.
[http://dx.doi.org/10.1016/S0015-0282(16)53644-9] [PMID: 2358076]

[68]   Fabbri R, Porcu E, Marsella T, Rocchetta G, Venturoli S, Flamigni C. Human oocyte cryopreservation: new perspectives regarding oocyte survival. Hum Reprod 2001; 16(3): 411-6.
[http://dx.doi.org/10.1093/humrep/16.3.411] [PMID: 11228204]

[69]   Stachecki JJ, Munné S, Cohen J. Spindle organization after cryopreservation of mouse, human, and bovine oocytes. Reprod Biomed Online 2004; 8(6): 664-72.
[http://dx.doi.org/10.1016/S1472-6483(10)61647-1] [PMID: 15169583]

[70]   Clark NA, Swain JE. Oocyte cryopreservation: searching for novel improvement strategies. J Assist Reprod Genet 2013; 30(7): 865-75.
[http://dx.doi.org/10.1007/s10815-013-0028-8] [PMID: 23779099]

[71]   Dobrinsky JR, Pursel VG, Long CR, Johnson LA. Birth of piglets after transfer of embryos cryopreserved by cytoskeletal stabilization and vitrification. Biol Reprod 2000; 62(3): 564-70.
[http://dx.doi.org/10.1095/biolreprod62.3.564] [PMID: 10684796]

[72]   Vajta G, Nagy ZP. Are programmable freezers still needed in the embryo laboratory? Review on vitrification. Reprod Biomed Online 2006; 12(6): 779-96.

[http://dx.doi.org/10.1016/S1472-6483(10)61091-7] [PMID: 16792858]

[73]   Loutradi KE, Kolibianakis EM, Venetis CA, Papanikolaou EG, Pados G, Bontis I, *et al.*
Cryopreservation of human embryos by vitrification or slow freezing: A systematic review and meta-
analysis. Fertil Steril [Meta-Analysis Review] 2008; 90(1): 186-93.
[http://dx.doi.org/10.1016/j.fertnstert.2007.06.010]

[74]   Fernandez Gallardo E, Spiessens C, D'Hooghe T, Debrock S. Effect of embryo morphology and
morphometrics on implantation of vitrified day 3 embryos after warming: a retrospective cohort study.
Reprod Biol Endocrinol 2016; 30;14(1): 40.
[http://dx.doi.org/10.1186/s12958-016-0175-8]

[75]   Liu SY, Teng B, Fu J, Li X, Zheng Y, Sun XX. Obstetric and neonatal outcomes after transfer of
vitrified early cleavage embryos. Hum Reprod 2013; 28(8): 2093-100.
[http://dx.doi.org/10.1093/humrep/det104] [PMID: 23569081]

[76]   Panagiotidis Y, Vanderzwalmen P, Prapas Y, *et al.* Open versus closed vitrification of blastocysts
from an oocyte-donation programme: a prospective randomized study. Reprod Biomed Online 2013;
26(5): 470-6.
[http://dx.doi.org/10.1016/j.rbmo.2013.01.016] [PMID: 23507132]

[77]   Chen Y, Zheng X, Yan J, Qiao J, Liu P. Neonatal outcomes after the transfer of vitrified blastocysts:
closed versus open vitrification system. Reprod Biol Endocrinol 2013; 11(1): 107.
[http://dx.doi.org/10.1186/1477-7827-11-107] [PMID: 24256633]

[78]   Hashimoto S, Amo A, Hama S, Ohsumi K, Nakaoka Y, Morimoto Y. A closed system supports the
developmental competence of human embryos after vitrification : Closed vitrification of human
embryos. J Assist Reprod Genet 2013; 30(3): 371-6.
[http://dx.doi.org/10.1007/s10815-012-9928-2] [PMID: 23315263]

[79]   Kang SM, Lee SW, Yoon SH, Kim JC, Lim JH, Lee SG. Comparison of clinical outcomes between
single and double vitrified-warmed blastocyst embryo transfer according to the day of vitrification. J
Assist Reprod Genet 2013; 30(6): 779-85.
[http://dx.doi.org/10.1007/s10815-013-0017-y] [PMID: 23715875]

[80]   Van Landuyt L, Van de Velde H, De Vos A, *et al.* Influence of cell loss after vitrification or slow-
freezing on further in vitro development and implantation of human Day 3 embryos. Hum Reprod
2013; 28(11): 2943-9.
[http://dx.doi.org/10.1093/humrep/det356] [PMID: 24014599]

[81]   Muthukumar K, Kamath MS, Mangalaraj AM, Aleyamma T, Chandy A, George K. Comparison of
clinical outcomes following vitrified warmed day 5/6 blastocyst transfers using solid surface
methodology with fresh blastocyst transfers. J Hum Reprod Sci 2013; 6(1): 59-64.
[http://dx.doi.org/10.4103/0974-1208.112384] [PMID: 23869154]

[82]   Roy TK, Brandi S, Tappe NM, *et al.* Embryo vitrification using a novel semi-automated closed system
yields in vitro outcomes equivalent to the manual Cryotop method. Hum Reprod 2014; 29(11): 2431-8.
[http://dx.doi.org/10.1093/humrep/deu214] [PMID: 25164022]

[83]   Murakami M, Egashira A, Tanaka K, Mine C, Otsubo H, Kuramoto T. Perinatal outcomes for transfer
of blastocysts vitrified and warmed in defined solutions with recombinant human albumin: 374 babies
born after 898 embryo transfers. J Assist Reprod Genet 2014; 31(12): 1605-10.
[http://dx.doi.org/10.1007/s10815-014-0343-8] [PMID: 25326854]

[84]   Levron J, Leibovitz O, Brengauz M, *et al.* Cryopreservation of day 2-3 embryos by vitrification yields
better outcome than slow freezing. Gynecol Endocrinol 2014; 30(3): 202-4.
[http://dx.doi.org/10.3109/09513590.2013.875995] [PMID: 24397397]

[85]   Reed ML, Said AH, Thompson DJ, Caperton CL. Large-volume vitrification of human biopsied and
non-biopsied blastocysts: a simple, robust technique for cryopreservation. J Assist Reprod Genet 2015;
32(2): 207-14.
[http://dx.doi.org/10.1007/s10815-014-0395-9] [PMID: 25464896]

[86]   Iwahata H, Hashimoto S, Inoue M, *et al.* Neonatal outcomes after the implantation of human embryos vitrified using a closed-system device. J Assist Reprod Genet 2015; 32(4): 521-6.
[http://dx.doi.org/10.1007/s10815-015-0431-4] [PMID: 25617086]

[87]   Cobo A, de los Santos MJ, Castellò D, Gámiz P, Campos P, Remohí J. Outcomes of vitrified early cleavage-stage and blastocyst-stage embryos in a cryopreservation program: evaluation of 3,150 warming cycles. Fertil Steril 2012; 98(5): 1138-46.e1.
[http://dx.doi.org/10.1016/j.fertnstert.2012.07.1107] [PMID: 22862909]

[88]   De Munck N, Petrussa L, Verheyen G, *et al.* Chromosomal meiotic segregation, embryonic developmental kinetics and DNA (hydroxy)methylation analysis consolidate the safety of human oocyte vitrification. Mol Hum Reprod 2015; 21(6): 535-44.
[http://dx.doi.org/10.1093/molehr/gav013] [PMID: 25833840]

[89]   Hiraoka K, Hiraoka K, Kinutani M, Kinutani K. Blastocoele collapse by micropipetting prior to vitrification gives excellent survival and pregnancy outcomes for human day 5 and 6 expanded blastocysts. Hum Reprod 2004; 19(12): 2884-8.
[http://dx.doi.org/10.1093/humrep/deh504] [PMID: 15347597]

[90]   Van Landuyt L, Verpoest W, Verheyen G, *et al.* Closed blastocyst vitrification of biopsied embryos: evaluation of 100 consecutive warming cycles. Hum Reprod 2011; 26(2): 316-22.
[http://dx.doi.org/10.1093/humrep/deq338] [PMID: 21138906]

[91]   Schoolcraft WB, Fragouli E, Stevens J, Munne S, Katz-Jaffe MG, Wells D. Clinical application of comprehensive chromosomal screening at the blastocyst stage. Fertil Steril Evaluation Studies Research Support, Non-U.S. Gov't 2010; 94(5): 1700-6.

[92]   Roque M, Lattes K, Serra S, *et al.* Fresh embryo transfer versus frozen embryo transfer in in vitro fertilization cycles: a systematic review and meta-analysis. Fertil Steril Comparative Study Meta-Analysis Review 2013; 99(1): 156-62.
[http://dx.doi.org/10.1016/j.fertnstert.2012.09.003] [PMID: 23040524]

[93]   Maheshwari A, Pandey S, Shetty A, Hamilton M, Bhattacharya S. Obstetric and perinatal outcomes in singleton pregnancies resulting from the transfer of frozen thawed versus fresh embryos generated through in vitro fertilization treatment: A systematic review and meta-analysis
[http://dx.doi.org/10.1016/j.fertnstert.2012.05.019]

[94]   Wennerholm UB, Söderström-Anttila V, Bergh C, *et al.* Children born after cryopreservation of embryos or oocytes: a systematic review of outcome data. Hum Reprod 2009; 24(9): 2158-72.
[http://dx.doi.org/10.1093/humrep/dep125] [PMID: 19458318]

[95]   Kato O, Kawasaki N, Bodri D, *et al.* Neonatal outcome and birth defects in 6623 singletons born following minimal ovarian stimulation and vitrified versus fresh single embryo transfer. Eur J Obstet Gynecol Reprod Biol 2012; 161(1): 46-50.
[http://dx.doi.org/10.1016/j.ejogrb.2011.12.005] [PMID: 22200255]

[96]   Shapiro BS, Daneshmand ST, Garner FC, Aguirre M, Hudson C, Thomas S. Evidence of impaired endometrial receptivity after ovarian stimulation for in vitro fertilization: a prospective randomized trial comparing fresh and frozen-thawed embryo transfer in normal responders. Gov't 2011; 96(2): 344-8.
[PMID: 21737072]

[97]   Shapiro BS, Daneshmand ST, Garner FC, Aguirre M, Hudson C, Thomas S. Evidence of impaired endometrial receptivity after ovarian stimulation for in vitro fertilization: a prospective randomized trial comparing fresh and frozen-thawed embryo transfers in high responders. Gov't 2011; 96(2): 516-8.
[PMID: 21737071]

[98]   Sanger F, Nicklen S, Coulson AR. DNA sequencing with chain-terminating inhibitors. Proc Natl Acad Sci USA 1977; 74(12): 5463-7.
[http://dx.doi.org/10.1073/pnas.74.12.5463] [PMID: 271968]

[99] Cobo A, Garrido N, Pellicer A, Remohí J. Six years' experience in ovum donation using vitrified oocytes: report of cumulative outcomes, impact of storage time, and development of a predictive model for oocyte survival rate. Fertil Steril 2015; 104(6): 1426-34.e1, 8.
[http://dx.doi.org/10.1016/j.fertnstert.2015.08.020] [PMID: 26353081]

[100] Venetis CA, Kolibianakis EM, Bosdou JK, Tarlatzis BC. Progesterone elevation and probability of pregnancy after IVF: a systematic review and meta-analysis of over 60 000 cycles. Hum Reprod Update 2013; 19(5): 433-57.
[http://dx.doi.org/10.1093/humupd/dmt014] [PMID: 23827986]

[101] Drakopoulos P, Blockeel C, Stoop D, *et al.* Conventional ovarian stimulation and single embryo transfer for IVF/ICSI. How many oocytes do we need to maximize cumulative live birth rates after utilization of all fresh and frozen embryos? Hum Reprod 2016; 31(2): 370-6.
[http://dx.doi.org/10.1093/humrep/dev316] [PMID: 26724797]

CHAPTER 10

Reproductive Cloning

Eman A. Hussien[1], Sara A. Mekkawy[1], Elham K. Eltahawy[2], Islam M. Saadeldin[3,4,5] and Mohamed M. Omran[2,*]

[1] *Biotechnology Program, Faculty of Science, Helwan University, Cairo, Egypt*

[2] *Chemistry Department, Faculty of Science, Helwan University, Cairo, Egypt*

[3] *Research Institute of Veterinary Medicine, Chungnam National University, 34134 Daejeon, Republic of Korea*

[4] *College of Veterinary Medicine, Chungnam National University; 34134 Daejeon, Republic of Korea*

[5] *Department of Physiology, Faculty of Veterinary Medicine, Zagazig University, 44519 Zagazig, Egypt*

Abstract: Hearing the name "Dolly" was and still stirs the minds of professionals and non-specialists towards the term "cloning", but the way of producing dolly is not the only aspect of cloning. Cloning is defined as the techniques through which identical or virtually identical individuals can be produced. Based on this definition, in this chapter, we are trying to clarify the different applications, aspects, and techniques of cloning such as gene cloning, therapeutic cloning, but to focus on reproductive cloning. Reproductive cloning is the method of making a genetically similar clone of a whole organism. Then it is needed to be discussed with all the scientific thoughts around it, advantages, disadvantages, legal or illegal, and comparing it to other aspects and this is our aim in this chapter.

Keywords: Cloning, Embryo splitting, Reproductive cloning, Somatic Cell Nuclear Transfer, Tetraploid complementation.

INTRODUCTION

The term "cloning" was and is still being widely used by biologists to express many concepts. Even though all these concepts depend on the usage of copies of the whole biological material, they vary in their precision. There are three common major applications of cloning; the first is gene cloning which means the cloning of DNA segments [1, 2], the second is cells cloning which is done to give rise to particular types of cells, and the third is individual cloning that seeks the

* **Corresponding author Mohamed Mostafa Omran**: Chemistry Department, Faculty of Science, Helwan University, Ain Helwan, 11795, Cairo, Egypt; Tel: 00201025585562;
E-mails: drmmomran@science.helwan.edu.eg and drmmomran@yahoo.com

mimicking of the natural process in the identical twin development specifically in multicellular organisms Table **1**. The first case to appear and to spread the concept of cloning to the public is the famously known case of Dolly, which was successfully produced by cloning after about 270 trials by the British scientist Ian Wilmut. After its success, trials have been done on many types of mammals such as mice, rats, pigs, sheep, cows, goats with an increasing success rate without reaching 100%. The spontaneous abortion was the fate of most of the cloning pregnancies [3]. In the case of fetus survival from death during pregnancies, if they did not face early death, they were seen to have serious health concerns such as distorted limbs, gross obesity, dysfunctional immune systems, and organs such as liver and kidneys, and also lungs and the progressive diseases and failures were one of the reasons of the early death of Dolly along with arthritis after living only 6 years after its development by cloning, causing its health to decay rapidly [4, 5]. With all these consequences, it was nearly impossible to apply to humans especially since there are more serious consequences that are abridged in devastating economic, health, and emotional ramifications due to the fetus's death [5].

Table 1. Different applications of cloning.

-	Gene Cloning (Molecular Cloning)	Therapeutic Cloning	Gene Cloning (Molecular Cloning)
Definition	The process of isolating a DNA sequence of interest to make multiple copies of it. The identical copies are clones	The technology used to produce tissues and organs from a single cell to replace damaged cells, tissues, and organs	The process of generating an organism that has the same DNA as another organism
Product of cloning	produces copies of genes or segments of DNA	produces embryonic stem cells of engineered tissues for experiments	produces copies of whole animals
Applications	1. **Main aim**: Studying the structure and function of genes in detail. 2. **Medical applications**: the synthesis of vitamins, hormones, and antibiotics 3. **Agricultural applications**: cloning in bacteria facilitate nitrogen fixation in plants	1. **Main aim**: To make use of the unique ability to generate virtually all types of cells in an organism. 2. **Medical applications:** Grow healthy tissues in the laboratory that can be used to replace damaged or diseased tissues. 3. Learn more about the molecular causes of diseases.	1. production of organs for transplantation to humans 2. Studies of human genetic diseases. 3. Basic research in the control of gene expression and function
Techniques	Recombinant DNA technology	Somatic cell nuclear transfer	1. Somatic cell nuclear transfer 2. Embryo splitting

-	Gene Cloning (Molecular Cloning)	Therapeutic Cloning	Gene Cloning (Molecular Cloning)
Potential drawbacks	Nature is an extremely complex, interrelated chain. It is believed by some scientists that the introduction of genetically-modified genes may have an irreversible effect with consequences not yet known.	There are striking similarities between stem cells and cancer cells. Both cell types can proliferate indefinitely and some studies show that after 60 cycles of cell division, stem cells can accumulate mutations that could lead to cancer.	Reproductive cloning is a very inefficient technique and most cloned animal embryos cannot develop into healthy individuals. For instance, Dolly was the only clone to be born live out of a total of 277 cloned embryos. This very low efficiency, combined with safety concerns, presents a serious obstacle to the application of reproductive cloning.

The characteristics of clones are not only genetically but also environmentally controlled, the nutritional inputs during pregnancy or from parents after birth, society, or life experience while growing up and other elements can be determinants of the clone's characteristics. So, if there are two genetically identical clones as the nucleus and the mitochondria donor are the same but are grown in different environments or at different times, they will not be identical in their physical or behavioral characteristics. Even monozygotic twins are not fully genetically identical or epigenetically identical as many elements contribute to their variation such as mutations, stochastic developmental variations, and varied imprinting effects (parent-specific chemical marks on the DNA) that differ in impact on each twin [6, 7]. If twin clones do not have the same mitochondria, as there may be multiple donors or the cell is obtained from an individual and the nucleus is obtained from another, there are additional variations that will exist especially in the high energy demanding body parts such as the heart, muscles, brain, and eye. Also, some differences exist in the body systems that are controlled by mitochondria to determine cell numbers by using mitochondrial control over cell death [8 - 10].

HUMAN CLONING *VS.* ANIMAL CLONING

In the context of human cloning, there is no scientific evidence that human embryos have been cloned (Table **2**). In 1998, there was a claim from South Korean scientists to be successful in cloning a human embryo till the phase of a four cells embryo, and then the experiment was interrupted. In 2002, a kind of religious group called Clonaid claimed the birth of a girl named eve to be the first cloned human and the first of another 12 cloned humans, but also without introducing any evidence to prove their claim. In 2004, a paper was published in the journal Science by a scientist called Woo-Suk Hwang and his group in Seoul National University in South Korea in which they claimed to be successful in cloning a human embryo in a test tube [11], then in 2006, Science had retracted Hwang's paper due to the absence of proofs that support their claim. Depending on all these claims, it cannot be admitted that humans had been cloned, but it also does not necessarily deny the possibility of human cloning. Scientifically, two reasons caused the failure of primates cloning. The first are two proteins called spindle proteins, which are essential to cell division, but their location differs in primates such as humans from other mammalians. In primate eggs, they are located very close to the chromosomes. This results in their removal during the removal of the egg's nucleus inhibiting the cell division. In contrast with primates, in the other mammalians, the spindle proteins are spread all over the egg, so the egg removal does not affect their existence and then does not affect the cell division. The second reason is the effect of some dyes and the ultraviolet light during the removal of the egg's nucleus, which can damage the cell and prevent it from growing [12].

Table 2. Major milestones in human and animal reproductive cloning.

	Human Cloning	**Animal Cloning**
Advantages	1. Help us find new ways to recover from trauma. 2. It could help couples resolve problems with fertility. 3. Lead to new advances in medical science. 4. Allow us to explore the potential benefits of modifying genes. 5. Reduce the impact of diseases in ways that vaccinations cannot. 6. Help us correct today's conditions for tomorrow's generation by eliminating defective genetic profiles and chromosomes. Allow us to create stem cells ethically. 7. Make people more resilient to disease. 8. Cloning humans could help us to be ready for global warming. 9. Allow our best and brightest to continue impacting the world. 10. Lead us toward organ development or regeneration.	1. Allow us to balance environmental habitats. 2. Create more security in the global food supply. 3. Advance scientific discoveries in other fields. 4. Help pet parents find greater comfort. 5. Cloning animals do not create an exact duplication 6. Allows us to preserve endangered species 7. Allows us to produce the most desirable traits 8. Help us to reduce human disease 9. Cloning animals would not impact the quality of the food supply

	Human Cloning	**Animal Cloning**
Disadvantages	1. Cloning humans might always be an imperfect science 2. Cloning humans would be a technology initially priced only for the wealthy 3. Creating a rapidly aging population 4. Alter our perceptions of individuality 5. Reduce the diversity of our genetic makeup. 6. Lead to the exploitation of women. 7. Turn people into potential commodities. 8. Change how we approach grief and unexpected loss.	1. The least effective way to produce offspring. 2. Cloning animals is expensive 3. Reduces the genetic diversity of that species 4. Eventually, slow the rate of reproduction 5. Cloning animals has a history of creating abnormal pregnancies 6. Create long-term health issues to manage 7. Result in sterility issues, unforeseen consequences, and more cancer-related issues 8. Eventually, this lead to the cloning of humans 9. Leads to higher levels of embryo destruction

While the future economic and social advantages of animal cloning are immense, certain facets of cloned animal processing need to be properly addressed before the large-scale application of this technology by proper testing. In the former, fetal growth tends to be stopped during conception at multiple stages, leading to abortions in the early and advanced stages of fetal development [13, 14]. Many facets of an embryo and fetal mortality appear to mean that the placenta does not naturally mature, likely due to an incorrect change from the yolk sac to the allantoic diet. Others also found that allantois growth is significantly retarded or even non-existent, as it is marked by a loss or decline in vascularization during early gestation, resulting in a breakdown in normal placental development [15, 16]. While the causes of poor placental growth remain unclear, many of the genes that regulate the growth and differentiation of both placental and fetal are' imprinted.' Loss or alteration of these genetic imprints in cloned embryos may be the source of the abnormalities observed. Placental instability, contributing to hydroallantois and the presence of fewer and swollen placentomes, enlarged umbilical vessels, and edematous placental membranes is the result of fetal deaths in bovine organisms at later gestation periods [17, 18]. It is of consequence that there were no equivalent fetal deaths in goats [19 - 21]. Elevated growth weight and high neonatal mortality are normal in cloned calves. Neonatal health issues [22], a phenomenon referred to as "large offspring syndrome," or LOS, was described in 922 May Vet J Volume41, December 2000. As LOS also happens in calves resulting from embryos that are in vitro fertilized (IVF), it is not yet clear if the primary cause is the nuclear transfer itself. Syndromes of respiratory distress have also been cited in cloned calves [23, 24], and lambs [25, 26], which may signify inadequate development and function of the adrenal gland, low levels of fetal cortisol, and, thus, insufficient lung surfactant. To accelerate the maturation of the fetal lungs along with oxygen therapy, corticosteroid treatment of the recipient before delivery may improve survival rates [27, 28]. The introduction of pulmonary surfactants to cloned calves has been studied with varying results, meaning the primary concern could be pulmonary hypertension. In infected cattle, pneumonia is widely found, requiring prompt antimicrobial treatment. As evidenced by a sudden drop in lymphocyte and red cell counts found in a cloned cow, vulnerability to infection of some type is also a typical feature and may result from immune dysfunction [29, 30].

REPRODUCTIVE CLONING

The method of making a genetically similar clone of a whole organism is reproductive cloning. While clones may be created close to what occurs spontaneously with identical twins by separating an early embryo. Adult animal cloning is usually achieved by a procedure called somatic cells nuclear transfer (SCNT) [31 - 33]. The implantation of a cloned embryo into a true or an artificial uterus is required for reproductive cloning where the egg is then brought to maturity and develops into a fetus. *Via* the embryo splitting process, reproductive cloning experiments have been conducted for more than 40 years, in which a single early-stage two-cell embryo is manually divided into two separate cells and then develops as two identical embryos [34]. In the 1990s, following the birth of Dolly, who was produced through the SCNT method, reproductive cloning techniques underwent major changes. This method involves extracting the whole nucleus from an organism's somatic (body) cell, followed by transferring the nucleus into an egg cell that has shed its nucleus [35, 36]. The egg is activated with a slight electric current until the somatic nucleus is within the egg and starts to separate. A cloned embryo is developed, and effectively an embryo of an exact twin of the original organism is created. Since the 1990s, the SCNT process has undergone considerable refinement, and techniques have been improved to avoid damage to eggs during nuclear extraction and nuclear injection of somatic cells [26]. The use of polarized light to imagine the nucleus of an egg cell, for instance, enables the extraction of the nucleus from the egg, leading to a stable, viable egg and thereby increasing the SCNT success rate [37]. Cloning for reproductive, research, or medicinal reasons 10. Scientific and ethical problems would emerge from the use of somatic-cell nuclear transfer for the reproduction of human beings. Unpredictable genetic and epigenetic disorders have arisen in all the species of mammals cloned so far, including rodents, rabbits, pigs, cattle, and sheep, which have not only contributed to a high incidence of defects and fetal mortality but have also created health problems for most animals born alive, problems that vary from one clone to another. Global research academies and other scientific experts from around the world have therefore voiced resistance to the reproductive application of the procedure in humans [10, 38].

Is Reproductive Cloning the Same as Stem Cell Research?

It cannot be admitted that either stem cell research is related to reproductive cloning or fully not related to it, but depending on scientific evidence, it can be said that they are not directly related [39]. What makes this conflict is following the common initial step in both the aspects. This step is called Somatic Cells Nuclear Transfer (SCNT), or Nuclear Transplantation. Stem cell research depends on the unique feature of stem cell's ability to divide repeatedly, to give rise to both specialized cells and more stem cells and we cannot overlook the advantage of giving us the ability to induce types of them such as the embryonic stem cells (ES cells) that can be obtained from pre-implanted embryos to give different types of mature cells [39 - 42]. ES cells are well known for their pluripotency, as they can generate many different types of specialized cells [30, 43 - 45]. Stem cell research nowadays is based on the idea of creating or producing cell lines, tissues, or maybe organs but it is not confirmed to be able to produce a whole developed organism [39, 46 - 51]. The reason for the importance of stem cell research is its usage in therapeutic transplantation to help people who suffer from many types of disease disorders. Despite the difficulty of obtaining adult stem cells, from a research point of view, they can be useful in therapy even in different tissue other than the original tissue of the cell [39, 40, 52 - 55]. Depending on the current research work, it can be deduced that adults can replace the pre-implanted embryo or the blastocyst as a source of stem cells that can be used for a therapeutic wise but it is still uncertain [39, 56 - 60]. One of the critical things to be taken into consideration is how far the riskiness of the therapeutic potential of the transplantation with stem cell-derived cells, tissue, or organ. The risk factor, in this case, is the extent of the body's ability to receive the new organ or tissue. The advantage of adult stem cells over ES cells in transplantation is that adult stem cells can be derived from the patient and also from the same organ so, the probability of rejection from the body's immune system is low. The main potential drawback is that if there is any genetic origin of the disease or the disorder, the adult stem cells derived from the patient would have the same defect unless they are genetically modified before transplantation [61].

To minimize the risk of transplant rejection, somatic cell nuclear transfer is applied or it can be called research cloning, therapeutic cloning, or non-reproductive cloning as it serves as an alternative way to get therapeutic transplantation stem cells. The application of SCNT allows scientists to generate pluripotent ES cells that are identical on the genetic level to the cells of a transplant recipient [62, 63] so, these cells are predicted to avoid the drawback of adult stem cells. The only adult stem cells in use currently for therapy are stem cells in the blood-forming bone marrow and skin stem cells. Researches introduced information about the work of the cell-derived from adult stem cells as they contribute directly to the tissue function, but most of these studies were done on a mouse, not humans. There is a dark side to using either adult or embryonic stem cells other than body resistance. Human ES cells are found to need a feeder to grow, which is a layer of mouse cells, which in many cases can be carrying viruses. Also, the ES cells can give rise to various types of cells in the same culture. It can also form benign tumors if they were introduced to a mouse before its differentiation [43, 64, 65]. Despite all the previous information on their risk, studies showed that mouse ES cells may be promising for treating Parkinson's disease [66, 67], diabetes [49, 68], and spinal cord injury [69, 70] and it also have a higher sustainability rate in the media more than adult stem cells.

The direction of ES cells to its specialization and making use of these cells to produce an organ requires the perfection of the techniques [71, 72]. The sequence of steps to reach the phase of ES cells line production starts with nuclear transplantation by the transfer of a somatic cell nucleus from a patient cell into an enucleated egg, then the embryo is cultured in vitro till the blastocyst stage, then from the inner mass of the blastocyst, the pluripotent ES cell line is gotten, then this cell line is directed to specialization in a laboratory culture to specific cell type then to an organ. Although this procedure is used to avoid transplantation rejection, there are many possible drawbacks of applying it as the animal model experiments illustrated that the divergent mitochondrial proteins may cause minor transplantation antigens to be produced, but this can be overlapped if the donated egg is from the patient's mother or the patient herself [73 - 76]. Transplantation of cloned cells originates from the patient's donated cells may be inappropriate in the case of autoimmune diseases as they can be targeted by the ongoing destructive process. It is quite similar to the disorders that have a genetic origin, as it is required to apply genetic modification in the adult stem cell before transplantation if the donated cell is from the patient him/herself which is already a challenging mission in the stem cell line that is derived from the nuclear transplantation of a self-cell. So, it would be appropriate in these cases to use foreign originating cloned stem cells with the necessity of immunosuppression [77, 78].

There are two methods other than the nuclear transfer that can be applied to drive less rejection likelihood ES cells. The first is ES cell lines bank but it is rated as difficult to conceive [39]. Engineered ES cells are the second method that can deliver or eliminate specific types of cell-surface proteins, so as not to be attacked by the recipient's immune system. But it must be mentioned that neither first nor the second methods promise success for adult stem cells [79, 80].

The difference between nuclear transplantation and reproductive cloning in the point of ES cells preparation is that there is no need for a uterus as embryos are not transplanted. Despite this difference, there is a need in all cases to surround the ES cells by cells derived from a host embryo such as trophoblast cells of the blastocyst which already participate in providing the essential patterning cues or signals to the embryo for the orientation of the embryo's head and rump (anterior-posterior) axis, and also to be surrounded by primitive right endoderm. Mouse ES cells must be provided with host embryonic cells to get the path of development, regardless these cells do not remain viable throughout gestation (Richard Gardner, personal communication) or it is combined with trophoblast vesicles from another embryo [26, 81].

There are two methods other than the nuclear transfer that can be applied to drive less rejection likelihood ES cells. The first is ES cell lines bank, but it is rated to be difficult to conceive [39]. Engineered ES cells are the second method that can deliver or eliminate specific types of cell-surface proteins, so as not to be attacked by the recipient's immune system. But it must be mentioned that neither first nor the second methods promise success for adult stem cells [26, 82].

In humans and primates, ES cells have been reported to have the ability to give rise to trophoblast cells in culture [83 - 86]. This trophoblast may lack positional cues that are normally demanded of its development from an egg. Researches admitted that either human ES cells clumps or mouse ES cells clumps produced disorganized forms which are called embryoid bodies [86 - 88].

One of the benefits of ES cells obtained by nuclear transfer is their contribution to human developmental biology research as this science cannot be studied using mouse or monkey ES cells, and also it is not possible to use the normally fertilized blastocysts to obtain ES cells from it. Such studies are like the study of genetic diseases and the role of mutated genes by using ES cells derived from nuclear transplantation of a nucleus from humans with genetic diseases. The need for a donor egg is the only disadvantage of this kind of study. In comparison between ES cells and other types of cells for the previous approach, other cell types could not drive primary cell lines from human tissues. So, we can depend on ES cells obtained from nuclear transplantation to produce genetically diverse cell lines without the need to obtain them from frozen embryos or that are more than clinical need in IVF clinics. So, it would be important to apply nuclear transplantation using nuclei from individuals who have diseases associated with both simple and complex (multiple-gene) heritable genetic predilections to produce stem cells for research-wise, such as in individuals with "Lou Gehrig's disease" (amyotrophic lateral sclerosis, or ALS) as the disease behavior must be studied as the disease appears only on some of the patients [80, 89].

REPRODUCTIVE CLONING METHODS

Embryo Splitting

The first human embryo cloning method was used by Jerry Hall and colleagues in 1993 and is called embryo splitting [90, 91]. This was done by separating cells from early phase embryos using micromanipulation techniques, then injecting them into "empty" natural zona pellucida (or an artificial one), and then the normal development is achieved producing an identical organism. The identical twins or monozygotic twins are developed from portions from the same embryo, and similar to that is the embryo splitting as the embryo can be split into two or more portions. And many animal embryos are successfully cloned using this technique [92].

Tetraploid Complementation

Tetraploid complementation is meant by the production of an embryo by a combination of two mammalian embryos [93]. The term "tetra" is due to the diploid nature of the somatic cell, as there are two copies of each chromosome, so if there are two embryos with two copies of genetic material the sum is tetra. The procedure starts with the fusion of two cells in the two-cell stage embryo to get the tetraploid cell by introducing it to an electric current. After several divisions, all the daughter cells in the new embryo will be also tetraploid till the phase of the blastocyst. This blastocyst is implanted in the wall of a uterus, and then the tetraploid cells can form extraembryonic tissue such as the placenta. The embryonic tissues are not developed by the tetraploid cells but are developed by the injection of ES cells into tetraploid embryos especially in the 4-cell stage tetraploid embryo. So, the final form of the embryo will be as follows; the fetus tissues are derived from the diploid ES cells, and the extraembryonic tissues are constructed by the tetraploid cells [94, 95].

Traditional and Alternative Methods of Cloning

The traditional method of cloning is the conventional method used to accomplish cloning in mammals. It depends on enucleating a mature oocyte and putting a somatic nucleus instead of the original nucleus. The previously mentioned procedure is known as nuclear transfer. It is somehow easy to identify the matured mammalian oocyte chromosomes as the cytoplasm is transparent especially in human oocytes as well as laboratory animals like mice and rats in contrast with the cytoplasm of large domestic animals cells as there is a need to use fluorescence staining such as Hoechst 33342 and ultraviolet (UV) exposure with cytochalasin B treatment during enucleation as the cells are darker and less clear because of the high lipid content. UV affects cells and embryonic development harmfully, so another procedure is needed to be done [96]. In 2013, Iuso and colleagues tried to avoid the drawbacks of UV usage by creating a new enucleation method in a large animal model as the sheep which showed an elevated blastocyst rate in SCNT If compared with the traditional method, without the exposure of the oocyte to UV [97]. The UV riskiness is not a concern in the traditional method of nuclear transfer, but also the need for expensive equipment and highly skilled personnel. The problem of the expensive equipment is covered by applying handmade cloning (HMC), which was done for the first time successfully by Peura and colleagues, and then it was widely done on buffalo, sheep, horses, and pigs [98 - 103]. Also, an enucleation of zona-free has been successfully done using a sharp microblade, and then the enucleated oocytes and donor cells are attached using phytohemagglutinin. Direct injection or electrofusion of the somatic cell to the cytoplasm of the oocyte is how the reconstruction of the enucleated oocytes can be done. Direct injection is divided into two major methods. The first is the traditional way, which in the repeated aspiration and ejection with a micro-capillary is used to lyse the donor cell membrane. The second way is called the alternative way Piezo device is used to apply the Piezo-driven technique or the "Honolulu technique", which was first used to produce the first cloned mouse by Wakayama at the University of Hawaii [104, 105].

Somatic Cell Nuclear Transfer (SCNT)

Somatic cell nuclear transfer is a cloning method that depends on the stimulation of the egg to get a blastocyst. The first step is the removal of the chromosomes in the form of the nucleus from an egg creating an enucleated egg. Then the enucleated cell is implanted with a nucleus separated from a somatic cell from the same individual or an embryo or the cell can be obtained from cells culture, or from frozen tissue to get the targeted chromosomes. Stimulation of the cell must be done to stimulate the cell division to get a pre-implantation embryo or the blastocyst that will be transferred to an animal's uterus for further development hoping to give birth to a new animal. The new cloned animal will have the genetic material of one parent whose nucleus was donated in the initial steps [26, 106 - 108]. If the donor of the somatic nucleus is the same individual whose egg cell was used from the beginning, the cloned embryo will have the whole genetic material of this individual either its genetic material or the mitochondrial genes. If the egg donor differs from the nucleus donor, then there will be little genetic variation from the nucleus donor results from the mitochondrial genes in the egg cell [26, 109 - 111]. Talking about the number of resulting clones, there are no limits related to the nucleus donor, but it depends on the number of eggs available and the number of females available to carry the cloned embryos.

The Source of the Donor Cell

One of the main critical aspects in reproductive cloning is the source of the nucleus donor cell especially if there is an aim to increase nuclear reprogramming efficiency [112, 113]. The cumulus cells are known for their ease of retrieval, so they are commonly used as donor cells [47, 104, 114]. There are other kinds of cells proved to be successfully used in cloning such as tail-tip fibroblasts [115, 116], Sertoli cell [117, 118], primordial germ cells [119, 120], fetal fibroblasts [121, 122], natural killer T cells [123, 124] and embryonic stem cells [120, 125]. Embryo-derived cell lines, adult and fetal fibroblasts, and cumulus cells were used from large mammals [126, 127]. The nuclear transfer embryos which are developed from embryonic cells are reported to be better developed to the blastocyst stage than the embryos developed from other cell types, but disappointedly, the percentage of developed embryos is still low. Cloned embryos show a broad variety on the level of epigenetic alternation, such as methylation, DNA methylation, non-coding RNA transcripts expression, and histone acetylation [108].

Induced Pluripotent Cells (iPS). Generation of Embryonic Stem Cell-like Cells

Induced pluripotent stem cells, usually abbreviated as iPS cells or iPSCs, are a type of artificially generated pluripotent cells from a non-pluripotent cell, usually an adult or neonatal somatic cell, by causing certain genes to be "forced" to express. were first produced from mouse cells in 2006 and from human cells in 2007. This has been cited as a crucial development in stem cell science because, without the problematic use of embryos, researchers can produce pluripotent stem cells, which are significant in research and potentially have medicinal applications. As iPSCs are formed from the somatic cells of the patient, it was assumed that iPSC therapy would prevent any immunogenic responses; however, this idea has recently been questioned [55, 128]. IPS cells are similar to normal stem cells in many ways, such as ES cells, but the full degree of this resemblance is still being researched. For gene expression, DNA methylation, and differentiation ability, variations between iPS-and blastocyst-derived embryonic stem cells were identified. Furthermore, iPS cell reprogramming tends to undermine genomic integrity, adding de novo mutations and changes in the number of copies. The ethical and legal controversies about human reproductive cloning have entered the "post-SCNT era" following the creation of iPS cells.

Until then, the moral position of the human embryo was the ethical focus of the issue. For those who claim human development starts at the moment of creation, the death of an actual fetus, which was inevitable for SCNT, was ethically unreasonable. Tetraploid complementation and iPS methods will now be used to build new species from the somatic cells of current humans, turning them back to the embryonic stem cell process first. Besides, a direct derivation of human sperm cells using iPS technology has recently been identified [129, 130], Mammalian oocytes and efforts to make them are already in the way [131, 132]. The death of an actual embryo could also no longer be sufficient for human reproductive cloning. However, these advances have not ended the ethical debates about the issue. In the sense of ethical and legal controversy regarding human reproductive cloning, two conclusions can be taken from the definition of these laboratory methods, which have been incredibly advanced over the last decade. First, SCNT is not the only potential human cloning technique; there are many other current and upcoming techniques. Second, there are alternate methods of reproductive cloning of mammals that do not need any current embryos to be killed [24, 133].

ADVANTAGES AND DISADVANTAGES OF REPRODUCTIVE CLONING

Advantages of Reproductive Cloning [42, 134, 135]

Satisfy the Desire to have Children

Reproductive cloning importance is evident for infertile individuals, who now have the chance to have children, and in some cases, they can have more than one child, or also they may have twins.

To Treat Life-threatening Diseases

The production of matching tissues by cloning can be an appropriate solution for the treatment of particular health conditions.

Genetic Modification

The allowance to choose particular traits like sex or any physical attribute in the cloning product, or also the elimination of undesired or defective genes by the cloning of healthy cells.

Speed up the Recovery Process

By lowering the recovery and healing time by cloned cells. This can help many categories of athletes and workers.

Compensation for Lost or Defected Organs

Reproductive cloning applications will provide us with live organs that can be transplanted and reduce the number of people waiting for transplantation.

Contribution to Genetic Research

For more understanding, studying and exploration of the genetic composition of every person if needed, this would be done by reproductive cloning technology.

Disadvantages of Reproductive Cloning [14, 133]

Medical Malpractice Appearance

As scientists and doctors could have the power to control a person's genetic makeup or to change it. Unfortunately, this opens the door for many setbacks and mistakes.

Lack of Diversity

Cloning duplicates the presence of an individual's genetic material by copying it into another individual without any change or even undergoing the natural evolution.

A Faster Aging Process

As the used cell in many cases is originally an adult cell, that already has its timeline, this enhances the speeding up of the aging process.

Endangering People's Life

Most of the reproductive cloning done faced the She faced the fate of life-threatening events such as miscarriage or stillbirth. Also, it is thought among the reproductive cloning experts that there is no fully healthy clone due to the need of an individual to complete the process, who can be put in danger.

Interference with Nature

This method disrupts nature because procreation cannot be tampered with, religious and other critics of this technology argue. They believe that it will have a dominant influence on the human race to interfere with how nature works. But one of the claims against artificial cloning that people have is that by producing life in unnatural forms, people are attempting to play God.

Making Division Between People

Some individuals could not consider cloned human beings to be humans. Division and unrest could cause this. Many are worried that this will only lead to more problems in the population.

Problems with no Solutions

DNA is something that people are not allowed to mess around with. If the process of cloning is not successful, there is a high likelihood that the host's DNA will be altered or, worse, harmed. If that happens, a whole new set of problems may be generated that we do not have any knowledge of.

Abuse Toward Reproductive Cloning

Many opponents believe that individuals with vested interests will misuse cloning technology. It may be used by others for criminal or illegal activities. Many individuals are also frightened that if reproductive cloning becomes commonplace, unscrupulous individuals could unwillingly clone individuals.

Fears on Many Levels

For one thing, it could take more than a hundred pregnancies, due to the current state of research, before a single successful clone is achieved. A lot of research, on the other hand, suggests that most cloned babies are not stable, leading to genetic abnormalities, premature aging, and, worse, death. It may be tainted with certain diseases or maybe too frail, even though a cloned child may live longer. There are several explanations for endorsing human cloning or for opposing it. For one, unlike a naturally conceived human being, a good clone would not be as perfect. It is also possible to bring it up differently, too. But if cloning is going to be used for a positive cause, since it can help cure incurable diseases, it can help improve humanity.

LEGALITY AND ILLEGALITY OF REPRODUCTIVE CLONING

It seems that there is widespread opposition among bioethicists, governments, and societies in general to the possibility of human reproductive cloning. A 2001 Pew Research Center for the People and the Press survey conducted in the United States found that 86 percent of survey respondents were strongly opposed or opposed to human cloning and only 14 percent supported or strongly favored cloning [136]. Respondents who self-identified as religious were more likely to be opposed to cloning [136]. A more recent survey conducted in Malaysia found that 58.2 percent of the 1920 respondents believed that reproductive cloning was contrary to their religious precepts [137]; nearly half of the Christian respondents and nearly two-thirds of the Muslim respondents accepted or strongly agreed that their religious beliefs were opposed to reproductive cloning. Problems of assigning and enacting parental responsibilities would be compounded. It places undue social and parental expectations on the cloned infant, resulting in psychological damage, violates the right to self-determination and an open future for cloned children, and affronts human dignity because it absorbs money required for basic healthcare for people who lack proper health. The parents, physicians, and scientists involved are dehumanized when reproductive activity falls beyond the natural model. Also the protection gaps, the potential for birth defects, fetal and neonatal deaths capitalize on infertility's social construction as a female affliction. Effects are created rather than engendered in children. This leads to objectification and racism towards cloned children. Discrimination and the second-class status of children conceived through sexual reproduction is possible. It contributes to the development of a metaphysical, utilitarian worldview that focuses on utility and market value. Can contribute to new eugenics and reduces the genetic diversity essential to the species' survival. Replace the supremacy of God over man with the dominion of man over man [10].

Reproductive Cloning: Legal or Illegal

Reproductive cloning objections have centered on the potential for the dehumanization of the procedure by parents, physicians, and scientists, the possibility of physical and/or psychological damage to the child caused by the cloning process, the potential adverse effects on public perception of parental roles and the existence of the family, and the potential use of cloning as the basis for new eugenics.

The increased choices that could arise from the procedure for infertile couples, same-sex couples, and single individuals and the need to uphold the right of individuals to autonomy in the reproductive process have been stressed by advocates of reproductive cloning. Determinations relating to whether and to what degree cloning should be allowed or prohibited for reproductive purposes and who should be in charge of those determinations are themselves subject to debate. Writers have cast the problem in different ways as a dispute between the right of parents to reproductive rights and the right of clones to self-determination [138, 139], between the interests of health researchers and practitioners as opposed to members of religious communities; between the person and the state; and, more generally, as a struggle to control the meaning to be attributed to the problem [140, 141].

Others have asked why it should be allowed to dominate the debate by the intractable beliefs of certain religious groups as opposed to both reproductive and scientific cloning in order to forbid either use of cloning except among those who do not adhere to those faiths. In reviewing the landscape of reasons for and against reproductive cloning, one astute author noted that many contradictory context assumptions derived from the key philosophical reasons for and against human reproductive cloning, which means that people who disagree on the concept or importance of harm or the essence and role of human dignity and autonomy will not be able to settle on any policy's universal validity [142, 143].

Opponents of reproductive cloning note that safety issues currently exist that can't be resolved without further study. These questions emerge from the observation of the adverse effects on health that have occurred in connection with animal cloning efforts. The consequences were liver, brain, and heart abnormalities, premature aging, and immune system functioning issues [10, 144]. Many of the anti-human cloning claims are conjectural, postulating what the uncertain future could bring. Some bioethicists have suggested that, because of the way they were born, the children resulting from cloning would face great social and parental expectations [24, 145]. Havstad and also Jones indicate that cloned children would be required to "recapitulate the personalities and lives" of the people from whom they were cloned, thus breaching their right and self-determination to an open future [138].

Sandel, for once, warned: the issue is not that parents usurp the child's autonomy that it is not as though the child might otherwise choose for herself her gender, height, and eye color. The issue lies in the hubris of the parents who design, in their drive to master the birth mystery. Even if this hubris does not make parents tyrants to their children, the relationship between parent and child is disfigured. It deprives parents of the modesty and expanded human sympathies that can be cultivated by an openness to our progeny's unbidden otherness [146]. Woloschak has proposed in a somewhat similar statement that humans will only like to be cloned for egotistical purposes. Parents will now, however, undergo testing to check for suspected genetic defects [147]. It will logically follow this line of reasoning that any such screening should be rejected in the hope that any resulting flaw, no matter how serious, would promote the creation of greater transparency for parents. It seems that this approach smacks of the hubris against which Sandel warns: that parents should be relying on their offspring to improve their qualities. Anti-cloning claims focused on the likelihood of a future influenced by parental expectations, and those who remember that one should not talk about a future individual being harmed by a choice that hurts his or her existence because he or she would not exist but for that choice have been opposed [148 - 151].

In addition, some scholars have noted that, regardless of the nature of their birth, parents may place expectations on their children and all children face a future that may be limited due to the nature of their circumstances, *e.g.* parental limitations and economic condition [152]. Other human cloning critics have argued that cloned children are likely to be objectified and/or experience prejudice that would lead to emotional or psychological damage in either case. It was proposed in the first scenario that cloned children whose genetic material is used to treat a sick child may be treated as an entity or tool rather than a unique human being [153]. In the second case, the cloned child would have the status of an illegitimate being if human cloning were to be carried out in a culture that has made the procedure illegal, resulting in their diminished independence and possibly prejudice. These claims have been countered by the observation that cloned children might be better genetically endowed than those born by sexual reproduction [154], increasing the likelihood that "egg and sperm" children will be given second-class status rather than cloned children [155]. Several scholars have raised concerns about the possible deterioration of family relationships that could result from cloning. However, this implies that the concept of family is confined to a family consisting of a male, a female, and their genetic offspring; that this idealized arrangement is the only family configuration that can succeed in a society, that the concept of family is permanently static and unchanging; and that within each structured family [154], the members are necessarily mutually supportive and caring. Commentators concerned with reproductive cloning, as well as those opposed to the procedure, pointed to the existence of alternative ways to tackle infertility and the urge to have children, including dependency and adoption on sperm or egg donors or surrogates [139, 154, 156]. However, in general, these writers have not discussed the logistical, legal, and ethical problems associated with such decisions, including the later ability of a child born from egg or sperm donation to discover the identity of the donor parent who wanted to remain anonymous; the potential for disruption of family relationships due to the identity of the donor parent; infringement of the right of the donor parent to privacy due to the discovery of their identity; the validity of surrogacy and the enforceability under a particular legal jurisdiction of surrogacy contracts [157 - 160], and the legal and ethical problems that may occur due to injuries sustained during pregnancy by the surrogate [160], the death or divorce of the commissioning parents during the pregnancy period, the refusal of the surrogate to give up the infant, or the child's birth defects [158]. Many of the ethical concerns associated with the donation and surrogacy of sperm and eggs are comparable to those posed in response to reproductive cloning, including the potential for poor women to be abused [158, 159], and the degree to which informed consent is informed [158 - 160]. Adoption is also rife with possible legal and ethical problems, including the legitimacy of the consent of the adoptive parents to recognize or revoke the rights and privacy concerns of the birth parents, confidentiality, paternalism, conflicts of interest, deceit, and truth-telling [161]. At least one bioethicist has expressly rooted his objections in terms commonly known as religious to reproductive cloning. Paul Ramsey in 1970 argued against reproductive cloning, with a claim that it restores God's dominion as the "dominion of man over man, facilitates the quest for unending emancipation and determinism, and is likely to lead to violations of human life, reproduction, parenthood, and determinism. While a prohibition against reproductive cloning may, by avoiding both foreseeable and unforeseeable harm, optimize the ethical principle of non-maleficence, it contravenes the ethical principle of autonomy, which suggests that individuals should be free to make their own reproductive decisions as long as they are in a position to give informed consent. But the as-yet-unborn cloned child does not give consent in the form of reproductive cloning, and has no say in the process, precisely because they are still unborn. Opponents contend, similar to the claims raised against in vitro fertilization, that no risk to cloned children can be justified because they have not decided to take such risks and do not have the right to be born [162]. Human cloning proponents point to the many advantages that the procedure can offer: the ability of parents of the same sex or a single parent to have a genetically identical child, the ability of infertile couples to have a genetically related child, and a source of genetic material for the care of a sick child [148]. Others have countered such arguments by claiming that these people's interests must be balanced against those of other people and society in general [163].

Both within and across religious traditions, the reaction of faith groups to the idea of cloning humans has varied. For instance, the Southern Baptist Convention (2001) passed a resolution opposing both reproductive and research cloning based on safety issues, the destruction of human embryos in research, and the conversion of procreation into development. The United Methodist Church opposed cloning because of embryo waste and the potential for interference with the nature of the family, the abuse of women, the consequent decline in human diversity, the violation of privacy, And the opportunity for the corporate benefit or personal profit regulation. Individuals claiming to speak on behalf of evangelicals have claimed that cloning results in the development of children rather than their begetting, a stance originating in the Nicene Creed from Son's relationship with the parent [164]. The United Church of Christ Committee on Genetics has also made its opposition to cloning public [165]. An editorial that appeared in Christianity Today opposed cloning as it violates human dignity, which for Christians is based on the biblical understanding that each person is an individual whose ultimate creator is the God of all Creation [166]. In the sense of Christian belief, Messer, writing regarding human cloning, implies that humans are accountable in and for the universe and must therefore behave in a way that is compatible with the redemptive work and intentions of God. In opposing reproductive cloning, he argues that it is not penultimate, *i.e.,* it does not pave the way for the redemptive word of God, but rather is ultimate, in that it will act as a replacement for the redemptive work of God and represent an unjustified attempt to control others' identities and thus overstep the limits of legitimate human responsibility for the creation of God [167], unrestrained ambition and the desire to be like God, and the desire to control and master is allowed to dominate, unrestricted ambition and the desire to be like God, and the desire to control and master is allowed to dominate [168]. From a Christian viewpoint, another author expressed opposition to reproductive cloning, arguing that manufacturing human clones makes people in the image of man-not God and equates the process with hideous idolatry [169].

In Jewish law, Golinkin described sources that indicate that cloning would be appropriate, but others that say it would be forbidden [170]. He states, as an example, that from Adam's rib, God created Eve; perhaps, then, since man was created in the image of God, man could also create clones. Judaism, however, he states, teaches that in life there are three partners: Heaven, man, and woman. Because cloning does not include these three spouses and distinguishes birth from marriage, it will result in the family being dissolved [171]. Golinkin relies on the creation of Eve from Adam's rib by God to argue not only in favor of cloning but also against it, meaning that God can clone humans, but this is a limit that cannot be crossed by man. King Hezekiah's story in the Mishnah provides the basis for Golinkin's claim that it is important to restrict the creation of inventions that are too dangerous; King Hezekiah hid the healing book because it contained recipes that could be used to cure or to kill. Finally, in weighing the Jewish explanations for and against cloning, Golinkin concludes that the claims against cloning are more convincing. Similarly, like Golinkin, Cohen relies on Genesis' creation story as the basis for his study, but his interpretation varies from Golinkin's. He suggests that the term "God created in the beginning" may stand as a declarative phrase describing what God did, implying that God created the world from nothing and that man was created by the plan of God. Creation was thus complete, and the man was not allowed to tamper with it. Alternatively, if the expression "when God created the heavens and the earth" is read as a positive clause, creation will be seen as not total but more transformative. If God is the Creator and man in the image of God is created, then the position of man is as a co-creator. Cohen notes that, thus, cloning itself is morally neutral; the moral meaning depends on how it is used. Breitowitz echoes the observation of Cohen that mankind was created in the image of God and, as such, each person has the right to respect, that is, individuality, sanctity, and singularity. He mentions the divergent accounts of Adam's creation of God that occur in Genesis' first two chapters. Adam is created in the image of God in the first verse, indicating that he is born with the characteristics of autonomy and strength and has the capacity to make moral decisions. As such, in changing the earth, he is a joint collaborator with Divinity. It is said that in a state of imperfection, God purposefully created man; God invites us to be a partner in changing the world for the better. However, Adam is created from dust in the second chapter of Genesis; his mission becomes one of preservation and defense.

From these two separate accounts of Adam's existence, Breitowitz deduces that: In the exercise of power, knowledge, and influence, human beings must live in a constant state of tension and inconsistency between understanding their divinity and acknowledging the need to submit to that which is greater and all-knowing [155]. He also points to other concepts in Judaism that contribute to the permissibility of reproductive cloning such as that the Jewish culture prefers non-existence to existence because potential is imbued with existence; that while cloned persons may be born with particular propensities because of the nature of their birth, he or she would maintain the right to make independent decisions and would be considered a special, autonomous individual by Judaism; and cloning as an effort to attain immortality would be counter to the Jewish teaching that reminds followers to remember the day of death so that every moment counts. He concludes his examination with the observation that the notion of the individual's individuality, in two directions. In the image of God, Tzelem Elokim calls upon people to be imaginative, to be independent, and to use their creativity and intelligence to relieve human suffering. At the same time, Tzelem Elokim demands that the dignity and uniqueness of each person be respected. These two elements literally cut in opposite directions in the case of reproductive cloning: the pro-cloning stance stresses the moral imperative of using human intelligence for the betterment of humanity, and yet we must use it at the same time in a way that does not undercut the singularity of every human being. This is the great question society must contend with [155].

Several references originate from Islamic legal and ethical teachings. These include the Quran, believed to be a revelation from Allah; hadith, the Prophet Mohammed's traditions and sayings; ijma ', referring to consensus; qiyas, meaning analogical reasoning; istihsan, a legal approach that chooses the community's best interests over analogical reasoning and maslahah mursalah, referring to public interests [172 - 175]. About the authority of some of these sources, there is a hierarchical order; the degree to which some of these sources are relied on varies throughout the different traditions within Islam. In their rejection of cloning for reproductive purposes, Islamic claims concerning reproductive cloning tend to be unanimous. Objections are focused on the potential loss of relationships of kinship in contravention of Sūrah an-Nisā', the possible social harms that could arise from reproductive cloning, the possible use of eugenic cloning, contrary to the provisions of the Quran, the controversy between reproductive cloning and the belief that Allah is the only creator, and the resulting restrictions on diversity [172].

Sekaleshfar has, however, countered these claims by noting that there is no text in Islam banning the development of fatherless children and that the introduction of adequate regulations and supervision would mitigate or prevent the likelihood of kinship-related uncertainty. Although not approving of reproductive cloning, Moeinifar and Ardebeli accept its possible advantage in circumstances in which couples are unable to conceive and stress the need to preserve the rights of the cloned kid [176]. In comparison, Dr. Nasr Farid Wasil, the former Grand Mufti of Egypt, has equated human cloning with a forbidden "satanic act" and suggested that it would lead to a rise in crime due to changes in personal identity [175]. Others have cloning for adultery analogized to [175].

The two key schools of Buddhism, the Theravada School, which predominates in Southeast Asian countries, and the Mahayana School, which is predominant in Central and East Asia, are very different in teaching [177]. Buddhist ethics is based on the concepts of non-harm (ahimsā), which apply to the motives of the actor; the effects or outcomes of the actions for the actor, as the wholesomeness or absence of them, may have consequences for the karma of the person in their next rebirth; the context of the actions; and the degree to which the actions encourage self-cultivation [177, 178]. Decision-making includes the original Buddhist scriptures to be reviewed, guidelines to be derived in a manner compatible with those texts, the opinions of esteemed teachers to be taken into account, and personal judgment and discretion [179].

Some Buddhist scholars, at least, say that reproductive cloning is not inconsistent with Buddhist ethics. Keown (2005) has indicated that cloning is seen as an alternate method for creating new life in Buddhism. Several scholars have stressed that whether reproductive cloning will be ethical is the purpose of the parties that is determinative [180]. Ratankul (1998) and Nimanong (n.d.) suggest that reproductive cloning would not be immoral if it were to satisfy a couple's desire for parent children and would not cause life to be lost or cause pain or suffering [181], Whereas Promta indicates that if it is to bring a new being into being, the act is optimistic [177].

Research into and/or financing of reproductive cloning in any number of countries has not been well discussed, whether or not the opinions of these denominations and theologians have affected the formulation of government decisions relating to prohibitions or moratoriums against them. However, it seems in the political and legislative realms, the religious values of key players have strongly shaped the direction that communities have chosen [24].

As an example, former President George Bush assembled a Bioethics Council of the President to advise the President on bioethical issues that may arise as a result of advances in biomedical science and technology. Their emphasis was to include: the fundamental investigation into the human and moral meaning of biomedical and behavioral science and technology innovations, the analysis of particular ethical and political issues relevant to these developments, the provision of a platform for a national debate on bioethical issues, the facilitation of a greater understanding of bioethical issues, the exploration of possibilities for the international collaboration on bioethical issues [182]. The President had previously expressed his conviction that the human embryo is sacred [183], a belief possibly centered on his beliefs of God or a Higher Power's creation and life. Leon Kass, the commission's named representative, had previously expressed his opposition to the in the vitro fertilization process, arguing that it would substitute procreation with development and cloning and other genetic innovations, characterizing them as repugnant [184, 185]. It was stated that both the vetting process for the commission appointment and the retention of the commission were based on aligning the ideological perspectives of individuals with those of George Bush [186]. An appeal to the expected negative social effects of enabling both reproductive and therapeutic cloning was included in the final report of the Council, arguing that it would minimize the respect toward human life [182]. It was argued that this view, adopted by not quite a majority of the members of the Council, represented themes that were religiously informed, and pervasive in conservative Christian biomedical writings [187]. No states within the United States expressly allowed reproductive cloning on 5 May 2014 [188]. Seven states banned human cloning for any reason by 2018, and two provided that healing is not forced against their conscience to engage in human cloning practice [188, 189]. In the international context, reproductive cloning has been met with comparable rejection.

By July 2004, 23 countries had introduced legislation expressly banning reproductive cloning and seven others had interpreted existing legislation prohibiting reproductive cloning activities [190]. Legislation that expressly forbids either the use of SCNT or the production of a genetically identical person has been enforced by a variety of jurisdictions outside the United States. As of 2010, Australia, Canada, China, Germany, Israel, Japan, Singapore, South Korea, and Sweden were among those [191].

Opposition to the idea of reproductive cloning has also been suggested by foreign bodies and organizations. In the various writings of the Universal Declaration on the Human Genome and Human Rights, promulgated in 1997 by the United Nations Educational, Scientific and Cultural Organization (UNESCO), it is provided that the underlying unity among all members of the human family, as well as the appreciation of their intrinsic integrity and diversity, underlies the human genome. Everyone has the right, regardless of their genetic characteristics, to respect their dignity and their freedom. Dignity makes it imperative not to reduce people to their genetic characteristics and to respect their individuality and diversity. Activities that are contrary to human dignities, such as human reproductive cloning, are not permitted. By July 2004, 23 countries had introduced legislation expressly banning reproductive cloning and seven others had interpreted existing legislation prohibiting reproductive cloning activities [190]. Legislation that expressly forbids either the use of SCNT or the production of a genetically identical person has been enforced by a variety of jurisdictions outside the United States. As of 2010, Australia, Canada, China, Germany, Israel, Japan, Singapore, South Korea, and Sweden were among those [191].

CONCLUSION

Reproductive cloning has made a revolution in the field of cloning, starting from the embryo splitting method which is considered the primitive method of reproductive cloning, passing through the tetraploid complementation method down to the somatic cell nuclear transfer (SCNT). This does not necessarily mean that there is only one method used for its preference so far, but rather that all methods are used based on the extent of their applicability and need. There has been a lot of research and scientific opinions about the benefit that accrues from its application. On the opposite side, some have called for not adopting it because of its potential harms and it is believed by some. Despite all these thoughts, there is still a clear argument between the scientists about its legality and illegality.

CONSENT FOR PUBLICATION

Not applicable.

k

CONFLICT OF INTEREST

The authors declare that there is no conflict of interest.

ACKNOWLEDGEMENT

Declared none.

REFERENCES

[1]   Lynch M. The Origins of Genome Architecture. Sunderland, MA: Sinauer 2007.

[2]   Jobling MA, Hurles ME, Tyler-Smith C. Human Evolutionary Genetics. Garland Science, NY: Origins, Peoples & Disease 2004.

[3]   Jabr F. Will cloning ever save endangered animals? Sci Am 2015; 2013 Available at: www.scientificamerican.com/article/cloning-endangered-animals/

[4]   Shiels PG, Kind AJ, Campbell KHS, *et al.* Analysis of telomere lengths in cloned sheep. Nature 1999; 399(6734): 316-7.
[http://dx.doi.org/10.1038/20580] [PMID: 10360570]

[5]   Highfield R. 2007 Dolly creator Prof. Ian Wilmut shuns cloning. Daily Telegraph. 2015.
www.telegraph.co.uk/news/science/science-news/3314696/Dolly-creator-Prof-Ian-Wilmut-shuns-cloning.html

[6]   Higham T, Douka K, Wood R, *et al.* The timing and spatiotemporal patterning of Neanderthal disappearance. Nature 2014; 512(7514): 306-9.
[http://dx.doi.org/10.1038/nature13621] [PMID: 25143113]

[7]   Pääbo S. Neanderthal Man: Search of Lost Genomes. New York: Basic Books 2014.

[8]   Boyd R, Richerson PJ. Culture and the Evolutionary Process. Chicago: Univ of Chicago Press 1985.

[9]   Ayala FJ. Darwin's Gift to Science and Religion. Washington, DC: Joseph Henry Press 2007.

[10]   Ayala FJ. Cloning humans? Biological, ethical, and social considerations. Proc Natl Acad Sci USA 2015; 112(29): 8879-86.
[http://dx.doi.org/10.1073/pnas.1501798112] [PMID: 26195738]

[11]   Hwang WS, Ryu YJ, Park JH, *et al.* Evidence of a pluripotent human embryonic stem cell line derived from a cloned blastocyst. Science 2004; 303(5664): 1669-74.
[http://dx.doi.org/10.1126/science.1094515] [PMID: 14963337]

[12]   Liu Z, Cai Y, Wang Y, *et al.* Cloning of Macaque Monkeys by Somatic Cell Nuclear Transfer. Cell 2018; 172(4): 881-887.e7.
[http://dx.doi.org/10.1016/j.cell.2018.01.020] [PMID: 29395327]

[13]   Wells DN. Animal cloning: current progress, challenges and future prospects. Rev Bras Reprod Anim 1999; 23: 86-97.

[14]   Ibtisham F, Niu Y, Wang Z, *et al.* Animal cloning drawbacks an-overview. J Dairy Vet Anim Res 2016; 3(4): 139-43.

[15]   Wrenzycki C, Wells D, Herrmann D, *et al.* Nuclear transfer protocol affects messenger RNA expression patterns in cloned bovine blastocysts. Biol Reprod 2001; 65(1): 309-17.
[http://dx.doi.org/10.1095/biolreprod65.1.309] [PMID: 11420254]

[16]   Drăgușin RC, Șorop-Florea M, Pătru CL, *et al.* Abnormalities of the Placenta. Congenital Anomalies - From the Embryo to the Neonate 2018.

[17]   Sharma D, Shastri S, Sharma P. Intrauterine Growth Restriction: Antenatal and Postnatal Aspects. Clin Med Insights Pediatr 2016; 10: 67-83.
[http://dx.doi.org/10.4137/CMPed.S40070] [PMID: 27441006]

[18]   Hanna CW. Placental imprinting: Emerging mechanisms and functions. PLoS Genet 2020; 16(4): e1008709.
[http://dx.doi.org/10.1371/journal.pgen.1008709] [PMID: 32324732]

[19]   Baguisi A, Behboodi E, Melican DT, *et al.* Production of goats by somatic cell nuclear transfer. Nat Biotechnol 1999; 17(5): 456-61.
[http://dx.doi.org/10.1038/8632] [PMID: 10331804]

[20]   Mitalipov S, Wolf D. Totipotency, pluripotency and nuclear reprogramming. In: Martin U, Ed. Engineering of Stem Cells Advances in Biochemical Engineering / Biotechnology. Berlin, Heidelberg: Springer 2009; Vol. 114.

[21]   Hajian M, Jafarpour F, Aghamiri SM, Rouhollahi Varnosfaderani S, Nasr Esfahani MH. Effects of ovary storage temperature and embryo vitrification on somatic cell nuclear transfer outcomes in goats. Reprod Fertil Dev 2019.
[PMID: 31816272]

[22]   Lopes JS, Alcázar-Triviño E, Soriano-Úbeda C, *et al.* Reproductive Outcomes and Endocrine Profile in Artificially Inseminated *versus* Embryo Transferred Cows. Animals (Basel) 2020; 10(8): 1359.
[http://dx.doi.org/10.3390/ani10081359] [PMID: 32781545]

[23]   Smith LC, Bordignon V, Babkine M, Fecteau G, Keefer C. Benefits and problems with cloning animals. Can Vet J 2000; 41(12): 919-24.
[PMID: 11143925]

[24]   Nabavizadeh SL, Mehrabani D, Vahedi Z, Manafi F. Cloning: A Review on Bioethics, Legal, Jurisprudence and Regenerative Issues in Iran. World J Plast Surg 2016; 5(3): 213-25.
[PMID: 27853684]

[25]   Keefer CL. Artificial cloning of domestic animals. Proc Natl Acad Sci USA 2015; 112(29): 8874-8.
[http://dx.doi.org/10.1073/pnas.1501718112] [PMID: 26195770]

[26]   Gouveia C, Huyser C, Egli D, Pepper MS. Lessons Learned from Somatic Cell Nuclear Transfer. Int J Mol Sci 2020; 21(7): 2314.
[http://dx.doi.org/10.3390/ijms21072314] [PMID: 32230814]

[27]   Tecirlioglu RT, Cooney MA, Lewis IM, *et al.* Comparison of two approaches to nuclear transfer in the bovine: hand-made cloning with modifications and the conventional nuclear transfer technique. Reprod Fertil Dev 2005; 17(5): 573-85.
[http://dx.doi.org/10.1071/RD04122] [PMID: 15907283]

[28]   Cortez JV, Vajta G, Valderrama NM, Portocarrero GS, Quintana JM. High Pregnancy and Calving Rates with a Limited Number of Transferred Handmade Cloned Bovine Embryos. Cell Reprogram 2018; 20(1): 4-8.
[http://dx.doi.org/10.1089/cell.2017.0024] [PMID: 29412737]

[29]   Hill JR. Incidence of abnormal offspring from cloning and other assisted reproductive technologies. Annu Rev Anim Biosci 2014; 2(1): 307-21.
[http://dx.doi.org/10.1146/annurev-animal-022513-114109] [PMID: 25384145]

[30]   Liu L. Nuclear Transfer and Cloning. In: Larson M, Ed. Transgenic Mouse Methods in Molecular Biology. New York, NY: Humana 2020; Vol. 2066.
[http://dx.doi.org/10.1007/978-1-4939-9837-1_10]

[31]   Tsai MC, Takeuchi T, Bedford JM, Reis MM, Rosenwaks Z, Palermo GD. Alternative sources of gametes: reality or science fiction? Hum Reprod 2000; 15(5): 988-98.
[http://dx.doi.org/10.1093/humrep/15.5.988] [PMID: 10783340]

[32]   Vajta G, Gjerris M. Science and technology of farm animal cloning: state of the art. Anim Reprod Sci 2006; 92(3-4): 211-30.
[http://dx.doi.org/10.1016/j.anireprosci.2005.12.001] [PMID: 16406426]

[33]   Simões R, Rodrigues Santos A Jr. Factors and molecules that could impact cell differentiation in the embryo generated by nuclear transfer. Organogenesis 2017; 13(4): 156-78.
[http://dx.doi.org/10.1080/15476278.2017.1389367] [PMID: 29020571]

[34]   Clapton G. Against All Odds? Birth Fathers and Enduring Thoughts of the Child Lost to Adoption. Genealogy 2019; 3(2): 13.
[http://dx.doi.org/10.3390/genealogy3020013]

[35]   Zhao M-T, Chen H, Liu Q, *et al.* Molecular and functional resemblance of differentiated cells derived from isogenic human iPSCs and SCNT-derived ESCs. Proc Natl Acad Sci USA 2017; 114(52): E11111-20.
[http://dx.doi.org/10.1073/pnas.1708991114] [PMID: 29203658]

[36]   Eguizabal C, Aran B, Chuva de Sousa Lopes SM, *et al.* Two decades of embryonic stem cells: a historical overview. Hum Reprod Open 2019; 2019(1): hoy024.
[http://dx.doi.org/10.1093/hropen/hoy024] [PMID: 30895264]

[37]   Brochard V, Beaujean N. Somatic Reprograming by Nuclear Transfer. In: Ancelin K, Borensztein M, Eds. Epigenetic Reprogramming During Mouse Embryogenesis Methods in Molecular Biology. New York, NY: Humana 2021; Vol. 2214.
[http://dx.doi.org/10.1007/978-1-0716-0958-3_8]

[38]   Nasrullah Iqbal. Ethical issues of human cloning. J Med Sci 2021; 40: 103-6.

[39]   Washington DC. Committee on Stem Cells and The Future of Regenerative Medicine, Board on Life Sciences and Board on Neuroscience and Behavioral Health. Stem Cells and the Future of Regenerative Medicine. Report of the National Academy of Sciences and the Institute of Medicine. 2001.

[40]   Baum CM, Weissman IL, Tsukamoto AS, Buckle AM, Peault B. Isolation of a candidate human hematopoietic stem-cell population. Proc Natl Acad Sci USA 1992; 89(7): 2804-8.
[http://dx.doi.org/10.1073/pnas.89.7.2804] [PMID: 1372992]

[41]   Rietze RL, Valcanis H, Brooker GF, Thomas T, Voss AK, Bartlett PF. Purification of a pluripotent neural stem cell from the adult mouse brain. Nature 2001; 412(6848): 736-9.
[http://dx.doi.org/10.1038/35089085] [PMID: 11507641]

[42]   Langlois A. The global governance of human cloning: the case of UNESCO. Palgrave Commun 2017; 3(1): 17019.
[http://dx.doi.org/10.1057/palcomms.2017.19] [PMID: 28382210]

[43]   Thomson JA, Itskovitz-Eldor J, Shapiro SS, *et al.* Embryonic stem cell lines derived from human blastocysts. Science 1998; 282(5391): 1145-7.
[http://dx.doi.org/10.1126/science.282.5391.1145] [PMID: 9804556]

[44]   Lee S-H, Lumelsky N, Studer L, Auerbach JM, McKay RD. Efficient generation of midbrain and hindbrain neurons from mouse embryonic stem cells. Nat Biotechnol 2000; 18(6): 675-9.
[http://dx.doi.org/10.1038/76536] [PMID: 10835609]

[45]   Weissman IL, Baltimore D. Disappearing stem cells, disappearing science. Science 2001; 292(5517): 601-1.
[http://dx.doi.org/10.1126/science.292.5517.601] [PMID: 11330301]

[46]   Gussoni E, Soneoka Y, Strickland CD, *et al.* Dystrophin expression in the mdx mouse restored by stem cell transplantation. Nature 1999; 401(6751): 390-4.
[http://dx.doi.org/10.1038/43919] [PMID: 10517639]

[47]   Lee MS, Liu CH, Lee TH, *et al.* Association of creatin kinase B and peroxiredoxin 2 expression with

age and embryo quality in cumulus cells. J Assist Reprod Genet 2010; 27(11): 629-39.
[http://dx.doi.org/10.1007/s10815-010-9459-7] [PMID: 20721618]

[48]   Wakayama T, Tabar V, Rodriguez I, Perry AC, Studer L, Mombaerts P. Differentiation of embryonic stem cell lines generated from adult somatic cells by nuclear transfer. Science 2001; 292(5517): 740-3.
[http://dx.doi.org/10.1126/science.1059399] [PMID: 11326103]

[49]   Lumelsky N, Blondel O, Laeng P, Velasco I, Ravin R, McKay R. Differentiation of embryonic stem cells to insulin-secreting structures similar to pancreatic islets. Science 2001; 292(5520): 1389-94.
[http://dx.doi.org/10.1126/science.1058866] [PMID: 11326082]

[50]   Shamblott MJ, Axelman J, Littlefield JW, *et al.* Human embryonic germ cell derivatives express a broad range of developmentally distinct markers and proliferate extensively in vitro. Proc Natl Acad Sci USA 2001; 98(1): 113-8.
[http://dx.doi.org/10.1073/pnas.98.1.113] [PMID: 11134532]

[51]   Kim J, Koo BK, Knoblich JA. Human organoids: model systems for human biology and medicine. Nat Rev Mol Cell Biol 2020; 21(10): 571-84.
[http://dx.doi.org/10.1038/s41580-020-0259-3] [PMID: 32636524]

[52]   Ferrari G, Cusella-De Angelis G, Coletta M, *et al.* Muscle regeneration by bone marrow-derived myogenic progenitors. Science 1998; 279(5356): 1528-30.
[http://dx.doi.org/10.1126/science.279.5356.1528] [PMID: 9488650]

[53]   Petersen BE, Bowen WC, Patrene KD, *et al.* Bone marrow as a potential source of hepatic oval cells. Science 1999; 284(5417): 1168-70.
[http://dx.doi.org/10.1126/science.284.5417.1168] [PMID: 10325227]

[54]   Negrin RS, Atkinson K, Leemhuis T, *et al.* Transplantation of highly purified CD34+Thy-1+ hematopoietic stem cells in patients with metastatic breast cancer. Biol Blood Marrow Transplant 2000; 6(3): 262-71.
[http://dx.doi.org/10.1016/S1083-8791(00)70008-5] [PMID: 10871151]

[55]   Yamanaka S. Pluripotent Stem Cell-Based Cell Therapy-Promise and Challenges. Cell Stem Cell 2020; 27(4): 523-31.
[http://dx.doi.org/10.1016/j.stem.2020.09.014] [PMID: 33007237]

[56]   Lanza RP, Caplan AL, Silver LM, Cibelli JB, West MD, Green RM. The ethical validity of using nuclear transfer in human transplantation. JAMA 2000; 284(24): 3175-9.
[http://dx.doi.org/10.1001/jama.284.24.3175] [PMID: 11135784]

[57]   Rodriguez-Polo I, Behr R. Non-human primate pluripotent stem cells for the preclinical testing of regenerative therapies. Neural Regen Res 2022; 17(9): 1867-74.
[http://dx.doi.org/10.4103/1673-5374.335689] [PMID: 35142660]

[58]   Winston R. Embryonic stem cell research. The case for..... Nat Med 2001; 7(4): 396-7.
[http://dx.doi.org/10.1038/86442] [PMID: 11283652]

[59]   Vogel G. Stem cell policy. Can adult stem cells suffice? Science 2001; 292(5523): 1820-2.
[http://dx.doi.org/10.1126/science.292.5523.1820] [PMID: 11397925]

[60]   Nielsen MEJ, Kongsholm NCH, Schovsbo J. Property and human genetic information. J Community Genet 2019; 10(1): 95-107.
[http://dx.doi.org/10.1007/s12687-018-0366-4] [PMID: 29713893]

[61]   Hansen M, von Lindern M, van den Akker E, Varga E. Human-induced pluripotent stem cell-derived blood products: state of the art and future directions. FEBS Lett 2019; 593(23): 3288-303.
[http://dx.doi.org/10.1002/1873-3468.13599] [PMID: 31520530]

[62]   Odorico JS, Kaufman DS, Thomson JA. Multilineage differentiation from human embryonic stem cell lines. Stem Cells 2001; 19(3): 193-204.
[http://dx.doi.org/10.1634/stemcells.19-3-193] [PMID: 11359944]

[63] Freel BA, Sheets JN, Francis KR. Corrigendum to "iPSC modeling of rare pediatric disorders" [J. Neurosci. Methods 332 (15) (2020) 108533]. J Neurosci Methods 2020; 339(15): 108739.
[http://dx.doi.org/10.1016/j.jneumeth.2020.108739] [PMID: 32381271]

[64] Studer L, Tabar V, McKay RD. Transplantation of expanded mesencephalic precursors leads to recovery in parkinsonian rats. Nat Neurosci 1998; 1(4): 290-5.
[http://dx.doi.org/10.1038/1105] [PMID: 10195162]

[65] Gorecka J, Kostiuk V, Fereydooni A, *et al.* The potential and limitations of induced pluripotent stem cells to achieve wound healing. Stem Cell Res Ther 2019; 10(1): 87.
[http://dx.doi.org/10.1186/s13287-019-1185-1] [PMID: 30867069]

[66] McDonald JW, Liu XZ, Qu Y, *et al.* Transplanted embryonic stem cells survive, differentiate and promote recovery in injured rat spinal cord. Nat Med 1999; 5(12): 1410-2.
[http://dx.doi.org/10.1038/70986] [PMID: 10581084]

[67] Jalali MS, Sarkaki A, Farbood Y, *et al.* Transplanted Wharton's jelly mesenchymal stem cells improve memory and brain hippocampal electrophysiology in rat model of Parkinson's disease. J Chem Neuroanat 2020; 110: 101865.
[http://dx.doi.org/10.1016/j.jchemneu.2020.101865] [PMID: 32991967]

[68] Godfrey KJ, Mathew B, Bulman JC, Shah O, Clement S, Gallicano GI. Stem cell-based treatments for Type 1 diabetes mellitus: bone marrow, embryonic, hepatic, pancreatic and induced pluripotent stem cells. Diabet Med 2012; 29(1): 14-23.
[http://dx.doi.org/10.1111/j.1464-5491.2011.03433.x] [PMID: 21883442]

[69] Lanza RP, Cibelli JB, West MD. Prospects for the use of nuclear transfer in human transplantation. Nat Biotechnol 1999; 17(12): 1171-4.
[http://dx.doi.org/10.1038/70709] [PMID: 10585712]

[70] Krabbe C, Zimmer J, Meyer M. Neural transdifferentiation of mesenchymal stem cells--a critical review. Acta Pathol Microbiol Scand Suppl 2005; 113(11-12): 831-44.
[http://dx.doi.org/10.1111/j.1600-0463.2005.apm_3061.x] [PMID: 16480453]

[71] Simpson E. Minor transplantation antigens: animal models for human host-*versus*-graft, graft-*versus*-host, and graft-*versus*-leukemia reactions. Transplantation 1998; 65(5): 611-6.
[http://dx.doi.org/10.1097/00007890-199803150-00002] [PMID: 9521192]

[72] Munsie MJ, Michalska AE, O'Brien CM, Trounson AO, Pera MF, Mountford PS. Isolation of pluripotent embryonic stem cells from reprogrammed adult mouse somatic cell nuclei. Curr Biol 2000; 10(16): 989-92.
[http://dx.doi.org/10.1016/S0960-9822(00)00648-5] [PMID: 10985386]

[73] Simpson E, Roopenian D. Minor histocompatibility antigens. Curr Opin Immunol 1997; 9(5): 655-61.
[http://dx.doi.org/10.1016/S0952-7915(97)80045-3] [PMID: 9368774]

[74] Bhuyan PK, Young LL, Lindahl KF, Butcher GW. Identification of the rat maternally transmitted minor histocompatibility antigen. J Immunol 1997; 158(8): 3753-60.
[PMID: 9103440]

[75] Amano T, Kato Y, Tsunoda Y. Comparison of heat-treated and tetraploid blastocysts for the production of completely ES-cell-derived mice. Zygote 2001; 9(2): 153-7.
[http://dx.doi.org/10.1017/S0967199401001162] [PMID: 11358322]

[76] Chung YG, Seay M, Elsworth JD, Redmond DE. Generation of pluripotent stem cells using somatic cell nuclear transfer and induced pluripotent somatic cells from African Green Monkeys. Stem Cells Dev 2020; 29(19): 1294-307.
[http://dx.doi.org/10.1089/scd.2020.0059] [PMID: 32715987]

[77] Rosa SB, Voltarelli JC, Chies JAB, Pranke P. The use of stem cells for the treatment of autoimmune diseases. Braz J Med Biol Res 2007; 40(12): 1579-97.
[http://dx.doi.org/10.1590/S0100-879X2006005000166] [PMID: 17713674]

[78]   Műzes G, Sipos F. Issues and opportunities of stem cell therapy in autoimmune diseases. World J Stem Cells 2019; 11(4): 212-21.
[http://dx.doi.org/10.4252/wjsc.v11.i4.212] [PMID: 31110602]

[79]   Byrne JA, Pedersen DA, Clepper LL, *et al.* Producing primate embryonic stem cells by somatic cell nuclear transfer. Nature 2007; 450(7169): 497-502.
[http://dx.doi.org/10.1038/nature06357] [PMID: 18004281]

[80]   Zhang ZP, Zhang JT, Huang SC, He XY, Deng LX. Double sperm cloning (DSC) is a promising strategy in mammalian genetic engineering and stem cell research. Stem Cell Res Ther 2020; 11(1): 388.
[http://dx.doi.org/10.1186/s13287-020-01907-0] [PMID: 32894201]

[81]   Thomson JA, Marshall VS. Primate embryonic stem cells. Curr Top Dev Biol 1998; 38: 133-65.
[http://dx.doi.org/10.1016/S0070-2153(08)60246-X] [PMID: 9399078]

[82]   Hwang W S, Lee B C, Lee C K, Kang S K. Cloned Human Embryonic Stem Cells for Tissue Repair and Transplantation. Stem Cell Reviews, 1(2), 099–110.
[http://dx.doi.org/10.1385/SCR:1:2:099]

[83]   Thomson JA, Kalishman J, Golos TG, *et al.* Isolation of a primate embryonic stem cell line. Proc Natl Acad Sci USA 1995; 92(17): 7844-8.
[http://dx.doi.org/10.1073/pnas.92.17.7844] [PMID: 7544005]

[84]   Itskovitz-Eldor J, Schuldiner M, Karsenti D, *et al.* Differentiation of human embryonic stem cells into embryoid bodies compromising the three embryonic germ layers. Mol Med 2000; 6(2): 88-95.
[http://dx.doi.org/10.1007/BF03401776] [PMID: 10859025]

[85]   Douglas GC, VandeVoort CA, Kumar P, Chang TC, Golos TG. Trophoblast stem cells: models for investigating trophectoderm differentiation and placental development. Endocr Rev 2009; 30(3): 228-40.
[http://dx.doi.org/10.1210/er.2009-0001] [PMID: 19299251]

[86]   Cinkornpumin JK, Kwon SY, Guo Y, *et al.* Naive Human Embryonic Stem Cells Can Give Rise to Cells with a Trophoblast-like Transcriptome and Methylome. Stem Cell Reports 2020; 15(1): 198-213.
[http://dx.doi.org/10.1016/j.stemcr.2020.06.003] [PMID: 32619492]

[87]   Rechitsky S, Strom C, Verlinsky O, *et al.* Accuracy of preimplantation diagnosis of single-gene disorders by polar body analysis of oocytes. J Assist Reprod Genet 1999; 16(4): 192-8.
[http://dx.doi.org/10.1023/A:1020312823155] [PMID: 10224562]

[88]   Bratt-Leal AM, Carpenedo RL, McDevitt TC. Engineering the embryoid body microenvironment to direct embryonic stem cell differentiation. Biotechnol Prog 2009; 25(1): 43-51.
[http://dx.doi.org/10.1002/btpr.139] [PMID: 19198003]

[89]   Sukoyan MA, Kerkis AY, Mello MRB, Kerkis IE, Visintin JA, Pereira LV. Establishment of new murine embryonic stem cell lines for the generation of mouse models of human genetic diseases. Braz J Med Biol Res 2002; 35(5): 535-42.
[http://dx.doi.org/10.1590/S0100-879X2002000500004] [PMID: 12011937]

[90]   Kolberg WC. Quick and Easy Optimal Approach Paths for Nonlinear Natural Resource Models. Am J Agric Econ 1993; 75(3): 685-95.
[http://dx.doi.org/10.2307/1243575]

[91]   Noli L, Ogilvie C, Khalaf Y, Ilic D. Potential of human twin embryos generated by embryo splitting in assisted reproduction and research. Hum Reprod Update 2016.
[http://dx.doi.org/10.1093/humupd/dmw041] [PMID: 27852683]

[92]   Omidi M, Khalili MA, Halvaei I, Montazeri F, Kalantar SM. Quality of blastocysts created by embryo splitting: a time-lapse monitoring and chromosomal aneuploidy study. Cell J 2020; 22(3): 367-74.
[PMID: 31863663]

[93] Kaufman MH, Webb S. Postimplantation development of tetraploid mouse embryos produced by electrofusion. 110 (4): 1121–1132.

[94] Hu J, Wang J. From embryonic stem cells to induced pluripotent stem cells-Ready for clinical therapy? Clin Transplant 2019; 33(6): e13573.
[http://dx.doi.org/10.1111/ctr.13573] [PMID: 31013374]

[95] Saragusty J, Ajmone-Marsan P, Sampino S, Modlinski JA. Reproductive biotechnology and critically endangered species: Merging in vitro gametogenesis with inner cell mass transfer. Theriogenology 2020; 155: 176-84.
[http://dx.doi.org/10.1016/j.theriogenology.2020.06.009] [PMID: 32702562]

[96] Gil MA, Maside C, Cuello C, *et al.* Effects of Hoechst 33342 staining and ultraviolet irradiation on mitochondrial distribution and DNA copy number in porcine oocytes and preimplantation embryos. Mol Reprod Dev 2012; 79(9): 651-63.
[http://dx.doi.org/10.1002/mrd.22071] [PMID: 22777700]

[97] Iuso D, Czernik M, Zacchini F, Ptak G, Loi P. A simplified approach for oocyte enucleation in mammalian cloning. Cell Reprogram 2013; 15(6): 490-4.
[http://dx.doi.org/10.1089/cell.2013.0051] [PMID: 24219576]

[98] Peura TT, Lewis IM, Trounson AO. The effect of recipient oocyte volume on nuclear transfer in cattle. Mol Reprod Dev 1998; 50(2): 185-91.
[http://dx.doi.org/10.1002/(SICI)1098-2795(199806)50:2<185::AID-MRD9>3.0.CO;2-G] [PMID: 9590535]

[99] George A, Sharma R, Singh KP, *et al.* Production of cloned and transgenic embryos using buffalo (Bubalus bubalis) embryonic stem cell-like cells isolated from in vitro fertilized and cloned blastocysts. Cell Reprogram 2011; 13(3): 263-72.
[http://dx.doi.org/10.1089/cell.2010.0094] [PMID: 21548826]

[100] Lagutina I, Lazzari G, Duchi R, *et al.* Somatic cell nuclear transfer in horses: effect of oocyte morphology, embryo reconstruction method and donor cell type. Reproduction 2005; 130(4): 559-67.
[http://dx.doi.org/10.1530/rep.1.00772] [PMID: 16183874]

[101] Du Y, Kragh PM, Zhang Y, *et al.* Piglets born from handmade cloning, an innovative cloning method without micromanipulation. Theriogenology 2007; 68(8): 1104-10.
[http://dx.doi.org/10.1016/j.theriogenology.2007.07.021] [PMID: 17889304]

[102] Vajta G. Handmade cloning: the future way of nuclear transfer? Trends Biotechnol 2007; 25(6): 250-3.
[http://dx.doi.org/10.1016/j.tibtech.2007.04.004] [PMID: 17434218]

[103] Zhang P, Liu P, Dou H, *et al.* Handmade cloned transgenic sheep rich in omega-3 Fatty acids. PLoS One 2013; 8(2): e55941.
[http://dx.doi.org/10.1371/journal.pone.0055941] [PMID: 23437077]

[104] Wakayama T, Perry AC, Zuccotti M, Johnson KR, Yanagimachi R. Full-term development of mice from enucleated oocytes injected with cumulus cell nuclei. Nature 1998; 394(6691): 369-74.
[http://dx.doi.org/10.1038/28615] [PMID: 9690471]

[105] Czernik M, Anzalone D A, Palazzese L, Oikawa M, Loi P. Somatic cell nuclear transfer: failures, successes and the challenges ahead. The International Journal of Developmental Biology, 63(3-4-5), 123–130.
[http://dx.doi.org/10.1387/ijdb.180324mc]

[106] Colman A. Somatic cell nuclear transfer in mammals: progress and applications. Cloning 1999-2000; 1(4): 185-200.
[http://dx.doi.org/10.1089/15204559950019825] [PMID: 16218819]

[107] Wilmut I, Bai Y, Taylor J. Somatic cell nuclear transfer: origins, the present position and future opportunities. Philosophical transactions of the Royal Society of London. Series B, Biological sciences, 370(1680), 20140366.

[http://dx.doi.org/10.1098/rstb.2014.0366]

[108] Matoba S, Zhang Y. Somatic Cell Nuclear Transfer Reprogramming: Mechanisms and Applications. Cell Stem Cell 2018; 23(4): 471-85.
[http://dx.doi.org/10.1016/j.stem.2018.06.018] [PMID: 30033121]

[109] Wolf E, Zakhartchenko V, Brem G. Nuclear transfer in mammals: recent developments and future perspectives. J Biotechnol 1998; 65(2-3): 99-110.
[http://dx.doi.org/10.1016/S0168-1656(98)00132-1] [PMID: 9828456]

[110] Chan AW, Dominko T, Luetjens CM, *et al.* Clonal propagation of primate offspring by embryo splitting. Science 2000; 287(5451): 317-9.
[http://dx.doi.org/10.1126/science.287.5451.317] [PMID: 10634789]

[111] Blesa JR, Tudela J, Aznar J. Ethical aspects of nuclear and mitochondrial DNA transfer. Linacre Q 2016; 83(2): 179-91.
[http://dx.doi.org/10.1080/00243639.2016.1180773] [PMID: 27833197]

[112] Wakayama T. Production of cloned mice and ES cells from adult somatic cells by nuclear transfer: how to improve cloning efficiency? J Reprod Dev 2007; 53(1): 13-26.
[http://dx.doi.org/10.1262/jrd.18120] [PMID: 17332696]

[113] Kfoury C. Therapeutic cloning: promises and issues. McGill journal of medicine: MJM: an international forum for the advancement of medical sciences by students, 10(2), 112–120.

[114] Hinrichs K. Assisted reproductive techniques in mares. Reprod Domest Anim 2018; 53 (Suppl. 2): 4-13.
[http://dx.doi.org/10.1111/rda.13259] [PMID: 30238661]

[115] Wakayama T, Yanagimachi R. Cloning of male mice from adult tail-tip cells. Nat Genet 1999; 22(2): 127-8.
[http://dx.doi.org/10.1038/9632] [PMID: 10369248]

[116] Kurd S, Zarei MA, Fathi F, Ghadimi T, Hakhamaneshi MS, Jalili A. Production of cloned mice by nuclear transfer of cumulus cells. Avicenna J Med Biotechnol 2013; 5(3): 186-92.
[PMID: 23919122]

[117] Ogura A, Inoue K, Ogonuki N, *et al.* Production of male cloned mice from fresh, cultured, and cryopreserved immature Sertoli cells. Biol Reprod 2000; 62(6): 1579-84.
[http://dx.doi.org/10.1095/biolreprod62.6.1579] [PMID: 10819758]

[118] Ogura A, Inoue K, Wakayama T. Recent advancements in cloning by somatic cell nuclear transfer. Philos Trans R Soc Lond B Biol Sci 2013; 368(1609): 20110329.
[http://dx.doi.org/10.1098/rstb.2011.0329] [PMID: 23166393]

[119] Miki H, Inoue K, Kohda T, *et al.* Birth of mice produced by germ cell nuclear transfer. Genesis 2005; 41(2): 81-6.
[http://dx.doi.org/10.1002/gene.20100] [PMID: 15712265]

[120] Rakha A. Cloning Efficiency and a Comparison between Donor Cell Types. Cloning Transgenes 2015; 4(3): 141.
[http://dx.doi.org/10.4172/2168-9849.1000141]

[121] Wakayama T, Yanagimachi R. Mouse cloning with nucleus donor cells of different age and type. Mol Reprod Dev 2001; 58(4): 376-83. a
[http://dx.doi.org/10.1002/1098-2795(20010401)58:4<376::AID-MRD4>3.0.CO;2-L]   [PMID: 11241773]

[122] Gómez NA, Ramírez MM, Ruiz-Cortés ZT. Primary fibroblast cell cycle synchronization and effects on handmade cloned (hmc) bovine embryos. Cienc Anim Bras 2018; 19(0)
[http://dx.doi.org/10.1590/1809-6891v19e-48555]

[123] Inoue K, Wakao H, Ogonuki N, *et al.* Generation of cloned mice by direct nuclear transfer from

natural killer T cells. Curr Biol 2005; 15(12): 1114-8.
[http://dx.doi.org/10.1016/j.cub.2005.05.021] [PMID: 15964276]

[124] Nair S, Dhodapkar MV. Natural Killer T Cells in Cancer Immunotherapy. Front Immunol 2017; 8: 1178.
[http://dx.doi.org/10.3389/fimmu.2017.01178] [PMID: 29018445]

[125] Wakayama T, Rodriguez I, Perry AC, Yanagimachi R, Mombaerts P. Mice cloned from embryonic stem cells. Proc Natl Acad Sci USA 1999; 96(26): 14984-9.
[http://dx.doi.org/10.1073/pnas.96.26.14984] [PMID: 10611324]

[126] Kato Y, Tsunoda Y. Role of the donor nuclei in cloning efficiency: can the ooplasm reprogram any nucleus? Int J Dev Biol 2010; 54(11-12): 1623-9.
[http://dx.doi.org/10.1387/ijdb.103203yk] [PMID: 21404183]

[127] Akagi S, Matsukawa K, Takahashi S. Factors affecting the development of somatic cell nuclear transfer embryos in Cattle. J Reprod Dev 2014; 60(5): 329-35.
[http://dx.doi.org/10.1262/jrd.2014-057] [PMID: 25341701]

[128] Zhao T, Zhang ZN, Rong Z, Xu Y. Immunogenicity of induced pluripotent stem cells. Nature 2011; 474(7350): 212-5.
[http://dx.doi.org/10.1038/nature10135] [PMID: 21572395]

[129] Yao L, Yu X, Hui N, Liu S. Application of iPS in assisted reproductive technology: sperm from somatic cells? Stem Cell Rev Rep 2011; 7(3): 714-21.
[http://dx.doi.org/10.1007/s12015-011-9236-8] [PMID: 21318596]

[130] Hendriks S. Towards the responsible clinical implementation of stem cell-based fertility treatments. 2017.

[131] Imamura M, Aoi T, Tokumasu A, *et al.* Induction of primordial germ cells from mouse induced pluripotent stem cells derived from adult hepatocytes. Mol Reprod Dev 2010; 77(9): 802-11.
[http://dx.doi.org/10.1002/mrd.21223] [PMID: 20722049]

[132] Fang F, Li Z, Zhao Q, Li H, Xiong C. Human induced pluripotent stem cells and male infertility: an overview of current progress and perspectives. Hum Reprod 2018; 33(2): 188-95.
[http://dx.doi.org/10.1093/humrep/dex369] [PMID: 29315416]

[133] Bach N, Bich Thuy L. Current Research, Challenges, and Perspectives of Biotechnology: An Overview. Vietnam Journal of Agricultural Sciences 2019; 1(2): 187-99.
[http://dx.doi.org/10.31817/vjas.2018.1.2.09]

[134] Mameli M. Reproductive cloning, genetic engineering and the autonomy of the child: the moral agent and the open future. J Med Ethics 2007; 33(2): 87-93.
[http://dx.doi.org/10.1136/jme.2006.016634] [PMID: 17264194]

[135] Austriaco RNPG. Dignity as a Faith-Based Consideration in the Ethics of Human Genome Editing. In: Trump B, Cummings C, Kuzma J, Linkov I, Eds. Synthetic Biology 2020: Frontiers in Risk Analysis and Governance Risk, Systems and Decisions. Cham: Springer 2020.
[http://dx.doi.org/10.1007/978-3-030-27264-7_17]

[136] Evans JH. Religion and Human Cloning: An Exploratory Analysis of the First Available Opinion Data. J Sci Study Relig 2002; 41(4): 747-58.
[http://dx.doi.org/10.1111/1468-5906.t01-1-00151]

[137] Kasmo MA, Usman AH, Said MM, Taha M, Aziz AA. The perception of human cloning: A comparative study between difference faiths in Malaysia. Rev Eur Stud 2015; 7(3): 178.
[http://dx.doi.org/10.5539/res.v7n3p178]

[138] Havstad JC. Human reproductive cloning: a conflict of liberties. Bioethics 2010; 24(2): 71-7.
[http://dx.doi.org/10.1111/j.1467-8519.2008.00692.x] [PMID: 19076121]

[139] Loue S. The New Frontier: Cloning. Case Studies in Society, Religion, and Bioethics. Cham: Springer

2020.
[http://dx.doi.org/10.1007/978-3-030-44150-0_9]

[140] Koenig HG. Religion, spirituality, and health: the research and clinical implications. ISRN Psychiatry 2012; 2012: 278730.
[http://dx.doi.org/10.5402/2012/278730] [PMID: 23762764]

[141] Yeary KHK, Alcaraz KI, Ashing KT, *et al.* Considering religion and spirituality in precision medicine. Transl Behav Med 2019.
[PMID: 31294809]

[142] Häyry M. Philosophical arguments for and against human reproductive cloning. Bioethics 2003; 17(5-6): 447-59.
[http://dx.doi.org/10.1111/1467-8519.00360] [PMID: 14959708]

[143] Hostiuc S. Procreative Autonomy *Versus* Beneficence in Assisted Reproductive Technologies. Balkan Journal of Philosophy 2020; 12(1): 63-74.
[http://dx.doi.org/10.5840/bjp20201214]

[144] National Human Genome Research Institute. 2017.Cloning fact sheet https://www.genome. gov/about-genomics/fact-sheets/Cloning-Fact-Sheet

[145] Camenzind S. On Clone as Genetic Copy: Critique of a Metaphor. NanoEthics 2015; 9(1): 23-37.
[http://dx.doi.org/10.1007/s11569-014-0218-6]

[146] Sandel MJ. The ethical implications of human cloning. Perspect Biol Med 2005; 48(2): 241-7.
[http://dx.doi.org/10.1353/pbm.2005.0063] [PMID: 15834196]

[147] American College of Obstetricians and Gynecologists. 2019.Prenatal genetic screening tests https://www.acog.org/Patients/FAQs/Prenatal-Genetic-Screening-Tests?IsMobileSet=false

[148] Parfit D. Future generations: Future problems. Philos Public Aff 1982; 11: 113-72.

[149] Parfit D. Reasons and persons. Oxford: Oxford University Press 1984.

[150] Harman E. Can We Harm and Benefit in Creating? Philos Perspect 2004; 18(1): 89-113.
[http://dx.doi.org/10.1111/j.1520-8583.2004.00022.x]

[151] Pummer T. Whether and Where to Give. Philos Public Aff 2016; 44(1): 77-95.
[http://dx.doi.org/10.1111/papa.12065]

[152] Jacobs AJ, Arora KS. Ritual male infant circumcision and human rights. Am J Bioeth 2015; 15(2): 30-9.
[http://dx.doi.org/10.1080/15265161.2014.990162] [PMID: 25674955]

[153] Staicu L. Human cloning and the myth of disenchantment. Journal for the Study of Religions and Ideologies 2012; 11(31): 148-69.

[154] de Melo-Martín I. On cloning human beings. Bioethics 2002; 16(3): 246-65.
[http://dx.doi.org/10.1111/1467-8519.00284] [PMID: 12211248]

[155] Breitowitz Y. What's so bad about human cloning? Kennedy Inst Ethics J 2002; 12(4): 325-41.
[http://dx.doi.org/10.1353/ken.2002.0023] [PMID: 12645610]

[156] Sparrow R. Cloning, parenthood, and genetic relatedness. Bioethics 2006; 20(6): 308-18.
[http://dx.doi.org/10.1111/j.1467-8519.2006.00509.x]

[157] Hansen M. As surrogacy becomes more popular, legal problems proliferate. ABA J 2011.

[158] Saxena P, Mishra A, Malik S. Surrogacy: ethical and legal issues. Indian J Community Med 2012; 37(4): 211-3.
[http://dx.doi.org/10.4103/0970-0218.103466] [PMID: 23293432]

[159] Burrell C, O'Connor H. Surrogate pregnancy: Ethical and medico-legal issues in modern obstetrics. Obstet Gynaecol 2013; 15(2): 113-9.

[http://dx.doi.org/10.1111/tog.12010]

[160] Finkelstein A, MacDougall S, Kintominas A, Olsen A. Surrogacy law and policy in the U.S.: A national conversation informed by global lawmaking. Report of the Columbia Law School Sexuality & Gender Law Clinic. 2016.https://web.law.columbia.edu/sites/default/files/ microsites/gender-sexuality/files/columbia_sexuality_and_gender_law_clinic__surrogacy_ law_and_policy_report_-_june_2016.pdf

[161] Reamer FG, Siegel DH. Ethical issues in open adoption: Implications for practice. Fam Soc 2007; 88(1): 11-8.
[http://dx.doi.org/10.1606/1044-3894.3587]

[162] Capron AM. Is it time to clone a bioethics commission? Hastings Cent Rep 1994; 24(1): 29-30.
[http://dx.doi.org/10.2307/3562384] [PMID: 8045766]

[163] McDougall R. A resource-based version of the argument that cloning is an affront to human dignity. J Med Ethics 2008; 34(4): 259-61.
[http://dx.doi.org/10.1136/jme.2006.018804] [PMID: 18375676]

[164] Meilaender G. Begetting and cloning. First Things 1997; (74): 41-3.
[PMID: 11656927]

[165] Cole-Turner R. Human cloning: Religious responses. Louisville, KY: Westminster John Knox Press. 9 The New Frontier. Cloning 1997; •••: 249.

[166] Anon ND. Glossary. United Nations Treaty Collection 2002.https://treaties.un.org/pages/Overview

[167] Messer NG. Cloning, creation, and control. Sci Christ Belief 2002; 16(1): 45-50.

[168] Jones DG. Human cloning: A watershed for science and ethics? Sci Christ Belief 2002; 14(2): 159-80.

[169] Brun RB. Cloning humans? Current science, current views, and a perspective from Christianity. Differentiation 2002; 69(4-5): 184-7.
[http://dx.doi.org/10.1046/j.1432-0436.2002.690411.x] [PMID: 11841476]

[170] Golinkin D. Cloning in Jewish law. The Schechter Institute.. 2001.https://schechter.edu/cloningin-Jewish-law/?print

[171] Rosner F. Test tube babies, host mothers and genetic engineering in Judaism. Traditio 1981; 19(2): 141-8.

[172] Al-Aqeel AI. Human cloning, stem cell research. An Islamic perspective. Saudi Med J 2009; 30(12): 1507-14.
[PMID: 19936411]

[173] Al-Hayani FA. Muslim perspectives on stem cell research and cloning. Zygon 2008; 43(4): 783-95.
[http://dx.doi.org/10.1111/j.1467-9744.2008.00960.x]

[174] Ghaly M. Human cloning through the eyes of Muslim scholars: The new phenomenon of the Islamic international religioscientific institutions. Zygon 2010; 45(1): 7-35.
[http://dx.doi.org/10.1111/j.1467-9744.2010.01056.x]

[175] Sekaleshfar FB. A critique of Islamic arguments on human cloning. Zygon 2010; 45(1): 37-46.
[http://dx.doi.org/10.1111/j.1467-9744.2010.01057.x]

[176] Moeinifar M, Ardebeli FA. Lineage and the rights of cloned child in the islamic jurisprudence. J Reprod Infertil 2012; 13(4): 183-92.
[PMID: 23926545]

[177] Schlieter J. 2004.Some aspects of the Buddhist assessment of human cloning.
[http://dx.doi.org/10.1007/978-94-017-6174-1_3]

[178] Promta S. 2005.Buddhism and human genetic research https://them.polylog.org/6/fps=en.htm

[179] Campbell CS. Religious perspectives on human cloning. Cloning human beings (Commissioned

papers). Rockville, MD: National Bioethics Advisory Commission 1997; Vol. 2: pp. D1-D64.

[180]   Keown D. Buddhist ethics: A very short introduction. Oxford: Oxford University Press 2005.
[http://dx.doi.org/10.1093/actrade/9780192804570.001.0001]

[181]   Ratankul P. Buddhism, prenatal diagnosis and human cloning. In: Fujiki N, Macer DRJ, Eds. Bioethics in Asia. Christchurch: Eubios 1998; pp. 405-7.

[182]   Human cloning and human dignity: An ethical inquiry. Washington, D.C.: Author 2002.

[183]   Lyon A. The cloning report: left of Bush but still a ban. Hastings Cent Rep 2002; 32(5): 7.
[PMID: 12360775]

[184]   Kass LR. Babies by means of in vitro fertilization: unethical experiments on the unborn? N Engl J Med 1971; 285(21): 1174-9.
[http://dx.doi.org/10.1056/NEJM197111182852105] [PMID: 5096639]

[185]   Kass LR. The wisdom of repugnance: why we should ban the cloning of humans. New Repub 1997; 216(22): 17-26.
[PMID: 11654974]

[186]   Dreifus C. A conversation with Elizabeth H. Blackburn: Finding clues to aging in the fraying tips of chromosomes. New York Times. 2007.

[187]   Green RM. The President's council on bioethics—Requiescat in pace. J Relig Ethics 2010; 38(2): 197-218.
[http://dx.doi.org/10.1111/j.1467-9795.2010.00426.x]

[188]   Bioethics Defense Fund. 2014.https://bdfund.org/wpcontent/uploads/2016/05/CLONINGChart-BDF2014.pdf

[189]   Macintosh KL. Human cloning: Stereotypes, public policy, and the law. In: Boonin D, Ed. Palgrave handbook of philosophy and public policy. New York: Palgrave 2018; pp. 637-47.
[http://dx.doi.org/10.1007/978-3-319-93907-0_48]

[190]   United Nations Educational, Scientific and Cultural Organization, 2004.

[191]   Lo B, Parham L, Alvarez-Buylla A, *et al.* Cloning mice and men: prohibiting the use of iPS cells for human reproductive cloning. Cell Stem Cell 2010; 6(1): 16-20.
[http://dx.doi.org/10.1016/j.stem.2009.12.004] [PMID: 20085739]

SUBJECT INDEX

www.ingramcontent.com/pod-product-compliance
Lightning Source LLC
Chambersburg PA
CBHW050817220326
41598CB00006B/233